Eucaryotic Microbes as Model Developmental Systems

MICROBIOLOGY SERIES

Series Editor
ALLEN I. LASKIN
EXXON Research and Engineering Company
Linden, New Jersey

Volume 1 Bacterial Membranes and Walls
edited by Loretta Leive

Volume 2 Eucaryotic Microbes as Model Developmental Systems
edited by Danton H. O'Day and Paul A. Horgen

Other Volumes in Preparation

Eucaryotic Microbes as Model Developmental Systems

edited by

DANTON H. O'DAY

Department of Zoology and Erindale College
University of Toronto
Mississauga, Ontario, Canada

PAUL A. HORGEN

Department of Botany and Erindale College
University of Toronto
Mississauga, Ontario, Canada

MARCEL DEKKER, INC. New York and Basel

Library of Congress Cataloging in Publication Data

Main entry under title:

Eucaryotic microbes as model developmental systems.

Includes indexes.
1. Developmental cytology. 2. Cell differentiation. 3. Micro-organisms. I. O'Day, Danton H. II. Horgen, Paul A.
QH491.E9 574.8'76 76-28079
ISBN 0-8247-6514-1

MARCEL DEKKER, INC.
270 Madison Avenue, New York, New York 10016

Current printing (last digit):
10 9 8 7 6 5 4 3 2 1

PRINTED IN THE UNITED STATES OF AMERICA

To our wives, Susan and Ilona

PREFACE

When we compiled this volume we were motivated by several stimuli. We felt that previous volumes on microbial development were often preoccupied with procaryotic cells and were of limited value to those interested in aspects of development, its control, and its maladies in eucaryotic cells. Not taking away from the inherent beauty of the organisms themselves, we also felt, in these times of relevance, that it was essential to reveal the merits of studying lower nucleated cells. The cellular systems discussed in this volume offer numerous advantages not afforded by higher plant and animal cellular systems, especially for approaching complex developmental questions. This volume focuses on many of the major problems of developmental biology and shows how the study of eucaryotic microbes is helping to elucidate these problems.

The book is divided into three major parts: Growth and Cellular Differentiation, Cell Communication and Morphogenesis, and Dormancy and Germination. Each part begins with an Editors' Introduction, which puts the contained articles into a general perspective but which is not intended as a comprehensive review. The volume will appeal mainly to senior undergraduates, graduate students, and scientists already working on the development of eucaryotic microbes. However, since each article begins with an introduction to the respective experimental organism, its morphology, and its life cycle, the book should have wider appeal to a more general readership.

The authors are indebted to Elinor Foden for her assistance in all phases of compilation of this volume and for her unending patience.

Danton H. O'Day
Paul A. Horgen

CONTRIBUTORS

G. B. Calleja,[*] Natural Science Research Center, University of the Philippines, Diliman, Quezon City, The Philippines

Linda E. Cameron, Department of Microbiology, University of Manitoba, Winnipeg, Manitoba, Canada

Randall L. Dimond, Department of Biology, University of California, San Diego, La Jolla, California

Allan Dingle, Department of Biology, McMaster University, Hamilton, Ontario, Canada

Larry D. Dunkle, Department of Plant Pathology, University of Nebraska, Lincoln, Nebraska

Antony J. Durston, Hubrecht Laboratory, International Embryological Institute, Universiteitscentrum "De Uithof," Utrecht, The Netherlands

Robert M. Eisenberg, Department of Biological Sciences, University of Delaware, Newark, Delaware

David W. Francis, Department of Biological Sciences, University of Delaware, Newark, Delaware

Stephen J. Free, Department of Biology, University of California, San Diego, La Jolla, California

Shelby N. Freer, Department of Plant Pathology, University of Nebraska, Lincoln, Nebraska

Swee H. Goh, Department of Microbiology, University of Manitoba, Winnipeg, Manitoba, Canada

Cheng-Shung Gong,[†] Department of Biological Sciences, Purdue University, West Lafayette, Indiana

James E. Haber, Department of Biology, Rosenstiel Basic Medical Sciences Research Center, Brandeis University, Waltham, Massachusetts

---

[*]Current affiliation: Division of Biological Sciences, National Research Council of Canada, Ottawa, Ontario K1A 0R6, Canada

[†]Current affiliation: School of Chemical Engineering, Purdue University, West Lafayette, Indiana

Paul A. Horgen, Department of Botany and Erindale College, University of Toronto, Mississauga, Ontario, Canada

Byron F. Johnson, Division of Biological Sciences, National Research Council of Canada, Ottawa, Ontario, Canada

Steven A. Johnson, Department of Biological Sciences, Purdue University, West Lafayette, Indiana

Glen R. Klassen, Department of Microbiology, University of Manitoba, Winnipeg, Manitoba, Canada

Gary Kochert, Department of Botany, University of Georgia, Athens, Georgia

Elaine Y. Lai, Department of Biology, Rosenstiel Basic Medical Sciences Research Center, Brandeis University, Waltham, Massachusetts

Herb B. LéJohn, Department of Microbiology, University of Manitoba, Winnipeg, Manitoba, Canada

Wallace M. LeStourgeon, Department of Molecular Biology, Vanderbilt University, Nashville, Tennessee

William F. Loomis, Department of Biology, University of California, San Diego, La Jolla, California

James S. Lovett, Department of Biological Sciences, Purdue University, West Lafayette, Indiana

Mark MacInnes, Department of Biological Sciences, University of Delaware, Newark, Delaware

David R. McNaughton, Department of Microbiology, University of Manitoba, Winnipeg, Manitoba, Canada

Renate U. Meuser, Department of Microbiology, University of Manitoba, Winnipeg, Manitoba, Canada

Danton H. O'Day, Department of Zoology and Erindale College, University of Toronto, Mississauga, Ontario, Canada

Richard P. Sutter, Department of Biology, West Virginia University, Morgantown, West Virginia

David A. Thomas, Department of Developmental Biology, Boston Biomedical Research Institute, Boston, Massachusetts

John E. Thompson, Department of Biology, University of Waterloo, Waterloo, Ontario, Canada

James L. Van Etten, Department of Plant Pathology, University of Nebraska, Lincoln, Nebraska

Peter J. Wejksnora, Department of Biology, Rosenstiel Basic Medical Sciences Research Center, Brandeis University, Waltham, Massachusetts

Sally S. White, Department of Biology, University of California, San Diego, La Jolla, California

Barbara E. Wright, Department of Microbiology and Molecular Genetics, Harvard Medical School, Boston Biomedical Research Institute, Boston, Massachusetts

Deborah D. Wygal, Department of Biology, Rosenstiel Basic Medical Sciences Research Center, Brandeis University, Waltham, Massachusetts

Bong Y. Yoo, Department of Biology, University of New Brunswick, Fredericton, New Brunswick, Canada

Shuhei Yuyama, Department of Zoology, University of Toronto, Toronto, Ontario, Canada

CONTENTS

Eucaryotic Microbes as Model Developmental Systems

PART I

Growth and Cellular Differentiation

EDITORS' INTRODUCTION

The majority of the current scientific investigations into the mechanisms that determine how a cell is transformed from one state into another have focused on the regulation of genomic expression and on the accumulation and function of specific gene products. Eucaryotic microbes lend themselves admirably to such studies because it is often possible to obtain large amounts of genetically identical cells which, by simple manipulation of environmental parameters, can be induced to grow and divide or to follow alternative pathways of development. For example: plasmodia of *Physarum* can either develop into sporangia or into thick-walled cysts; amoebas of *Naegleria* can differentiate into flagellates or into cysts; and cellular slime mold amoebas may embark on either a multicellular developmental pathway to form fruiting bodies or macrocysts or undergo unicellular differentiation to form microcysts. Thus, like the fertilized egg of animals or plants which develop into the multifarious cells of the mature organism, the cells of lower eucaryotes may undergo diverse kinds of differentiation. The advantage these microbes provide, however, is that each differentiation may be studied alone or, at most, in the presence of but a few other simultaneous events rather than in the complex, heterogeneous cellular environment that characterizes the multicellular embryo. The contributors to this part reveal how eucaryotic microbes are being used to solve some of the most perplexing problems of cellular differentiation.

GENETIC ANALYSIS OF DEVELOPMENTAL EVENTS

The discovery of the sexual cycle (macrocyst formation) of cellular slime molds [1] allows for the possibility of precise genetic mapping in this important group of eucaryotic microbes. David W. Francis and Robert M. Eisenberg review the history behind this discovery and show how developmental genetics, using both the parasexual and sexual cycle, is beginning to yield important information on the program of development in cellular slime molds.

In the yeasts, genetic analysis has been realized for many years. Since the mating-type locus has been shown to play a critical role in controlling sporulation, it has been the subject of much investigation. James E. Haber's group is pursuing the problem by employing conditional mutants which affect the mating-type locus. Their work with temperature-sensitive mutants suggests that an amber mutation in one (α) allele converts one mating type (α) to the other mating phenotype (a).

THE GENOME AND TRANSCRIPTION

If we are to understand the way the information that is stored in the genome is retrieved and utilized during the differentiation process, we must understand the organization of the eucaryotic chromosome. In the eucaryotic chromosome, the histone proteins are implicated as general repressors of gene function while certain nonhistone (acidic) proteins have been suggested as specific gene regulators [2,3]. Wallace M. LeStourgeon's early work with *Physarum* provided the first correlation between changes in the complement of nonhistone chromosomal proteins and altered patterns of genetic activity during cellular differentiation [4]. In this part, LeStourgeon reviews the current model for the organization of the basic chromatin fiber of the chromosome [5] and shows why *Physarum* is an excellent system for characterizing the residual nonhistone proteins of chromatin. His work demonstrates a remarkable correlation between the developmental changes in the nonhistone proteins of *Physarum* and mammalian cells.

In addition to regulatory proteins, the complex group of nonhistone proteins also includes enzymes such as DNA and RNA polymerases. There are three major species of RNA polymerase in eucaryotic cells, and these may contain a couple of subspecies [6]. How so few enzymes can selectively produce the specific RNA transcripts that characterize a specific kind of cellular differentiation is a perplexing problem. In procaryotes it is suggested that highly phosphorylated nucleotides may regulate RNA polymerase function [7]. Herb B. LéJohn and his co-workers demonstrate the appearance of polyphosphate compounds during the development of *Achlya*. The three polyphosphates of *Achlya* show complex activating and inhibiting effects on the various RNA polymerase species isolated from this organism, suggesting a role for these compounds in eucaryotic development.

Since the first products of gene action are various classes of RNA, the developmental appearance of these molecules has received much attention. Shuhei Yuyama has examined the patterns of RNA synthesis during heat-synchronized cell division in *Tetrahymena*. By starving synchronized cells, he was able to dramatically reduce RNA synthesis to one-fiftieth of that of control cells without significantly altering the time of the first synchronous cell division. Characterizing the species of RNA synthesized, he revealed that the synthesis of rRNA is not essential for cell division while the synthesis of certain species of mRNA is essential.

During sporulation in *Saccharomyces*, which occurs under conditions of pseudostarvation, rRNA synthesis appears to be significant and important. Using temperature-sensitive mutants for rRNA synthesis, James E. Haber's group provides data that suggest that different controls regulate rRNA synthesis during vegetative growth and sporulation. They also provide evidence indicating that the length of the poly(A) moiety of mRNA is shorter in sporulating cells as compared to vegetatively growing cells.

ENZYME ACCUMULATION AND FUNCTION

A cell is transformed from one state to another as a consequence of the accumulation of new, specific gene products. Generally, the intracellular accumulation of specific enzymes is accepted as the driving force of cellular differentiation. In this part, William F. Loomis and his co-workers show how the selection of enzyme-deficient mutants of *Dictyostelium* is providing information on the physiological roles of certain stage-specific enzymes. Through the accumulation of large amounts of information on developmental mutants, they are also able to show that certain biochemical events are independent of previous events while others are dependent on previous biochemical differentiations. This concept of sequence of events or timing sequences is also pursued by David W. Francis and Robert M. Eisenberg in this part.

The work of Byron F. Johnson, G. B. Calleja, and Bong Y. Yoo is concerned with the role enzymes play in the morphogenesis of fission yeasts. They propose a model of coordinated enzyme activities involving both autolytic and synthetic enzymes, which can explain the cell extension in yeast. With certain modifications, this model has also been used to explain cell division, conjugation, and spore liberation.

Barbara E. Wright and David A. Thomas acknowledge the importance of enzymes but indicate that the mechanism of enzyme accumulation is less important than the functional role of the enzyme in the differentiation process. The accumulation of activity of an enzyme is only important if that enzyme plays a key role in the developmental process and is rate-limiting. (Of five enzymes that have been shown to accumulate during the development of *Dictyostelium*, only the accumulation of glycogen phosphorylase seems to be the product of differential gene activation.) Their data reveal the value of employing kinetic models which integrate many different kinds of information relevant to enzyme action and substrate availability.

THE PRIMARY CONTROL OF CELLULAR DIFFERENTIATION

One of the primary questions of developmental biology is exemplified by the following simple question: what is the initial stimulus that tells a cell how it should differentiate? Current models of cellular differentiation suggest pivotal roles for ions and cyclic AMP [8,9]. In this part, Allan Dingle shows how *Naegleria* is a useful organism for pursuing such problems. Of special interest is the synchrony of the developmental sequence and the precision with which the independent events of the amoeba-flagellate transformation can be timed.

REFERENCES

1. Clark, M. A., D. Francis, and R. Eisenberg, 1973. Biochem. Biophys. Res. Comm. 52: 672-678.
2. Allfrey, V. G., 1970. Fed. Proc. 29: 1447-1460.
3. Cameron, I. L., and J. R. Jeter (eds.), 1974. Acidic Proteins of the Nucleus, Academic Press, New York.
4. LeStourgeon, W. M., and H. P. Rusch, 1971. Science 174: 1233-1235.
5. Kolata, G. B., 1975. Science 188: 1097-1099.
6. Kedinger, C., P. Nuret, and P. Chambon, 1971. FEBS Letters 15: 169-174.
7. Aboud, M., and I. Pastan, 1975. J. Biol. Chem. 250: 2189-2195.
8. Barth, L. G., and L. J. Barth, 1974. Develop. Biol. 39: 1-22.
9. McMahon, D., 1974. Science 185: 1012-1021.

MACROMOLECULAR SYNTHESIS AND CELL DIVISION IN *TETRAHYMENA*

Shuhei Yuyama

Department of Zoology
University of Toronto
Toronto, Ontario

I. INTRODUCTION

In studying any cellular morphogenetic events, four categories of problems may be discerned: (1) the problem of the triggering (stimulation or activation) mechanism; (2) the problem of control of macromolecular synthesis which is involved in that particular developmental event; (3) the problem of structural changes (assembly process) that are brought about by the products of macromolecular synthesis; and (4) the problem of relationships between cell growth (cell cycle) and developmental events. The present paper will deal mainly with the problem of the fourth category, i.e., the relationships between growth and division in heat-synchronized *Tetrahymena*, with emphasis on RNA synthesis. Other aspects of heat-synchronized *Tetrahymena* have been comprehensively reviewed by Zeuthen [1-3].

Tetrahymena pyriformis GL is a hymenostome ciliate protozoan (70 × 40 μm) that is known only to grow and divide, although some other species of *Tetrahymena* are known to be polymorphic or to differentiate into cysts [4]. The GL strain lacks micronuclei and does not undergo sexual differentiation [5]. The ciliates divide transversely during cytokinesis, in spite of the fact that all ciliates have a distinct polarity with quite different structures associated with the posterior and anterior halves. Complex morphogenetic events take place during the cell cycle in order for a cell to generate two essentially identical daughter cells with their full complement of organelles. One marked feature of these morphogenetic events is the development of a new oral apparatus [6]. A diagrammatic representation of morphogenetic changes in heat-synchronized *Tetrahymena* is shown in Fig. 1.

T. pyriformis possesses many advantages for studying the relationships between macromolecular synthesis and morphogenetic events, i.e., either cortical morphogenesis or cell division, or both. The cells can be grown axenically in simple [5] or defined [7] medium, have a short generation time (2.5 h), take up nutrients, labeled precursors, and metabolic inhibitors readily, and can be

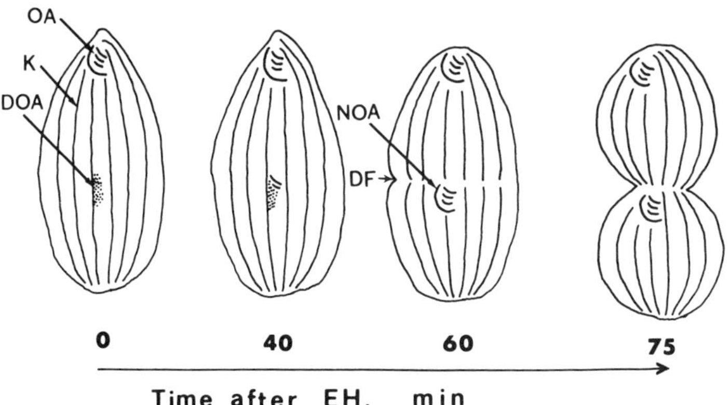

FIG. 1. Diagrammatic representation of heat-synchronized *Tetrahymena pyriformis* GL cells after the end of the last heat shock treatment. EH = end of the last heat shock; OA = oral apparatus; DOA = developing oral apparatus; K = kinetics; DF = division furrow; note that no detectable change in cell size occurs in the starvation medium.

synchronized by a G_2-sensitive event [8]. They exhibit a complex but clearly defined cortical pattern and are ideally suited for studying pattern formation of the cell surface [6]. The surface patterns can also be used as a marker to identify different regions of the cell. In addition, we already possess diversified and detailed information on the taxonomy, morphology, nutrient requirements, metabolic pathways, and growth characteristics (for reviews see Refs. 9 and 10), which enable us to continue investigating important contemporary issues without being hampered by numerous technical problems.

For studying biochemical aspects of developmental events, synchronous cultures must be available. Synchrony in *Tetrahymena* can be obtained by certain selection methods or by various induction methods (see Ref. 11). For studying a normal growth process, the use of a selection synchrony system is imperative, while various induction synchrony systems have advantages for answering other kinds of questions. The classic heat shock-induced synchrony [11] is well suited for studying the relationships between macromolecular synthesis and cell division. In cell cycle studies, elucidation of such relationships, in the simplest possible model system, is the first step toward our understanding of normal developmental processes occurring during the cell cycle. Such a model system has been established recently in our laboratory [12].

Zeuthen and his associates first established that cell division can be artificially synchronized [11,13]. Logarithmically growing *T. pyriformis* GL cells were subjected to alternate temperatures of 28° (optimal growth temperature) and 34° for 30-min intervals. This was repeated five to ten times. The cells then divided synchronously with a maximum division index of 70 to 80% approximately 72 min after the last heat shock treatment (EH) was completed. The 72-min interval between EH and division is shorter than the normal G_2 period. The induction of division synchrony can be explained by the fact that single cells isolated at different stages of the cell cycle display an increasing duration of division delay with increasing cell age (especially during the G_2 phase). When treated by a single heat shock at 34° [14], the cells display an age-dependent "excess delay" pattern. Since each heat shock delays division longer at the late G_2 phase than at the early G_2 phase, a population of log cells can be induced to divide synchronously by giving multiple heat shock treatments. The "excess delay" phenomenon is compatible with the following concepts: (1) certain temperature-sensitive events which are involved in division occur during the G_2 phase; (2) the heat shock damages these events; and (3) cells must reinitiate these events from the start during the recovery after EH.

Pulse treatments of heat-synchronized cells at different times after EH with para-fluorophenylalanine [15], an amino acid analog, or with cycloheximide [16], an inhibitor of protein synthesis, induce the age-dependent "excess delay" phenomenon similar to that observed with heat shock treatments. Based on these observations, a "division protein" model has been proposed by Zeuthen and his associates [13,15,17,18]. The essence of this model assumes that the substance of the temperature-sensitive events is a protein (division protein). When a certain threshold level of the division protein has been synthesized in the G_2 phase, cell division will be triggered. The stage when this threshold level is reached is the "transition point," after which time division is not inhibited either by a heat shock or by protein synthesis inhibitors. The division protein model assumes that a temporary inhibition of protein synthesis results in the functional disappearance of division protein so that cells must begin resynthesizing the protein. This assumption can also explain the "excess delay" phenomenon induced by many different types of antimitotic agents, since these agents conceivably affect protein synthesis (directly or indirectly). One problem yet to be resolved in this model is the lack of a theoretical basis for linking the temporary inhibition of protein synthesis to the hypothetical disappearance (inactivation or utilization for purposes unrelated to division) of division protein, for there exists a priori no obvious reason for this assumption.

The division protein model is important in that the "excess delay" phenomenon has been observed in a wide variety of organisms including procaryotes (for a review see Ref. 19). It is the first theoretical attempt to explain the timing mechanism of the cell division cycle with good experimental evidence. In addition, it is generally accepted that certain protein species must be synthesized after the completion of the S phase in order for a cell to enter mitosis (for reviews see Refs. 19 and 20). During the preparation phase of cell division, a cell must coordinate synthesis and assembly of many different molecules, leading to the complex but organized mitotic events [21]. It is possible that synthesis of "division protein" during the G_2 phase serves as a central clock around which various cellular processes related to division are coordinated; the division protein acts as the "ubiquitous relaxation oscillator" to measure time through each cell cycle [22]. The problem of timing mechanisms has been theorized using various biological rhythms [22-25]. Although the division patterns of *Tetrahymena* appear unrelated to circadian rhythms, the problem of intracellular timing mechanisms may share close relations to the time-sensing mechanisms manifested in circadian rhythms. In many systems the observable biological events occurring cyclically are the culmination of many coordinated intracellular activities.

Various workers have investigated the identity of the division protein in *Tetrahymena* without success [26-29]. This failure is likely a result of not knowing the exact role that this putative protein plays in cell division.

Although, in the "division protein" model, the synthesis of protein is closely linked to the "excess delay" phenomenon, certain exceptions to the model do exist. Both dividing sea urchin eggs

[30] and *Tetrahymena* [14] display "excess delay" pattern by heat shocks and by β-mercaptoethanol [31-33]. However, in sea urchin eggs the recovery from the mercaptoethanol treatment does not require protein synthesis [32]. There is a dissociation in timing of the protein synthesis required for division and of the period when the "excess delay" occurs. This suggests that the "excess delay" pattern observed in sea urchin eggs is not directly linked to the synthesis of protein. Dan and co-workers [34-37] suggested that the fluctuations of sulfhydryl (-SH) groups that occur in dividing sea urchin eggs and heat-synchronized *Tetrahymena* may act as the mitotic clock. While it is conceivable that such changes in -SH groups in *Tetrahymena* are closely linked in timing with the synthesis of "division protein," this doesn't appear to be the case in dividing sea urchin eggs.

One important aspect of heat-synchronized *Tetrahymena* is that nuclear DNA synthesis is not synchronized. Although the division delay is induced initially by interfering with the G_2 events, a certain percentage of cells still incorporate [^{3}H]thymidine into nuclear DNA asynchronously during the interval between EH and division [1,38]. This suggests that the preparation for cell division that normally occurs following the completion of nuclear DNA synthesis is uncoupled from nuclear DNA synthesis during the multiple heat shock regime (normally seven or eight heat shocks requiring about 7 h). Studies on the effects of multiple heat shocks on DNA synthesis using a selection synchrony system revealed that cells do undergo one or two rounds of DNA replication with replicating and nonreplicating periods occurring rhythmically during the heat shock regime [39,40]. Thus it appears that signals to initiate DNA replication reach the nucleus as though the nuclear and cell division had taken place, while actually nuclear and cell division are inhibited during the entire heat shock regime.

II. CURRENT RESEARCH

A. Aspects of RNA Synthesis

RNA synthesis in *Tetrahymena* has been studied in many laboratories (for reviews see Refs. 2,9,41). Like log cells, heat-synchronized *Tetrahymena* cells synthesize tRNA, rRNA, and DNA-like RNA. This is clearly demonstrated by methylated albumen Kiesulguhr (MAK) chromatographic fractionations of the RNA [42,43]. Heat-synchronized cells also respond, like log cells [44], to nutritional shift down or starvation by reducing rRNA synthesis [45-47]. Cells which are transferred to the starvation medium after the seventh heat shock treatment and are then given one additional heat shock in this starvation medium synthesize less rRNA [45-47] and less DNA-like RNA [47] although, interestingly, the division schedule is not significantly affected. However, close examination of the starved cells shows that they divide with a delay of 12 min and that this delay disappears when fresh nutrient medium is added immediately after EH or when the starvation medium contains 1% of the normal strength nutrient medium [47]. These cells also exhibit an increased incorporation of labeled precursors into RNA. These observations suggest that the synthesis of RNA, especially that of DNA-like RNA, is involved in the timing mechanisms of cell division.

Inhibitors of RNA synthesis (or metabolism) (fluorinated pyrimidines [48,49] or actinomycin D [49-54]) affect cell division as does enucleation [55]. However, the effects of inhibitors are less pronounced when cells are in the starvation medium (for a review see Ref. 2). In addition, the degree of inhibition of RNA synthesis by actinomycin D is not correlated with the inhibition of cell division [51,52]. Thus the stepdown experiments and actinomycin D experiments have shown a similar phenomenon, i.e., the marked reduction in the synthesis of RNA is not correlated with a similar degree of inhibition of cell division.

Several problems remained to be resolved on the relationships between the synthesis of RNA after EH and cell division. First, is the RNA synthesis that occurs in various degrees at different nutritional conditions required for division? Second, if the answer to the first question is yes, then what are these species of RNA? Third, do cells regulate growth-related RNA synthesis differently from division-related RNA synthesis? Finally, what are the relationships between the synthesis of RNA and protein required for division?

In order to answer these questions it is necessary to reduce the synthesis of RNA unrelated to division to a minimal level so that the requirement for RNA synthesis can be more concretely established. It appeared possible to achieve this condition, since starvation even at the end of the heat-synchronizing regime reduced the synthesis of RNA markedly without affecting the division schedule. The following series of experiments was performed. The cells were transferred to the starvation medium after the 1st, 2nd, 3rd, 4th, 5th, 6th, or 7th heat shock treatment and were given an additional 7, 6, 5, 4, 3, 2, or 1 heat shocks, respectively (i.e., the cells were given a total of 8 heat shocks). The mitotic indices of each group of cells were analyzed by taking samples at different times after EH. The cells transferred to the starvation medium after the 1st heat shock (and given 7 additional heat shocks in the starvation medium) exhibited a maximum division index (MDI) of 25%. The MDI of cells transferred to the starvation medium after the 2nd, 3rd, 4th, 5th, 6th, and 7th heat shocks were 42, 65, 80, 81, 74, and 65%, respectively (average of 2 experiments) (Fig. 2A). The time of maximum division occurred at about 72 mins after EH in groups of cells transferred to the starvation medium after the 2nd, 3rd, 4th, or 5th heat shock. The cells transferred to the starvation medium after the 1st, 6th, or 7th heat shock divided with a delay of about 10 to 15 min. These data show that the cells transferred to the starvation medium earlier divided without delay whereas those transferred at the end of the heat-synchronizing

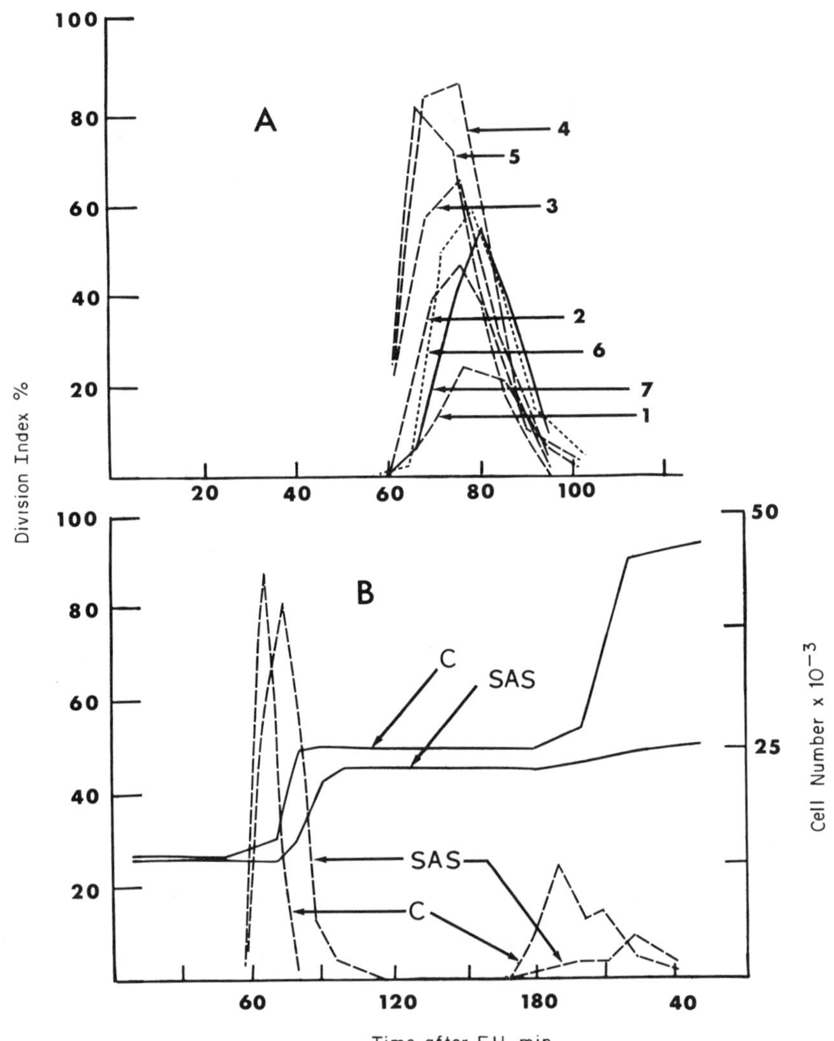

FIG. 2. Division patterns of *Tetrahymena* starved for various periods during the heat shock synchronization regime. (A) Cells were washed, transferred to a starvation medium (inorganic medium) after the 1st, 2nd, 3rd, 4th, 5th, 6th, or 7th heat shock treatment and were given additional 7, 6, 5, 4, 3, 2, or 1 heat shocks, respectively. Division indices of each group of cells have been plotted against the time after EH. Numbers associated with arrows indicate the number of heat shocks given in the nutrient medium. (B) Comparison of division patterns between the control and synchronized and starved (SAS) cells. SAS cells were transferred to the starvation medium after the 4th heat shock and were given 4 additional heat shocks in the starvation medium. C = control; SAS = SAS cells; solid line = cell number per ml of culture; dashed line = division index.

regime divided with delay. This indicates that starvation alone is not the cause of the small delay observed with these cells. A simple net incorporation study of pulse labeling (10 min) with [^{3}H]uridine after EH in different groups of cells revealed that those cells starved for a longer period incorporate less radioactivity into TCA-insoluble material.

Using cells which were given 4 heat shocks in the nutrient medium and 4 additional shocks in the starvation medium, further analysis was conducted. These cells will be referred to as synchronized and starved (SAS) cells. The reason for choosing this synchronization-starvation scheme was that with fewer heat shock treatments in the nutrient medium (1, 2, or 3 heat shocks), the division synchrony was inadequate; with more heat shock treatments in the nutrient medium (5, 6, and 7 heat shocks), the starvation periods were too short for an adequate reduction of RNA synthesis. First, the schedule of division was compared with that of control cells. The control cells referred to in this paper are cells washed and suspended in the starvation medium in the same manner as SAS cells but supplemented with nutrient medium whose strength is one-tenth of the normal nutrient medium. The control cells were treated this way to keep the ionic environment similar to SAS cells and to facilitate the incorporation of precursors. In the full-strength medium, the incorporation of [^{3}H]uridine or [^{3}H]adenosine into RNA was much less, due to the dilution of labeled precursors by unlabeled precursors in the culture medium. Fig. 2B shows that SAS cells and control cells start to divide at the same time, although it takes longer for SAS cells to complete the furrowing process. In addition, most of the SAS cells do not undergo a second synchronous division. A net incorporation study has shown that SAS cells incorporate about one-fiftieth the amount of label as compared to control cells when pulsed with [^{3}H]adenosine for 45 min starting at 5 min after EH. This pulse labeling will be expressed as a 5- to 50-min EH pulse.

In the second series of experiments, the classes of RNA synthesized after a 5- to 50-min EH pulse with [^{3}H]adenosine and [^{14}C]uridine have been characterized using either MAK chromatography or sodium dodecyl sulfate (SDS) sucrose gradient centrifugation. This pulse interval was chosen since the transition point of heat-synchronized *Tetrahymena* occurs at about 50 min after EH. If the RNA synthesis which occurs after EH is required for the first synchronous division, this RNA synthesis should be detected using this pulse-labeling period. The MAK fractionation pattern of the [^{3}H]adenosine-labeled RNA extracted from SAS cells is shown in Fig. 3A. Bulk nucleic acids are eluted in the order of tRNA, 5S RNA, DNA, 17S rRNA, and 25S rRNA. Tritium-labeled RNA is eluted as tRNA, Q_2RNA, and TD-RNA. Q_2 RNA and TD-RNA are both DNA-like RNA and are considered to contain mRNA, since they cosediment with polyribosomes [43]. The rRNA is not labeled significantly and labeling of ribosomal precursor RNA is not detectable [43]. In the average of four experiments, the radioactivity accumulated is 12.8 ± 1.4% for tRNA, 12.6 ± 7.3% for Q_2 RNA, and 75.2 ± 4.7% for TD-RNA. Thus the predominant species of RNA is DNA-like and is eluted mainly as TD-RNA. The ^{14}C labeling pattern was almost identical. Additional experiments using [^{14}C-methyl]methionine verified the previous contention that no rRNA is synthesized in SAS cells, as no detectable radioactivity was associated with 17S and 25S rRNA (Fig. 3A). The first conclusion that was drawn from the above observation was that after EH, new rRNA synthesis is not required for the first synchronous division. Similar pulse-labeling experiments have been conducted using control cells. With [^{3}H]adenosine and [^{14}C-methyl]methionine labeling, rRNA synthesis is clearly detectable (Fig. 3B). The extent of tritium incorporation into TD-RNA of control cells is much greater than that of SAS cells (Fig. 3B). For an additional comparison, the RNA synthesis of cells starved for a much longer period was examined. Cells were transferred to the starvation medium, starved for 24 h and were then given eight heat shocks. Some of these cells synchronously develop oral apparatus and round up after EH at about the time of cell division of control cells [56]. In addition, inhibitors of RNA and protein synthesis block the oral development and rounding up with a similar transition point [57]. Thus the RNA species synthesized in these cells during the 5- to 50-min EH pulse period appears to be responsible for the oral development and rounding-up process. RNA ex-

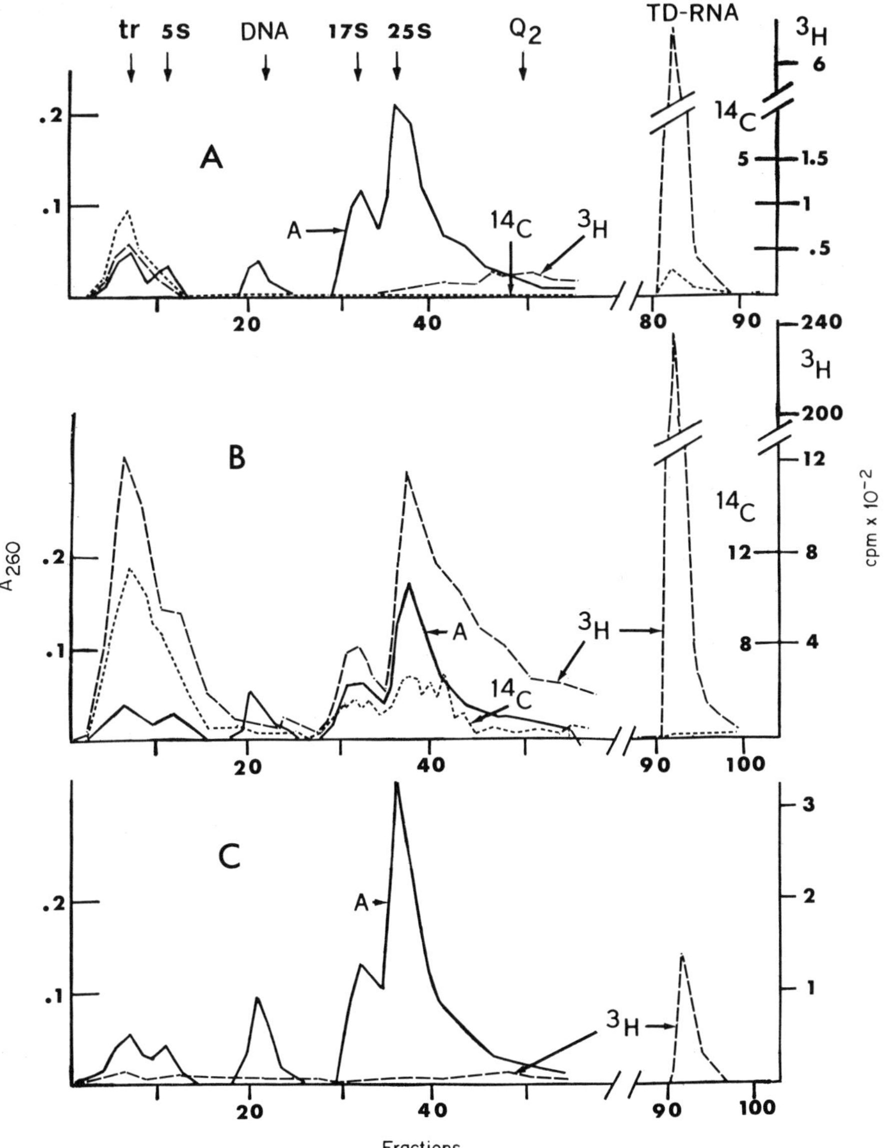

FIG. 3. MAK chromatographic profiles of RNA extracted from heat-synchronized *Tetrahymena*. (A) SAS cells (1,200,000 cells/ml) were pulsed with [^{3}H]adenosine (5 μCi/ml) and [^{14}C-methyl]-methionine (2.5 μCi/ml) for 45 min starting at 5 min after EH. (B) Control cells (870,000 cells/10 ml) were pulsed with [^{3}H]adenosine and [^{14}C-methyl]-methionine under the same conditions as (A). (C) Starved and synchronized cells (synchronously rounding cells) (5,050,000 cells/20 ml) were pulsed with [^{3}H]adenosine (5 μCi/ml) for the same period as (A) and (B). Counting efficiency of ^{3}H in (C) is identical to those of (A) and (B). 0.4 ml aliquot from each fraction (5.8 ml) was used for determining cpm. (For detailed methods see Ref. 12). A = absorbance at 260 nm.

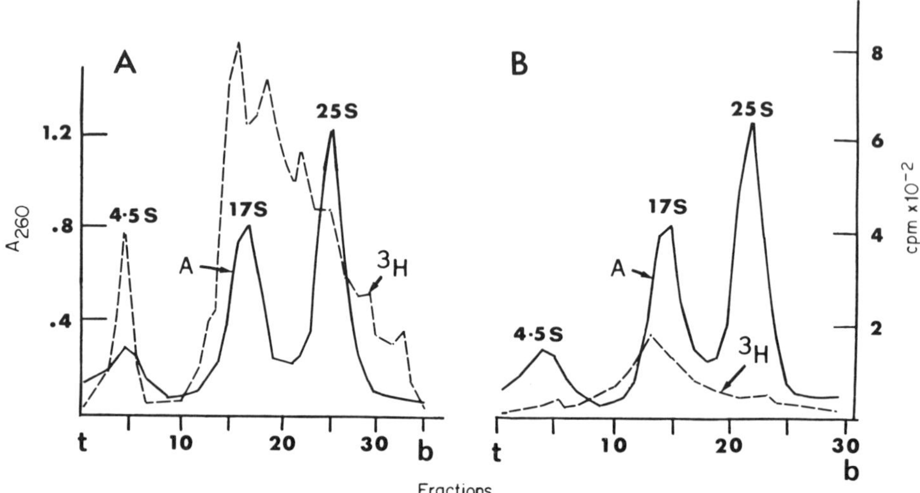

FIG. 4. SDS-sucrose gradient centrifugation profiles of nucleic acids from SAS cells. Cells (1,980,000 cells/20 ml) were pulsed with [^{3}H]adenosine (5 μCi/ml) for 45 min starting at 5 min after EH. Nucleic acids were extracted and divided into two parts and fractionated. (A) With each fraction, A_{260} and TCA precipitable radioactivity were determined. (B) With each fraction, A_{260} and RNase-resistant radioactivity were determined. For RNase digestion, to each fraction (1 ml) was added 2 ml of 3x SSC, and the fractions digested with RNase A (10 μg/ml) and T_1 (20 units/ml) for 16 h. After the digestion, TCA precipitable radioactivity was determined. Similar experiments using [^{3}H]uridine have shown that no TCA precipitable radioactivity was present after the RNase digestion. A = absorbance at 260 nm. (For detailed methods see Ref. 12.)

tracted from these cells shows that little tRNA, no rRNA, and about one-twentieth of TD-RNA (compared with the SAS cells) was labeled (Fig. 3C).

In the third experiment, RNA extracted from SAS cells was fractionated utilizing SDS-sucrose gradient centrifugation. The results show that RNA extracted from SAS cells after a 5- to 50-min EH pulse sediment as tRNA and heterogeneous high-mol wt RNA [presumably mRNA and possibly its precursor, heterogeneous nuclear RNA (HnRNA)] with a peak fraction sedimenting at about 17S (Fig. 4A). It should be noted that HnRNA has not been demonstrated in *Tetra-*

hymena. When each fraction was digested with ribonuclease (RNase) A and T_1 at high ionic strength, RNase-resistant radioactivity was found. The peak fraction contained RNase-resistant RNA sediments at about 15S (Fig. 4B). When TD-RNA was extracted and analyzed using SDS-sucrose gradient centrifugation, the resulting pattern was quite similar to Fig. 4A except for the nonprominence of low mol wt RNA (tRNA).

B. Synthesis of Poly(A)+ RNA

In order to fractionate DNA-like RNA, RNA extracted from SAS cells was separated on oligo-deoxythymidine cellulose (oligo-dT) columns. The RNA was separated into RNA containing polyadenylic acid [poly(A)+] which binds to oligo-dT columns and RNA deficient in polyadenylic acid [poly(A)-] which does not bind. These classes of RNA were then subjected to SDS-sucrose gradient centrifugation. We found that there was no significant binding of 25S or 17S RNA to the affinity columns. With SAS cells, the poly(A)+ RNA sediments as high-mol wt RNA have two peaks at about 17S and 22S (Fig. 5A) whereas the nonbinding RNA sediments as tRNA and high-mol wt RNA have the greater proportion of RNA sedimenting over 17S rRNA (Fig. 5B). This nonbinding [poly(A)-] RNA presumably represents more HnRNA and less mRNA [58-61]. Further analysis of binding and nonbinding RNA utilizing formalin sucrose gradient centrifugation after heating the RNA samples in 3% formalin indicates that the binding RNA sediments as one peak (about 11S), while nonbinding RNA sediments as tRNA and one broad peak (with a peak fraction about 10S and skewed to the higher-mol wt side).

In the fourth series of experiments, the rates of synthesis of poly(A)+ RNA were compared between SAS cells and controls. In order to do this, it was necessary to study the kinetics of incorporation of [^{3}H]adenosine into adenosine triphosphate (ATP). [^{3}H]adenosine was added to cultures of both SAS cells and control cells at 5 min after EH. Samples (80 ml) were taken at intervals up to about 120 min after EH and the specific activity of ATP was

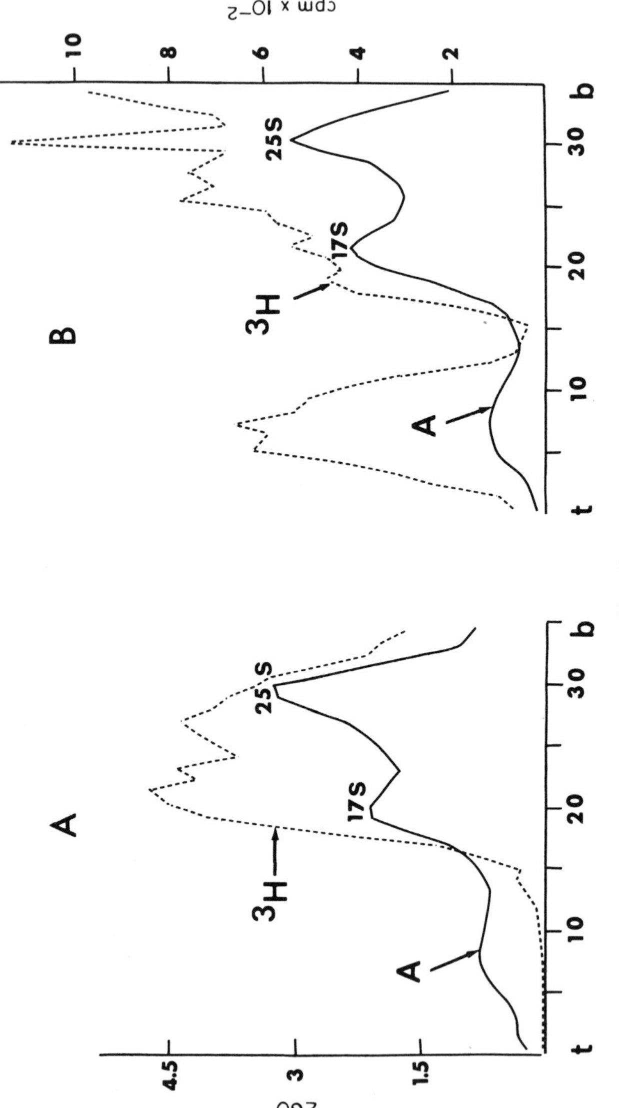

FIG. 5. SDS-sucrose gradient centrifugation profile of oligo-dT cellulose binding and nonbinding RNA extracted from SAS cells. Cells were pulse-labeled with $[^3H]$adenosine (0.5 μCi/ml) for 45 min starting at 5 min after EH, RNA extracted, fractionated into oligo-dT binding [poly(A)+] and oligo-dT nonbinding [poly(A)-] RNA and fractionated with SDS-sucrose gradient centrifugation. (A) Poly(A)+, DNA-like RNA; nonradioactive RNA was added as a marker. (B) Poly(A)-, DNA-like RNA; no marker was added. A = absorbance at 260 nm 63 .

determined after fractionating the free nucleotides using diethylaminoethyl (DEAE) cellulose columns. With SAS cells, the specific activity of ATP increases gradually within the period studied (Fig. 6A). This continuous increase in the specific activity of ATP in SAS cells presumably reflects the slow uptake of [^{3}H]adenosine by SAS cells. In cells in the nutrient medium, the kinetics of ATP labeling is quite different from that in SAS cells; there is a rapid rise of the specific activity within 30 min after the addition of [^{3}H]adenosine, followed by a gradual decline (Fig. 6B). The rapid rise obviously reflects a rapid uptake and phosphorylation of [^{3}H]adenosine in control cells. Since the pool size of ATP stays fairly constant in both SAS and control cells after EH, the slow decline of specific activity suggests a restricted entry of [^{3}H]adenosine coupled with the increased turnover of nonradioactive RNA at this time of the cell cycle in heat-synchronized cells. This confirms the earlier observation by Stocco [62]. The mechanisms of adenosine uptake and phosphorylation are complex in *Tetrahymena* [11], and are discussed elsewhere [63]. Using the data on the kinetics of the specific activity of ATP, it was calculated that the net rate of synthesis of poly(A)+ RNA in SAS cells is 11.8% of the controls. Thus both groups of cells divide at about the same time with a similar percentage of the cells dividing, in spite of the fact that the net synthesis of poly(A)+ RNA is vastly different. Obviously, net rate of synthesis of mRNA is not proportional to the schedule of cell division in heat-synchronized *Tetrahymena*.

In the final series of experiments, the requirement of RNA synthesis for the first division in heat-synchronized *Tetrahymena* was studied using actinomycin D as an inhibitor of RNA synthesis. SAS cells were treated with actinomycin D (5, 20, or 50 μg/ml) starting at 20 min after EH, and [^{3}H]adenosine was added 10 min later. RNA was extracted after a 15-min pulse and the amount of incorporation into TD-RNA (after MAK column fractionation) was determined for each sample. With this pulse labeling, the radioactive RNA is mainly localized in the nucleus, although polyribo-

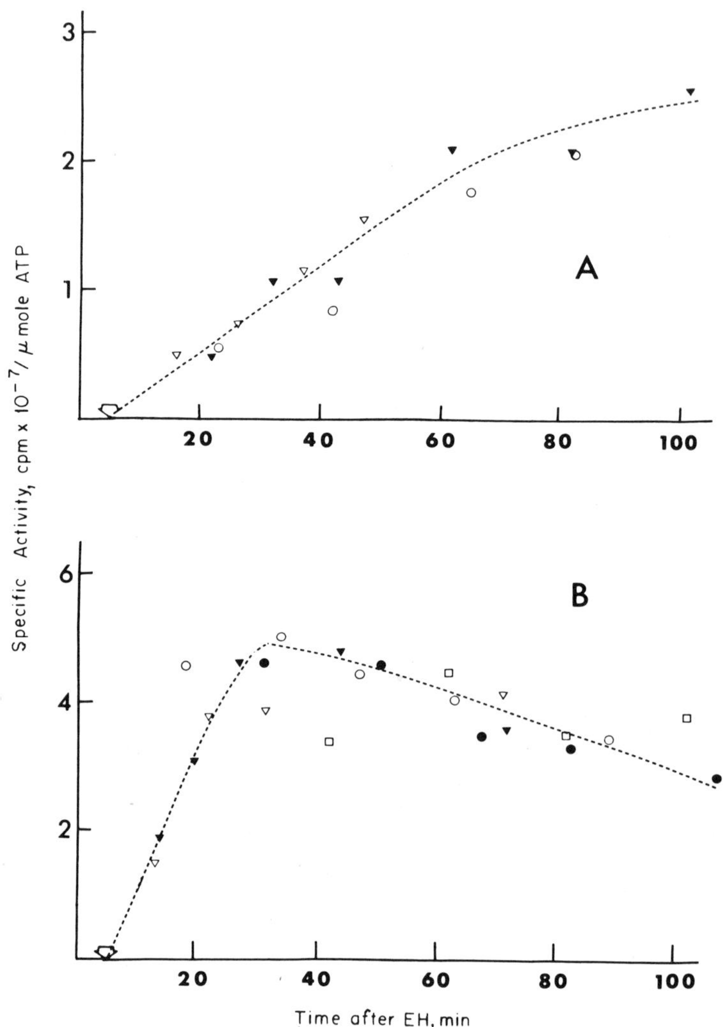

FIG. 6. Kinetics of [^{3}H]adenosine incorporation into ATP in SAS and control cells. [^{3}H]adenosine was added at 5 min after EH into the culture and 80-ml samples were taken at intervals. ATP was isolated with DEAE-cellulose chromatography [64] and the specific activities were determined. (A) Changes in the specific activity of ATP in SAS cells. (B) Changes in the specific activity of ATP in control. Arrows indicate the time of addition of ATP. Different symbols indicate different series of experiments [63].

somal-bound RNA is also detectable [43,64]. The percentage of cells divided in each group of actinomycin D-treated samples was determined using duplicate cultures. In this way RNA synthesis and division can be correlated. The results are shown in Fig. 7A and B. There is a good correlation between the synthesis of TD-RNA and the percentage of cells divided. Similar experiments were also conducted to establish the correlation between the extent of poly(A) synthesis [or poly(A) addition to RNA] and the percentage of cell division. The data are presented in Fig. 7C. There is a fairly good correlation between the RNase-resistant radioactivity [presumably poly(A)] and the percentage of division. These observations strongly indicate that the synthesis of mRNA (or at least DNA-like RNA) is required for the first synchronous division. These are the first such data presented for any developmental system [12].

Because of various unsolved problems, the significance of the present observation is complex. First, it is not known whether the mRNA synthesis that is directly related to division is actually enriched in SAS cells. If one assumes that all cells synthesize mRNA in redundant quantities, it is possible that synthesis of all classes of mRNA is uniformly repressed in SAS cells. Sedimentation studies of poly(A)+ RNA extracted from SAS cells and controls under denaturing conditions have shown that the size classes are about the same, shedding no light on the problem. In procaryotes, recent studies have demonstrated that cells under amino acid deprivation do not synthesize rRNA or mRNA for ribosomal proteins [65-67]. However, comparable information for eucaryotes is not available. At present it is tempting to speculate that the synthesis of mRNA for ribosomal proteins is repressed in SAS cells. This problem is important and warrants further investigation. Second, there is no theoretical basis for believing that the established correlation between the synthesis of mRNA and cell division (as presented here) in a population of cells reflects a causal relationship between the two. The present data are only consistent with the concept that

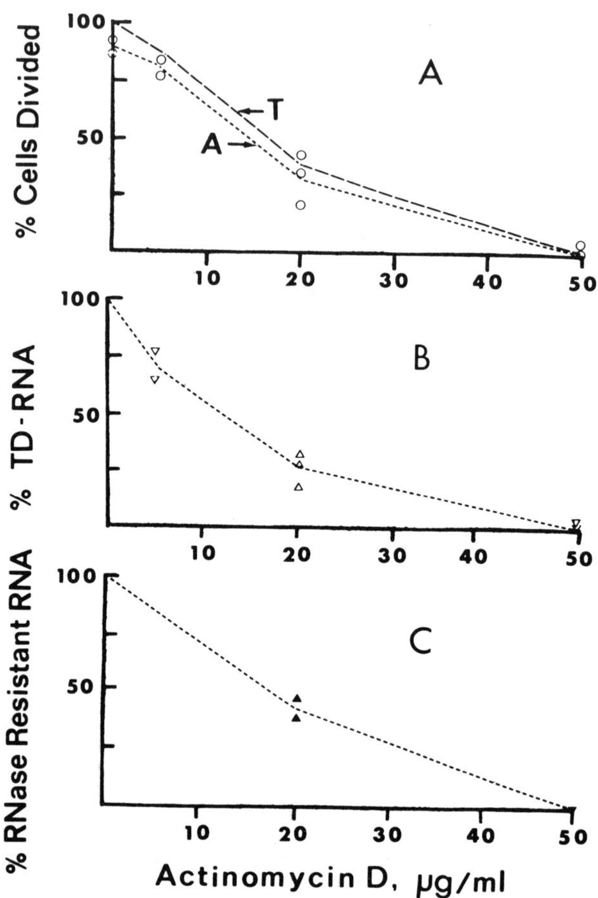

FIG. 7. Relationships between the percentage of SAS cells divided and the amount of radioactivity either in TD-RNA or RNase A- and T_1-resistant RNA. (A) Percentage of cells divided between 40 min and 120 min after EH in different concentrations of actinomycin D added at 20 min after EH. A = actual percentage of division; T = theoretical line calculated using the % of division of untreated SAS control cells as 100. (B) Percentage of relative radioactivity in TD-RNA accumulated in SAS cells treated with actinomycin D at different concentrations. (C) Percentage of relative radioactivity in RNase-resistant RNA accumulated in SAS cells treated with actinomycin D at different concentrations [12].

inherent relationships between the mRNA synthesis and cell division in individual cells are expressed statistically in a cell population as a correlation between the decrease in the extent of RNA synthesis and that of cell division. This problem is difficult to solve when a population of cells is used. Third, the effects of actinomycin D on cell division are not clear. In general, effects of antimitotic agents on cell division have been expressed either in terms of division delay or percentage of inhibition of cell division in a population. In addition, many publications do not distinguish between the delay in the initiation of the division process and the delay in the completion of the division process. When actinomycin D (50 μg/ml) was added at different times after EH, the effects of the drug were expressed either in terms of a complete lack of division or in terms of delay in the furrowing process [12]. Actinomycin D at this concentration does not cause delay in the initiation of the furrowing process [12]. Since this type of analysis can only examine the terminal transcriptional events in relation to cell division, whether or not the earlier transcriptional events are related to the initiation of the furrowing process--and thus timing mechanisms--cannot be assessed. One attractive hypothesis of timing mechanisms involves the linear (or sequential, but not necessarily causally related) transcription of genes [23]. Synthesis of step enzymes during the cell cycle appears to be linked to the linear transcription of DNA [19,68]. Whether or not such a linear transcriptional scheme controls the timing mechanisms of heat-synchronized *Tetrahymena* remains to be studied. An additional problem relating to the effects of actinomycin D on division is the specificity of its action. Although actinomycin D is known to affect nontranscriptional activities of cells [69], there is no such evidence for this effect on the cell division of *Tetrahymena* (for a review see Ref. 2). Fourth, there is a question on the role of terminal transcriptional events on cell division. Since actinomycin D added at the critical time prolongs the furrowing process, it is tempting to speculate that the last mRNA synthesis is related to the furrowing process. Since the furrowing process during the cell division of animal cells requires the formation of contractile machinery [70-73], it is possible that the last mRNAs synthesized during cell division code for contractile proteins. Finally, there is a question on the role of low-mol wt RNA in cell division. In the present study, no clear correlation was established between the synthesis of tRNA and cell division, although this does not exclude the possible involvement of tRNA synthesis with cell division. There is evidence in mammalian cells that low-mol wt RNA (smaller than tRNA) participates in the translational process [74,75]. In heat-synchronized *Tetrahymena*, synthesis of low-mol wt RNA species has been reported [76]. It is possible that the inhibition of RNA synthesis which interferes with the cell division process is related to the inhibition of synthesis of such low-mol wt RNA.

C. Polyribosomes and Messenger RNA

Polyribosomes of heat-synchronized *Tetrahymena* have been studied by various workers [77-80]. Antimitotic agents such as pressure [77], heat or cold shocks [79], or x-ray [80] reduce the total polyribosomal contents, suggesting that these antimitotic agents reduce the rate of protein synthesis. In heat or cold shock-treated cells, reduction in the polyribisomes appears to result from the inhibition of the initiation of translation [79]. However, these treatments do not abolish the polyribosomes completely.

Fewer polyribosomes were recoverable from SAS cells as compared to control cells (Fig. 8A and B). A net incorporation study, as well as the amounts of radioactivity in nascent proteins of polyribosomes, using [^{3}H]leucine or [^{3}H]phenylalanine pulses, have shown that SAS cells incorporate radioactivity at a rate of about one-third of the controls. Considering the fact that the labeled amino acids are vastly diluted in control cells in the amino acid-rich PPL medium, lowering the specific activity of cytoplasmic

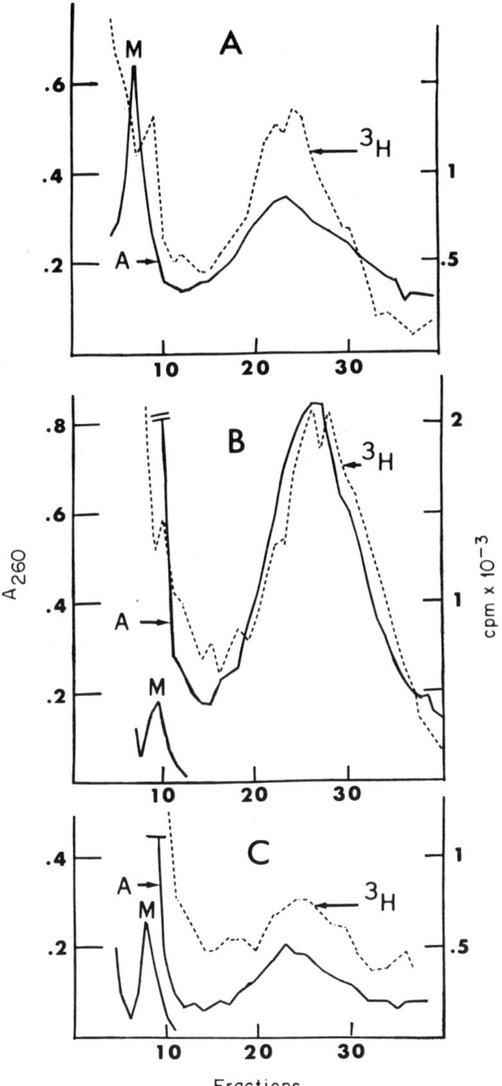

FIG. 8. Polyribosomes of heat-synchronized *Tetrahymena*. (A) SAS cells (1,760,000 cells/20 ml) were pulsed with [^{3}H]phenylalanine (5 μCi/ml) for 45 min starting at 5 min after EH. Polyribosomes were isolated and fractionated by sucrose gradient centrifugation. Postmitochondrial supernatant (4 ml) was layered on top of linear sucrose gradient (15 to 40%) and centrifuged for 3 h at 25,000 rpm. For each fraction, A_{260} and TCA-precipitable radioactivity was determined. (B) Control cells (1,760,000 cells/20 ml) were pulsed with [^{3}H]phenylalanine in the same manner as (A). The monoribosome profile was plotted in a scale one-tenth of polyribosomes. (C) SAS cells (1,800,000 cells/20 ml) were pulsed with [^{3}H]phenylalanine in the same manner as (A). Additional heat shock was given starting at 30 min after EH. Polyribosomes were isolated and fractionated. The monoribosomes profile was plotted in a scale one-tenth of the polyribosomes. M = monoribosomes; A = absorbance at 260 nm [81].

pools, the actual rate of protein synthesis in SAS cells must be much lower than that of control cells. These observations indicate that the schedule of cell division is not altered by the lower rate of protein synthesis occurring in SAS cells. Here the situation is quite analogous to the synthesis of poly(A)+ RNA in SAS cells. It appears likely that synthesis of ribosomal protein does not occur in SAS cells because no rRNA is synthesized and no labeled ribosomes appear after 2 h of pulse labeling with [^{3}H]leucine [81]. Experiments involving the fractionation of soluble proteins with SDS gels after a 5- to 50-min EH pulse with [^{3}H]phenylalanine revealed quite different labeling patterns between SAS and fed control cells, although no detectable difference was apparent in the protein-staining patterns [82].

When SAS cells are heat shock-treated (one additional heat shock started at 30 min after EH), there is a further reduction in the amount of polyribosomes; the amount of radioactivity in nascent polypeptides is also reduced when compared with untreated SAS cells (Fig. 8C). If the reduction in the rate of protein synthesis, which is reflected in the reduction of either the amount of recoverable polyribosomes or the rate of net incorporation of labeled amino acids, is responsible for the excess delay phenomenon, this inhibition of protein synthesis must be specific. A reduction in protein synthesis by starvation does not show inhibition of cell division, whereas heat shock-induced reduction of protein synthesis does inhibit division. Whether the "excess delay" phenomenon in *Tetrahymena* induced by treatments with the heat shocks, or with a pulse treatment with p-fluorophenylalanine, is due to the inhibition of protein synthesis or due to the inactivation of division protein is by no means established. p-Fluorophenylalanine is not a very effective inhibitor of protein synthesis [28] and is actually incorporated into various proteins of *Tetrahymena* [83,84].

At present little is known about the stability of mRNA in heat-synchronized *Tetrahymena*. Byfield and Scherbaum [85,86], using an actinomycin D-chasing method, studied the decay of RNA under normal and heat shock treatment conditions and found that RNA during the heat shock treatment decays more rapidly. Whether this decay is actually due to the mRNA, HnRNA, rRNA, or ribosomal precursor RNA remains to be solved. In SAS cells, there is no measurable decay of RNA at 28° in the presence of actinomycin D (50 μg/ml), while at 34° about 50% of total TCA-insoluble material decays with a half-life of about 25 min under similar chasing conditions.

III. FINAL COMMENTS

The present study has demonstrated that prolonged starvation during the heat shock synchronization regime of *Tetrahymena* abolishes the synthesis of rRNA and reduces the net rate of synthesis of DNA-like RNA to one-eighth of that of control cells. These cells (SAS cells) retain the capacity to divide synchronously without delay and the schedule of the first division is comparable to that of control cells. Protein synthesis is also markedly reduced in these cells. It is conceivable that the synthesis of mRNA and protein required for division is considerably enriched in SAS cells. One hypothesis that can be proposed from the present study is that synthesis of mRNA involved in division is separated from the mRNA synthesis which is regulated by the availability of nutrients [47]. A major candidate belonging to the latter category is mRNA coding for ribosomal proteins. Another group of proteins that may belong to the latter category is lysosomal enzymes which are involved in feeding activities. In SAS cells, cellular RNase activity does not increase at all after EH whereas the RNase activity increases after EH in fed control cells with the levels of the increase depending upon the degree of nutrient concentrations [87]. Proteins present in cilia and membranes may also fit into this category. As mentioned earlier, some cells synthesize nuclear DNA after EH in heat-synchronized *Tetrahymena*. Autoradiographic analysis revealed that some SAS cells also incorporate [^{3}H]thymidine into nuclei and cytoplasm although both the percentage of labeled cells and the level of radioactivity are much less than those in fed control cells [81]. Thus histone synthesis may not be regulated similarly.

Another problem with SAS cells is that correlations between mRNA or protein synthesis and division using inhibitors do not prove that cell division in SAS cells requires synthesis of any specific mRNA and protein after EH. To prove such requirements, two lines of research are currently being conducted. One is to establish a bioassay system by microinjecting isolated macromolecules into *Tetrahymena* whose own synthetic activities are inhibited at the G_2 phase. The other is to determine the synthesis of specific proteins which are known to be associated with contractile activities, on the assumption that a contractile ring is formed during the cell division of *Tetrahymena*. The protein we are studying at present is actin. However, an important question is what information is present in the cell cortex which enables the actin fiber to assemble at the furrowing region. In *Tetrahymena* the division furrow is formed in the predesignated area in every successive division. In sea urchin eggs, the positions of the cleavage furrow during the initial cleavages follow a well-defined pattern. Thus answers to these questions will bring us to one of the most important problems of cell biology, that of cortical differentiation.

REFERENCES

1. Zeuthen, E., 1971. *In* Advances in Cell Biology, Vol. 2, D. M. Prescott, L. Goldstein and E. McConkey (eds.), Appleton-Century-Crofts, New York, p. 111.
2. Zeuthen, E., and L. Rasmussen, 1972. *In* Research in Protozoology, Vol. 4, T. T. Chen (ed.), Pergamon Press, New York, p. 9.
3. Zeuthen, E., 1974. *In* Cell Cycle Controls, G. M. Padilla, I. L. Cameron and A. M. Zimmerman (eds.), Academic Press, New York, p. 1.
4. Corliss, J. O., 1974. *In* Biology of *Tetrahymena*, A. M. Elliott (ed.), Dowden, Hutchinson and Ross, Stroudsburg, Pa, p. 1.
5. Lwoff, A., 1923. C. R. Acad. Sci. Paris 176: 928-930.
6. Frankel, J., and N. E. Williams, 1974. *In* Biology of *Tetrahymena*, A. M. Elliott (ed.), Dowden, Hutchinson and Ross, Stroudsburg, Pa, p. 375.

7. Kidder, G. W., and V. C. Dewey, 1951. *In* Biochemistry and Physiology of Protozoa, Vol. 1, A. Lwoff (ed.), Academic Press, New York, p. 323.

8. Zeuthen, E., 1971. Exp. Cell Res. 68: 49-60.

9. Hill, D. L., 1972. The Biochemistry and Physiology of *Tetrahymena*, Academic Press, New York.

10. Elliott, A. M. (ed.), 1974. Biology of *Tetrahymena*, Dowden, Hutchinson and Ross, Stroudsburg, Pa.

11. Scherbaum, O. H., and E. Zeuthen, 1954. Exp. Cell Res. 6: 221-227.

12. Yuyama, S., 1975. Exp. Cell Res. 90: 381-391.

13. Zeuthen, E., 1964. *In* Synchrony in Cell Division and Growth, E. Zeuthen (ed.), Interscience, New York, p. 99.

14. Thormar, H., 1959. CRT Lab. Carlsberg 31: 207-226.

15. Rasmussen, L., and E. Zeuthen, 1962. CRT Lab. Carlsberg, 32: 333-358.

16. Frankel, J., 1969. J. Cell Physiol. 74: 135-148.

17. Cann, J. R., 1963. CRT Lab. Carlsberg 33: 431-453.

18. Zeuthen, E., and N. E. Williams, 1969. *In* Nucleic Acid Metabolism, Cell Differentiation and Cancer Growth, E. V. Cowdry and S. Seno (eds.), Pergamon Press, New York, p. 203.

19. Mitchison, J. M., 1971. The Biology of the Cell Cycle, Cambridge Univ. Press, London, England.

20. Tobey, R. A., D. F. Petersen and E. C. Anderson, 1971. *In* The Cell Cycle and Cancer, R. Gaserga (ed.), Marcel Dekker, New York, p. 309.

21. Mazia, D., 1961. *In* The Cell, Vol. 3, J. Brachet and A. Mirsky (eds.), Academic Press, New York, p. 77.

22. Winfree, A. T., 1975. Nature 253: 315-319.

23. Ehret, C. F., and E. Trucco, 1967. J. Theor. Biol. 15: 240-262.

24. Pavlidis, T., 1969. J. Theor. Biol. 22: 418-436.

25. Njus, D., F. M. Sulzman, and J. W. Hastings, 1974. Nature 248: 116-120.

26. Watanabe, Y., and M. Ikeda, 1965. Exp. Cell Res. 39: 443-452.

27. Watanabe, Y., and M. Ikeda, 1965. Exp. Cell Res. 39: 464-469.

28. Crocket, R. L., P. B. Dunham, and L. Rasmussen, 1965. CRT Lab. Carlsberg 34: 451-485.

29. Lowe-Jinde, L., and A. M. Zimmerman, 1971. J. Protozool. 18: 20-23.

30. Geilenkirchen, W. L. M., 1964. Biol. Bull. 127: 370.

31. Yuyama, S., 1965. Biol. Bull. 129: 429.

32. Yuyama, S., 1971. Biol. Bull. 140: 339-351.

33. Mazia, D., and E. Zeuthen, 1965. CRT Lab. Carlsberg 35: 341-361.

34. Sakai, H., and K. Dan, 1959. Exp. Cell Res. 16: 24-41.

35. Ikeda, M., and Y. Watanabe, 1965. Exp. Cell Res. 39: 584-590.

36. Dan, K., 1966. *In* Cell Synchrony, I. L. Cameron and G. M. Padilla (eds.), Academic Press, New York, p. 307.

37. Dan, K., and M. Ikeda, 1971. Dev. Growth Diff. 13: 285-301.

38. Hjelm, K. K., and E. Zeuthen, 1967. CRT Lab. Carlsberg 36: 127-160.

39. Jeffery, W. R., J. Frankel, L. C. DeBault and L. M. Jenkins, 1973. J. Cell Biol. 59: 1-11.

40. Cleffmann, G., 1975. J. Cell Biol. 66: 204-209.

41. Conner, R. L., and M. J. Koroly, 1974. *In* Biology of *Tetrahymena*, A. M. Elliott (ed.), Dowden, Hutchinson and Ross, Stroudsburg, Pa, p. 123.

42. Leick, V., and P. Plesner, 1968. Biochim. Biophys. Acta 169: 398-408.

43. Yuyama, S., and A. M. Zimmerman, 1972. Exp. Cell Res. 71: 193-203.

44. Leick, V., 1967. CRT Lab. Carlsberg 36: 113-126.

45. Byfield, J. E., and O. H. Scherbaum, 1967. Exp. Cell Res. 49: 202-206.

46. Byfield, J. E., and O. H. Scherbaum, 1968. Nature 218: 1271-1273.

47. Yuyama, S., and M. A. Tarnowka, 1973. Cytobios 7: 113-126.

48. Cerroni, R. E., and E. Zeuthen, 1962. Exp. Cell Res. 26: 604-605.

49. Frankel, J., 1965. J. Exp. Zool. 159: 113-148.

50. Lazarus, L. H., M. R. Levy, and O. H. Scherbaum, 1964. Exp. Cell Res. 36: 672-676.

51. Cleffmann, G., 1965. Z. Zellforsch. 67: 343-350.

52. Mita, T., 1965. Biochim. Biophys. Acta 113: 182-185.

53. Moner, J. G., 1967. Exp. Cell Res. 45: 618-630.

54. Nachtwey, D. S., and W. J. Dickinson, 1967. Exp. Cell Res. 47: 581-595.

55. Nachtwey, D. S., 1965. CRT Lab. Carlsberg 35: 25-35.

56. Watanabe, Y., 1963. Jap. J. Med. Sci. Biol. 16: 107-124.

57. Frankel, J., 1970. J. Exp. Zool. 173: 79-100.

58. Darnell, J. R., W. R. Jelinek, and G. R. Molloy, 1973. Science 181: 1215-1221.

59. Weinberg, R. A., 1973. Ann. Rev. Biochem. 42: 329-354.

60. Brawerman, G., 1974. Ann. Rev. Biochem. 43: 621-642.

61. Greenberg, J. R., 1975. J. Cell Biol. 64: 269-288.

62. Stocco, D. M., 1972. Ph.D. Thesis, Univ. of Toronto, Toronto.

63. Yuyama, S., and J. Lipman. Manuscript in preparation.

64. Hermolin, J., 1971. Ph.D. Thesis, Univ. of Toronto.

65. Dennis, P. P., and M. Nomura, 1974. Proc. Nat. Acad. Sci. USA 71: 3819-3823.

66. Dennis, P. P., and M. Nomura, 1975. Nature 255: 460-465.

67. Ryan, A. M., and E. Borek, 1971. Progr. N.A. Res. Mol. Biol. 11: 193-228.

68. Tauro, P., E. Schweizer, R. Epstein, and H. O. Halvorson, 1969. *In* The Cell Cycle: Gene Enzyme Interactions, G. M. Padilla, G. L. Whitson, and I. L. Cameron (eds.), Academic Press, New York, p. 101.

69. Sawicki, S. G., and G. C. Godman, 1971. J. Cell Biol. 50: 746-761.

70. Rappaport, R., 1971. Int. Rev. Cytol. 31: 169-213.

71. Perry, M. M., H. A. John, and N. S. T. Thomas, 1971. Exp. Cell Res. 65: 249-253.

72. Forer, A., and O. Behnke, 1972. Chromosoma 39: 175-190.

73. Schroeder, T., 1973. Proc. Nat. Acad. Sci. USA 70: 1688-1692.

74. Heywood, S. M., D. S. Kennedy, and A. G. Bester, 1974. Proc. Nat. Acad. Sci. USA 71: 2428-2431.

75. Reichman, M., and S. Penman, 1973. Proc. Nat. Acad. Sci. USA 70: 2678-2682.

76. Hellung-Larsen, P., S. Frederiksen, and P. Plesner, 1971. Biochim. Biophys. Acta 254: 78-90.

77. Hermolin, J., and A. M. Zimmerman, 1969. Cytobios 3: 247-256.

78. Hartman, H., and R. W. Dowben, 1970. Biochem. Biophys. Res. Commun. 40: 964-967.

79. Klemperer, H. G., and V. A. Rose, 1974. Nature 248: 443-446.

80. Paul, I., and A. M. Zimmerman, 1975. Radiat. Res. 61: 144-157.

81. Yuyama, S. Manuscript in preparation.

82. Grant, N. J., and S. Yuyama, unpublished results.

83. Zeuthen, E., and L. Rasmussen, 1966. J. Protozool. 13 (suppl.): 29.

84. Leick, V., 1969. Eur. J. Biochem. 8: 215-220.

85. Byfield, J. E., and O. H. Scherbaum, 1967. Proc. Nat. Acad. Sci. USA 57: 602-606.

86. Byfield, J. E., and Y. C. Lee, 1970. J. Protozool. 17: 445-453.

87. Tarnowka, M. A., and S. Yuyama, unpublished results.

THE NONHISTONE PROTEINS OF CHROMATIN DURING GROWTH AND DIFFERENTIATION IN *PHYSARUM POLYCEPHALUM*

Wallace M. LeStourgeon

Department of Molecular Biology
Vanderbilt University
Nashville, Tennessee

I. INTRODUCTION

A. Aspects of Growth and Differentiation

Physarum polycephalum is one of over 425 morphologically and physiologically linked protozoan-like organisms frequently referred to as the acellular, plasmodial, or "true" slime molds (in contrast to the cellular slime molds typified by *Dictyostelium discoideum*). The term slime mold erroneously suggests that *Physarum* and its relatives are fungi. It is true that the mycologists are largely responsible for characterizing and classifying these organisms, and because of this they are generally described as being members of the fungal class Myxomycetes [1-3]. Interestingly, however, DeBary (1887)--one of the founders of the science of mycology [1]--considered the slime molds to be more animal than fungus and for developmental and biochemical reasons most of today's mycologists and protozoologists agree with this interpretation. For a more complete review of the history, biology, and taxonomy of these organisms, the reader may consult refs. 1 to 4.

Physarum polycephalum, like other myxomycetes yet unlike the saprophytic true fungi, is a motile free-living organism which obtains nutrients by phagocytosis (bacteria and organic material) or, if cultured in axenic liquid media, by pinocytosis. The free-living assimilative growth phase of the life cycle is a true plasmodium (Fig. 1). For a mental picture of the plasmodium, one can consider the hypothetical case of an amoeba having lost its ability to undergo cell division following mitosis. In such a case, nuclear number and protoplasmic volume increase exponentially but cell number remains constant. Depending on the age and size of the plasmodium, only a few, hundreds, or millions of diploid nuclei are encased in a common cytoplasm. Should the need arise, single plasmodia of *Physarum* can be cultured to a diameter of several feet in the laboratory and some species may in nature reach 15 ft in diameter. Plasmodial growth is indeterminate and dependent only on nutrient supply and suitable ambient conditions. Unlike fungi, the plasmodium is not encased in a chitinous or cellulose wall but, like animal cells, is bounded only by a delicate plasma membrane. When plasmodia are cultured in axenic liquid media on a gyratory platform in baffled flasks, the plasmodial spheres continuously fragment into "microplasmodia" as they grow to a critical mass (ca 0.3 to 0.5 mm in diameter). Today in most laboratories where *Physarum* is used as

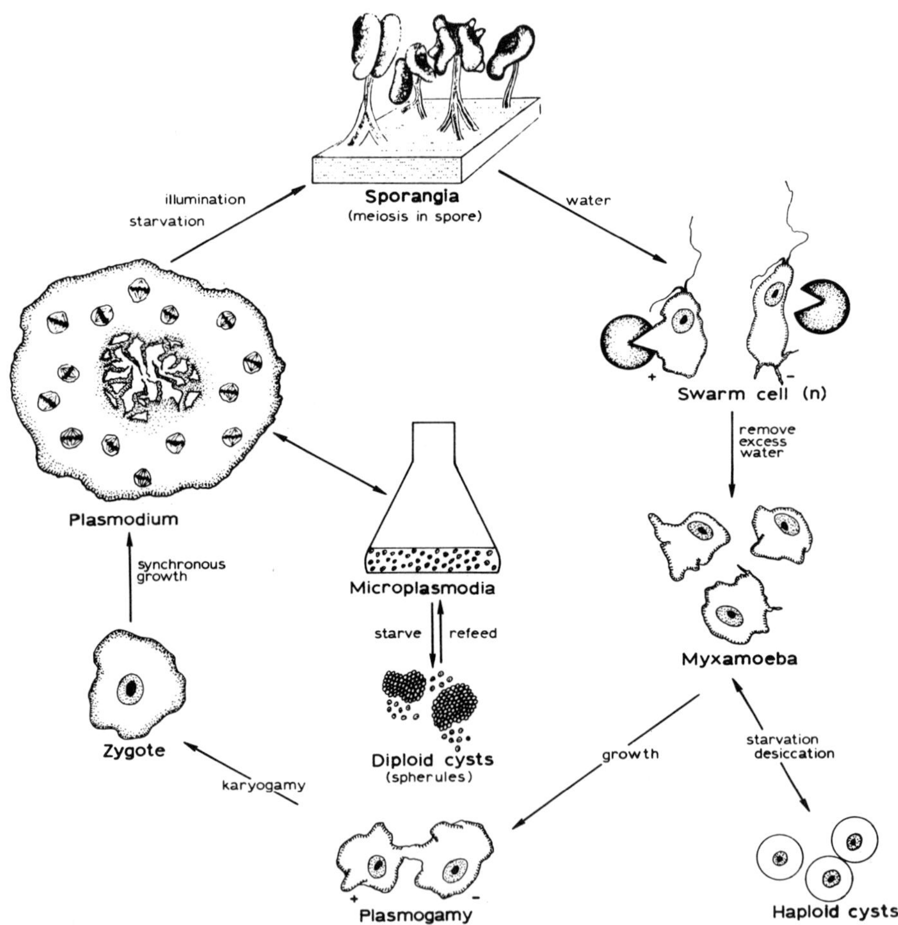

FIG. 1. Life cycle of *Physarum polycephalum*: nuclei in plasmodium shown in synchronous mitosis, structures not drawn to scale.

an experimental tool for biochemical studies, culture stocks are maintained as microplasmodia in liquid media [5].

Although large amounts of plasmodia can be obtained from shake flask cultures [6], the microplasmodia are usually spread on nutrient-saturated filter paper disks and allowed to fuse into a single "macroplasmodium" [7]. Depending on the inoculation size, the newly formed macroplasmodium--through new growth and active migration--can reach an overnight diameter of 12 cm and have a wet wt of 5 g. Since the nuclei in a single macroplasmodium are encased in a common cytoplasm and undergo naturally synchronous mitosis, biochemically speaking the investigator has in hand a single cell weighing 5 g. As will be discussed in more detail elsewhere, the phenomenon of naturally synchronous mitosis can be credited with attracting many cancer researchers and molecular biologists to *Physarum* as a model eucaryote for studies on growth, cell cycle, and differentiation.

Growing plasmodia can undergo two morphologically and physiologically distinct forms of differentiation (Fig. 1). If macroplasmodia growing on nutrient-saturated filter paper disks are transferred to a nonnutrient medium and incubated in darkness for 4 to 7 days, then exposed to light [8], the plasmodium will transform itself over a 12- to 14-h period into hundreds of elevated sporangia [5,9]. A final synchronous mitosis occurs in each newly formed sporangium [9] and each daughter nucleus (about 5×10^4 per sporangium in some species [10]) is then encased in a highly pigmented and resistant spore wall by a process of cytoplasmic cleavage [11]. Meiosis occurs inside the spore over a period of 24 to 48 h following cleavage [12,13]. Three of the meiotic daughter nuclei degenerate leaving an uninucleate haploid spore, which may remain viable for many years. The haploid nucleus contains 20 to 25 chromosomes and a total DNA content of 0.6 pg/nucleus [14].

A second form of synchronous differentiation (diploid encystment) can be induced by starvation of microplasmodia in liquid shake culture [5,15] or by desiccation of macroplasmodia growing on nutrient substrates (Fig. 2). Diploid encystment has been described as scleritization by mycologists [1] and "spherulation" by bio-

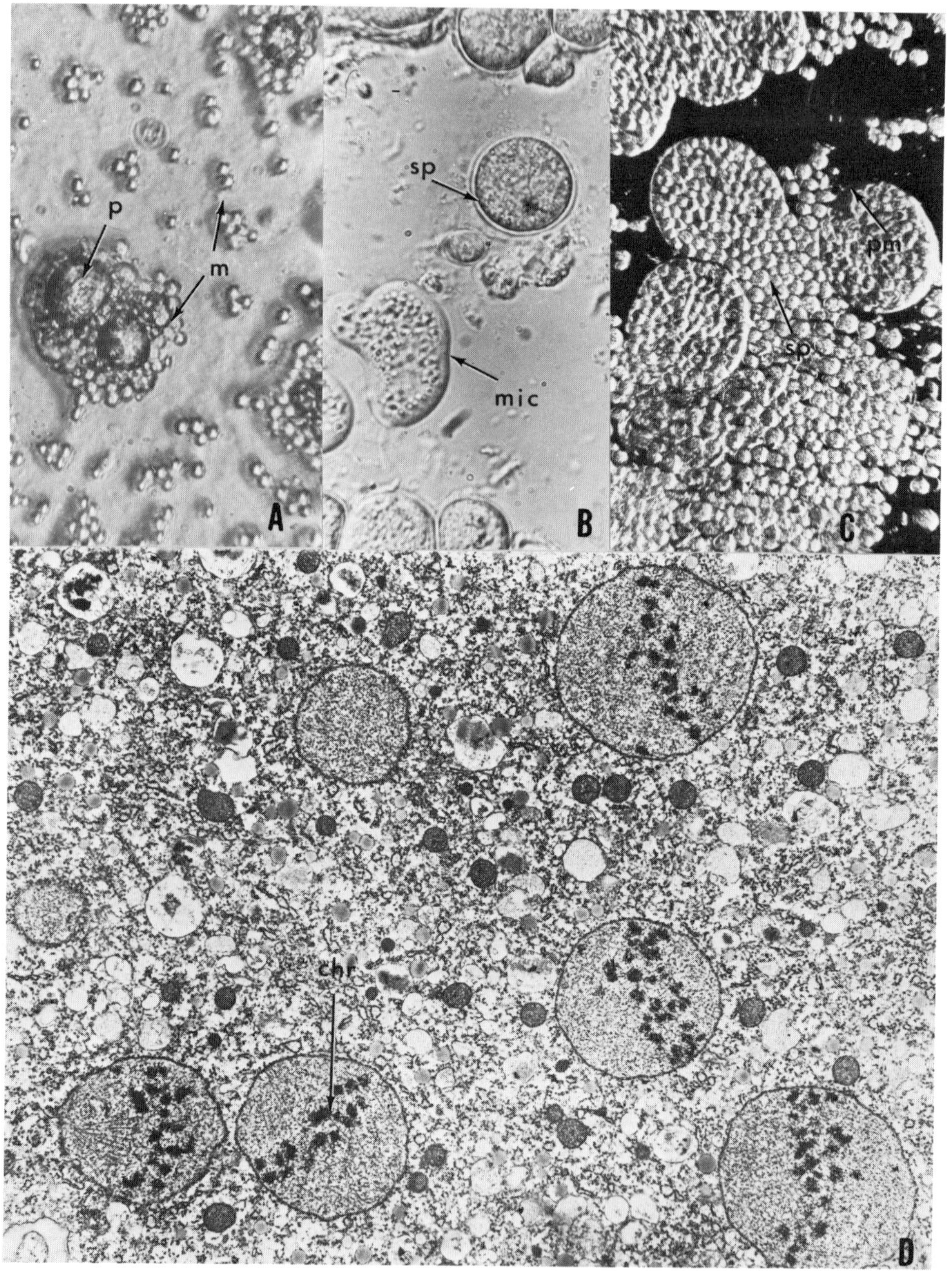

FIG. 2. (A) Myxamoeba (M) and newly formed plasmodia (P) on agar substrate. (B) Diploid cysts "spherules" (sp), and a recently excysted spherule or microplasmodium (mic). (C) Diploid cysts "spherules" spilling out of original plasma membrane of microplasmodia; "spherules" (sp), plasma membrane (pm). (D) Electron micrograph of synchronous mitosis in a myxomycete; note chromosomes (chr) in metaphase and microtubules in some nuclei. Note also that the nuclear membrane does not disappear during mitosis.

chemists [5,15]. Briefly, in response to adverse stimuli the cytoplasm of growing plasmodia is partitioned into hundreds of thick-walled cysts (Fig. 2B and C), usually containing between one and five diploid nuclei per cyst. For experimental purposes diploid encystment is usually induced to occur synchronously by transferring actively growing microplasmodia in shake flasks to a salts-citrate solution [15]. Depending on the incubation temperature, cyst formation may begin 16 to 21 h after the initiation of stepdown culture and be essentially complete in 4 to 6 h. If actively growing plasmodia are placed in a refrigerator at 4°, then a process of differentiation perhaps fundamentally identical to encystment occurs; however, the cysts induced by refrigeration are not encased in a resistant thick wall as are the cysts induced by starvation. Unlike the haploid spores, the diploid cysts are not melanized, and simply placing the cysts in distilled water or fresh nutrient solution induces excystment and the formation of new plasmodia (Fig. 2B). Nutrient-induced excystment represents a classic example of eucaryote cell dedifferentiation leading to renewed cell proliferation. As will be documented below, this process of dedifferentiation is physiologically and biochemically not unlike the nutrient and dilution-induced proliferation of contact-inhibited mammalian cells in tissue culture.

As diagrammed in Fig. 1, the uninucleate haploid spores germinate when placed in water for several hours. In the presence of excess liquid the emerging protoplasts are biflagellate and highly motile. In the absence of excess water, such as on a moist agar surface, the flagella are resorbed and the gametes become amoeboid (myxamoeba). The amoeboid gametes are themselves free-living and usually obtain nutrients by phagocytizing bacteria. More recently it has been possible to mass culture myxamoeba axenically [16]. In response to nutrient depletion or desiccation the gametes can also encyst and remain viable for periods of well over a year. In the normal scheme of events, myxamoeba proliferate by mitosis and cell division and, on the establishment of a suitable cell population, gametes compatible for the mating-type allele undergo plasmogamy and karyogamy with the establishment of a uninucleate diploid protoplast. In the diploid state mitosis is uncoupled from cell division and a new multinucleate plasmodium is formed (Fig. 2A). Depending on the strain and incubation temperature, mitosis in the plasmodium occurs synchronously every 8 to 10 h [17-19].

The actively growing plasmodium of *Physarum*, existing as a naked syncytial mass of migrating protoplasm, is perhaps the simplest and most undifferentiated form of eucaryotic life known. In nature plasmodia are found in the spring and summer months on moist forest floors and on decaying logs. The various forms of differentiation displayed by the plasmodium are clearly survival mechanisms in times of nutrient depletion, drought, or rapid change in ambient temperature. To the researcher interested in understanding the most fundamental genetic mechanisms which direct and regulate growth and differentiation, *Physarum* is particularly attractive due to its natural mitotic synchrony, nuclear identity, inducible synchronous differentiation, and the ability to obtain large quantities of physiologically and biochemically homogeneous material. Novel procedures for experimentally manipulating growth and differentiation in *Physarum* can be found in an excellent review by Mohberg [19] and in many of the references cited in this article.

B. Studies on Genome Function

Perhaps the most primitive and fundamental form of eucaryote cell differentiation is the ability of actively dividing cells to differentiate into quiescent states and the ability of cells in nonproliferative states to dedifferentiate into forms committed to active proliferation. This basic pattern of cell differentiation holds not only for microbes but for many highly evolved mammalian cells in complex organs. Today, of course, much attention is focused on this most basic form of cell differentiation, since neoplastic disease seems to result from aberrations in the normal gene regulatory mechanisms which control cell proliferation and quiescence.

To the molecular biologist researching the biochemical mechanisms which regulate genetic activity, several fundamental problems exist. First, current technology dictates that some minimum amount of tissue must be available before specific constituents can be purified and characterized. Even in the most sophisticated radioactive tracer techniques or in situ hybridization experiments, still the researcher is frequently faced with the problem of obtaining sufficient amounts of purified known reactants. In studies on the mechanisms which regulate genetic activity, this problem is especially real since in most cases nonstatic differentiating tissues must be used. A second frequently encountered problem is related to and in fact compounds the first; namely, many experimental systems used in studying differentiation processes are not adequately amplified for the phenomenon under study. As a case in point, attention can be focused on perhaps the most widely used system for studying the biochemistry of new cell proliferation, namely, the induction of cell proliferation in liver by surgically removing a portion of the liver (partial hepatectomy). Through autoradiography it has been demonstrated that at best only about 30% of the liver cells newly synthesize DNA and go on to divide about 20 to 50 h after hepatectomy [25]. While the great majority of liver is comprised of parenchymal hepatocytes, still methods for isolating cell organelles yield some material from connective, vascular, and duct cells, as well. In essence, new fluxes in the constituents of replicating cells may be masked by the natural constituents of the more predominant and varied nonproliferating cells. In the case of posttranscriptional or posttranslational events such as enzyme synthesis or activation, diagnostic assays frequently have sufficient sensitivity to detect trauma-induced changes. However, it now seems apparent that gene-regulatory macromolecules are present at normally undetectable levels even in cells where new patterns of genetic expression have been induced. A further handicap in an experimental system such as liver or cells in tissue culture is that the cells exist in a random distribution with respect to their point in the cell cycle at any given time.

This brief example emphasizes the need for an experimental system where all of the tissue responds to a specific stimulus with absolute synchrony. As stated previously, the plasmodium of *Physarum* can be defined as a single cell, since the lack of a cellular substructure facilitates biochemical homogeneity throughout the cytoplasm, a fact proven by the synchronous division of millions of nuclei at predictable times.

In the case of studies on the proteins of eucaryote chromatin, many experiments (to be summarized later) have demonstrated that at least some of the nonhistone proteins apparently function in altering the patterns of genetic expression. This evidence is as yet only correlative, and in eucaryotes a specific gene-regulatory protein has not been characterized. In order to determine the function of a specific chromosomal protein, to understand its mechanism of action, and ultimately to gain manipulatory control over specific genes, it will be necessary to have in hand purified native macromolecules for use as experimental probes. In *Physarum* or in mammalian cells there are hundreds of nonhistone proteins associated with the genetic complex (chromatin). Some of these proteins are present at very high intranuclear concentrations (1×10^6 copies) while others just fall into detectable ranges (5×10^4 copies) even when high-resolution electrophoretic procedures are used [26]. Even if as many as 30,000 copies of a single regulatory protein were required for an altered gene function, still frighteningly large amounts of nuclei would be needed before isolation and purification could be achieved. Through simple calculations it can be shown that in *Physarum* one would need approximately 13 g dry wt nuclei (6×10^{11} nuclei) to obtain (assuming 100% recovery) only 1 mg of a 34,000-mol wt protein if the protein were present at an intranuclear concentration of 30,000 copies per nucleus. This amount of dry wt nuclei extrapolates to about 240 ml wet-packed nuclear volume. Since purification procedures are rarely highly efficient, realistically the amount of nuclei would need to be several times the above figure. Using mammalian cells in tissue culture or laboratory animals, the cost of obtaining the above quantities of nuclei would be in the thousands of dollars whereas in *Physarum* the cost would be only a few hundred dollars. A routine daily nuclear isolation from *Physarum* can yield 20 ml wet-packed nuclei. More importantly all these nuclei are physiologically and biochemically identical with respect to their point in interphase, mitosis, or differentiation.

C. Chromosomal Proteins and the Structure and Regulation of the Genome

A fundamental concept in molecular biology and biophysics is the direct correlation between the structure of macromolecules and their function. Clearly, knowledge of the structure of macromolecules is a prerequisite to thoroughly understanding their function and mechanism of action. It follows then that if the structure of the chromatin fiber were known we would be imminently closer to elucidating the mechanisms regulating which genes will or will not be transcribed during particular developmental events.

Largely through this reasoning much excitement has recently been generated, for within the past year several laboratories researching the structure of chromatin have published findings which discount the widely accepted DNA "supercoil" and establish a new model for the organization of DNA and histone in chromatin [27-36]. As an outgrowth of the recent discoveries and adding to their significance, it now seems that a fundamental structural role for most of the basic histone proteins has finally been established. Briefly, the new view of the chromatin fiber depicts a "beads on a string" configuration, the string being either a short stretch of naked duplex DNA or DNA complexes with histone H1 and the beads formed by 200 to 230 base pairs of DNA packed in an ordered histone complex containing two molecules each of H2A, H2B, H3, and H4. The beads are about 70 to 100 Å in diameter and the distance between the beads is about 140 Å. The packing ratio of DNA in the beads is between 6.5:1 and 7.5:1.

While there are many problems yet to be solved, having a preliminary knowledge of chromatin organization will help in planning the most informative experiments. With the knowledge that histones do not recognize and bind specific gene sequences and that they are primarily involved in a DNA-packing mechanism, it still seems apparent that specific macromolecules are required to initiate the transcription of unique DNA sequences perhaps by destabilizing or preventing the histone DNA-packing phenomenon.

Unlike the histones which are few in number, the nonhistone proteins are highly heterogeneous. The nonhistone proteins which are present at high intranuclear concentrations (2×10^5 to 1×10^6 copies per nucleus) appear to be involved in nuclear processes other than the regulation of specific genes [37-39]. At least two of the major nonhistone proteins of mammalian chromatin are components of nuclear ribonuclear protein particles (see Fig. 3) which are thought to function in the processing of heterogeneous nuclear RNA (HnRNA) [40]. Other of the major nonhistones probably comprise portions of ribosomal precursor particles, the nuclear pore complex, and some are proteins involved in metaphase chromosome structure and motility (i.e., α and β tubulin, actin, kinetochore complex, etc.). A wide range of intranuclear enzymes have been described and at least some of the proteins present at moderate concentrations (6×10^4 to 5×10^5 copies per nucleus) may well be enzymes. Clearly the DNA and RNA polymerases are nonhistone proteins.

It has been suggested, however, that some of the proteins which can be observed as major bands in stained gels may--through plieotropic mechanisms (perhaps structural alterations of chromatin)--quantitatively regulate genetic activity [37-39]. In addition to the major nuclear proteins of *Physarum* and other eucaryotes there are many polypeptides whose existence can only be detected through tracer techniques [41] (less than 30,000 copies per nucleus). These proteins may function and bind specific regulatory sequences in the genome and alter the patterns of RNA synthesis in growing and differentiating cells [42].

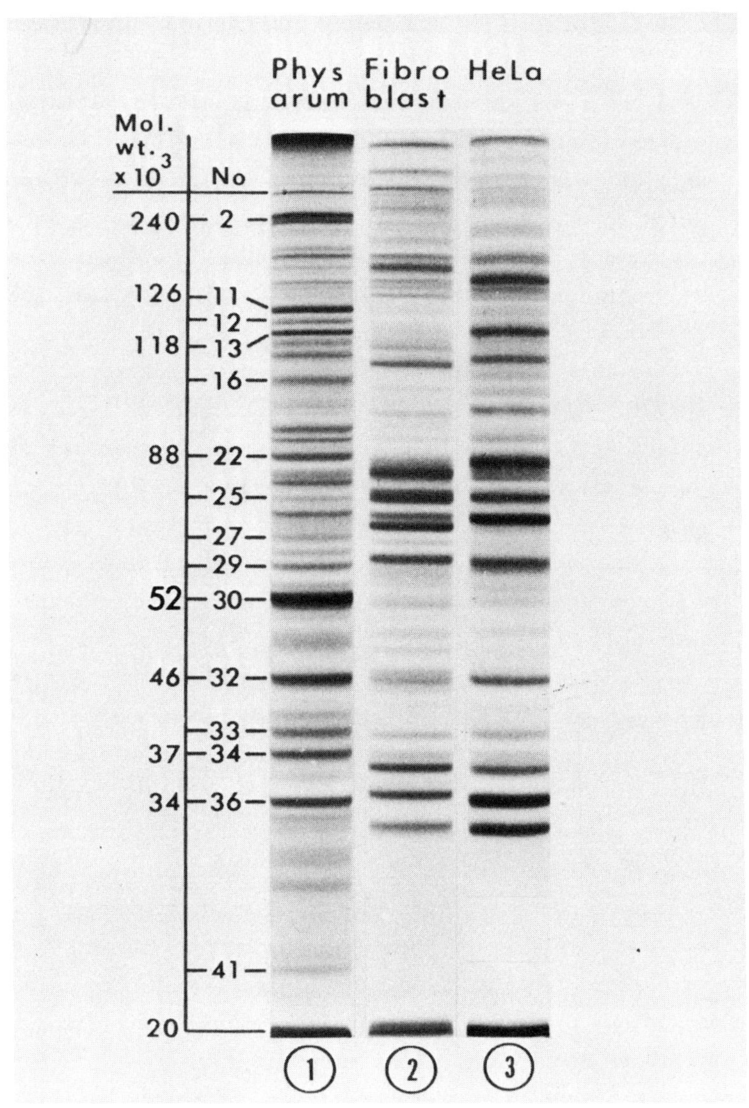

FIG. 3. Electrophoretic profiles of the residual nonhistone proteins of *Physarum*, mouse embryo fibroblasts, and HeLa cells. Note that in the higher-mol wt regions 52,000 daltons and above, numerous proteins are present at lower concentrations. Note that most of the species-specific differences occur in the higher-mol wt proteins. The two major lower-mol wt proteins (32,000 and 34,000 daltons) in fibroblasts and HeLa cells are the major constituents of 30- to 40-S nuclear ribonucleoprotein particles--"informosomes." Band 36 from *Physarum* has an amino acid composition identical to the two "informosome" proteins from the mammalian cells. The 46,000 dalton protein (band 32) is actin.

Perhaps the first evidence which established a correlation between the nonhistone proteins and chromatin activity came from the early quantitative studies on chromatin from various cell types [43,44]. The recurring observation was that the inactive chromatin from nondividing or metabolically quiescent cells was deficient in nonhistone proteins and rich in histones. Again, the inactive avian erythrocyte nucleus is almost entirely DNA and histone, yet the actively dividing stem cells are rich in nonhistone protein [45,46]. In more recent studies where interphase chromatin has been separated into fractions active and inactive in RNA synthesis, it has been shown that the active (euchromatin) fractions are rich in nonhistone proteins whereas the inactive (heterochromatin) fractions contain fewer nonhistone proteins and significantly more histones [48-51].

The observation that the eucaryote nucleus contains hundreds of different nonhistone proteins first suggested that some may regulate specific genes. Perhaps more suggestive was the observation that the complement of nonhistone proteins varied among different species and even among differentiated tissue within an organism (species and tissue specificity) [52-58]. Clearly different patterns of gene expression must be responsible for the formation and maintenance of differentiated cell states. In *Physarum*, for example, the complement of nonhistone proteins is distincly different in plasmodia and actively growing amoeboid gametes [59]. Also, during periods of differentiation in *Physarum*, significant changes occur in the nonhistone proteins [26,37,38,52,59-64]. The finding that numerous changes occur in the complement of nonhistone proteins during periods of cellular differentiation established a more direct correlation between specific proteins and altered patterns of genetic activity. Of particular relevance to this discussion is the fact that *Physarum* was the first experimental system in which the above was demonstrated to occur [60]. More recently it has been found that changes in the nonhistone proteins also occur during embryogenesis in higher eucaryotes [65-68].

Additional evidence that the nonhistone proteins play a role in gene activation came from studies demonstrating that the synthesis of specific nonhistone proteins is significantly increased at times of gene activation induced by hormones such as cortisol [69], estradiol [70], and glucagon [71]. Also in various cells which can be stimulated to proliferate--including *Physarum*--there is an early increase in nonhistone protein synthesis [62,72-75].

A further characteristic of the nonhistone proteins is that many are phosphorylated [41,42, and as reviewed elsewhere,76] and several lines of evidence indicate that phosphorylation may be a key to the activity of some of these macromolecules. For example, in cells induced to divide by plant mitogens [77] or by hormones [78-81], there is an early increase in protein phosphorylation, and in several cases this precedes the new rounds of RNA synthesis induced by the various stimuli for cell division.

If specific nonhistone proteins do function in a positive control mechanism, then as a prerequisite to their activity they should bind DNA. While mechanisms other than direct DNA binding might be envisioned, there are several reports that nonhistone proteins do bind DNA in a selective manner [53,78,82-88]. The binding of nonhistone protein to DNA alters the pattern of in vitro RNA synthesis when either purified DNA or reconstituted chromatin is employed [89]. More recently, it has been demonstrated that specific nonhistone proteins bind selectively to fractions of DNA differing in their incidence of reiteration in the genome [82,90]. If, as has been suggested [91], reiterated sequences of DNA function in a regulatory capacity in the genome, then proteins which selectively bind DNA of differing C_0t values may influence the function of the regulator sequences. The observations summarized above represent only a small portion of the work on the nonhistone proteins of eucaryotes. For more complete reviews the interested reader is referred to refs. 38, 39, 64, 76, and 90.

II. CURRENT RESEARCH

A. Characteristics of the Nonhistone Proteins

An important characteristic of the nonhistone proteins is that most have isoelectric points considerably below neutrality. As a further distinction from the histones, in the literature the nonhistone proteins are often referred to as the acidic nuclear or chromosomal proteins. Since proteins with neutral or acidic isoelectric points are frequently insoluble or stabilized in acidic solutions, the basic histones can almost selectively be extracted from nuclei or chromatin with mild acid (0.25 M HCl). Portions of the proteins remaining can be solubilized in various salt solutions while still others (the residual nonhistone proteins) require more powerful protein denaturants such as urea, guanidine-HCl, phenol, or sodium dodecyl sulfate (SDS). A second procedure for obtaining fractions of nonhistone proteins is to extract chromatin in high-ionic strength solutions and then selectively remove the histone using ion exchange chromatography [92]. An early premise in studies on chromatin proteins was that proteins not easily solubilized in low-salt solution were directly bound to DNA or firmly associated in some unique fashion with the chromatin fiber.

Unless otherwise specified, the observations reviewed here will focus on a class of nonhistone proteins which are not quantitatively solubilized in physiological saline. Whether all of these "residual proteins" are in fact bound to DNA directly remains to be proven. Detailed methods for extraction, characterization, electrophoresis, and purification can be found elsewhere [37,38,64].

In *Physarum* the residual nonhistone proteins which remain associated with chromatin following removal of the saline-soluble and histone proteins comprise 45% of total nuclear protein [61]. The ratio of residual protein to DNA is 0.67:1 [61]. These values are very similar to those reported for mammalian cells [93]. The electrophoretic profiles show a high degree of heterogeneity and, like similar fractions from mammalian cells, there is a certain polarity of polypeptide distribution in high-resolution gels (Fig. 3). Briefly, in the lower-mol wt regions (30,000 to 50,000 daltons) several proteins can be resolved which comprise almost half of the total residual nonhistone fraction. In the higher-mol wt regions (50,000 to 250,000 daltons) numerous polypeptides are resolved and these proteins are generally present in much lower intranuclear concentrations. Although photographic procedures do not record all of the polypeptides separated through SDS-polyacrylamide gel electrophoresis, on visual inspection well over 100 protein bands can be detected. Quantitative or qualitative differences in the complement of residual chromatin proteins cannot be detected whether isolated from intact nuclei or from nuclei which have been stripped of their nuclear membranes [64], an observation indicating that nuclear membrane proteins constitute a very small percentage of total residual nuclear protein. Isoelectric point (pI) determinations [64] have demonstrated that almost all of the residual chromatin proteins have isoelectric points considerably below neutrality (pI 3.5 to 6.5) and, as will be discussed in more detail below, several of the residual proteins are phosphorylated [41,42]. In *Physarum* maximum amino acid incorporation in the residual nonhistone proteins occurs during S period [61].

B. Nonhistone Proteins and Cell Cycle

With respect to studies on the nonhistone proteins of eucaryotes, especially those proteins thought to tightly bind DNA, an initial objective was to determine their role in cell cycle events (i.e., initiation of DNA synthesis, mitosis, etc.). The naturally synchronous cell cycle and the cytoplasmic homogeneity in *Physarum* suggested that plasmodia represent an ideal system for cell cycle studies on nuclear proteins. Interestingly, however, in studies where protein was extracted from nuclei isolated every 20 min during the 8-h interphase period and compared electrophoretically, no quantitative or qualitative differences could be detected even in the most minor protein species [60,61]. In these studies well over 100 protein species were compared and, considering the rather significant metabolic differences between the chromatin of S and G_2 periods, it seemed initially surprising not to see changes during interphase periods. While there is no precise explanation for this phenomenon, several points deserve consideration. First, during growth only a small fraction of the genome is functional and there is no evidence that the battery of genes expressed during growth are regulated in a temporal on-off fashion. Second, proteins which may function to activate specific gene transcription may be present at intranuclear concentrations too low to be observed directly through current "screening" techniques (fewer than 30,000 copies). The presence of many such proteins has been demonstrated in residual chromatin fractions from *Physarum* [41]. Finally, the mechanism of gene-regulatory molecules may not be in their presence or absence in nuclei but rather their activity may be a function of phosphorylation, methylation, or perhaps activation through specific hormone binding. For example, there is some evidence that the estrogen-receptor complex enters the nucleus and binds to preexisting nonhistone protein associated with regulatory gene sequences [94,95].

In *Physarum*, actively growing plasmodia have no G_1 period [19]. Since biochemical events occurring during G_1 set the stage for DNA synthesis and cell division, it might be argued that changes in the residual nonhistone proteins may be detected in cells with true G_1 periods. Indeed, early reports [93] suggested that quantitative changes (although minor) do occur in mammalian chromatin during the G_1 period. However, more recent studies have failed to confirm even minor cell cycle changes in synchronized cells [96]. In this context, then, it appears that no fundamental or mechanistic difference exists between the major nonhistones of *Physarum* and mammalian cells.

While quantitative changes in the residual nonhistone proteins have not been detected during interphase, in *Physarum* significant changes do occur during mitosis [37]. As is evident in Fig. 4, the complement of protein in interphase chromatin and metaphase chromosomes differs significantly with respect to several key proteins. Chromosomes are discrete packets of highly organized interphase chromatin. Some of the nonhistones may function to bring about chromosome formation during prophase while others function in chromosome structure and motility. A recent finding has been the fact that the contractile proteins myosin and actin are major

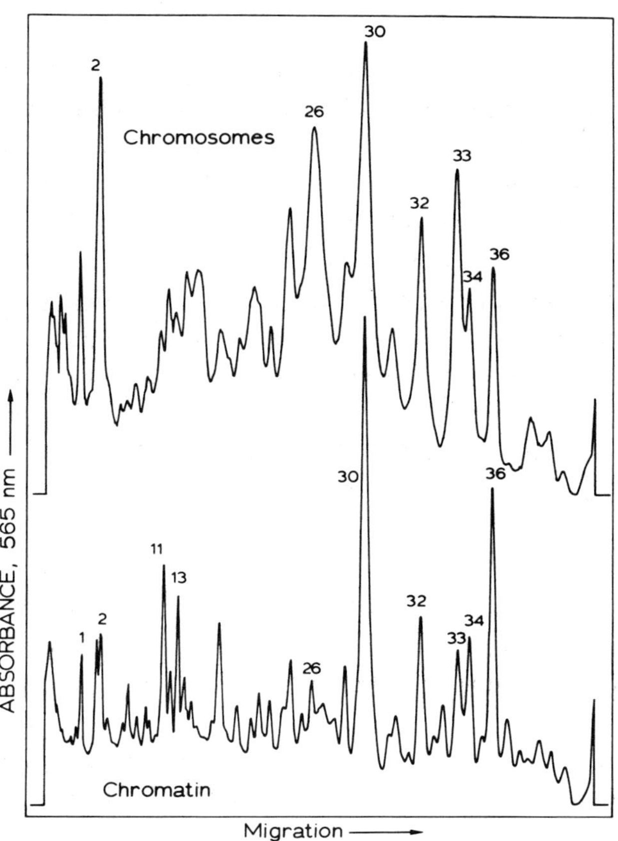

FIG. 4. Densitometer tracings of electrophoretically separated residual nonhistone protein from interphase chromatin and metaphase chromosomes. Note the relative increase in chromosomes in the peaks corresponding to myosin (band 2) and actin (band 32) as well as the significant increase in a protein with mol wt (55,000 daltons; band 26) similar to tubulin from higher eucaryotes. The lower tracing is unchanged during interphase periods.

components of interphase chromatin in *Physarum* (bands 2 and 32, Figs. 3 to 5) and apparently in mammalian cells as well [37]. Of apparent significance is the observation that isolated metaphase chromosomes of *Physarum* are enriched in both myosin and actin (bands 2 and 32, Fig. 4) as well as a major protein with a mol wt very similar to higher eucaryote tubulin (band 26, Fig. 4). The presence of most of the proteins of interphase chromatin in metaphase chromosomes suggests that these proteins are in fact associated with the genetic complex.

The absence of changes in the major nonhistone proteins during interphase periods in *Physarum* does not (for reasons previously discussed) prove a nonregulatory role for some of these macromolecules. Studies, where the degree and turnover of protein-bound phosphate was followed during the cell cycle, indicate that phosphorylation may be an important aspect of the metabolic activity of specific polypeptides [41,42]. Approximately 15 polypeptides present in the residual nonhistone fractions from *Physarum* incorporate ^{32}P following the addition of labeled orthophosphate to the culture medium [41]. Interestingly, most of the major nonhistone proteins are not phosphorylated (band 30, Fig. 5, being the most notable exception). In some cases proteins which cannot be detected as stained bands due to their very low intranuclear concentration can be detected in gel autoradiograms due to their incorporation of radioactive phosphate [42]. The phosphorylated proteins contain varying amounts of phosphate (0.02 to 0.12% phosphorus by weight or one to four phosphate groups per polypeptide) [42]. By comparing the results of short-term pulse-labeling experiments, pulse-labeling cold chase experiments, and experiments in which plasmodia were labeled over several generations, it has been clearly shown that some proteins rapidly "turn over" phosphate while still others accumulate phosphate more slowly and retain label for much longer periods of time (half-lives of phosphorylated proteins range from 1.5 to 3 h for some polypeptides). The evidence so far available indicates that

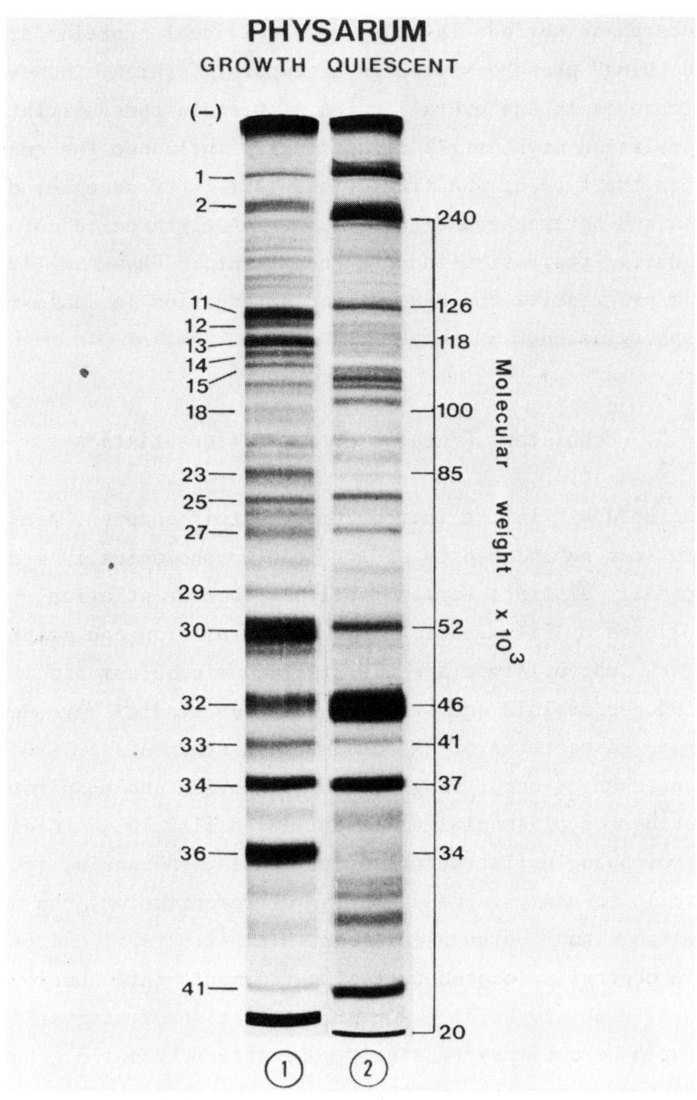

FIG. 5. Electrophoretic profiles of the residual nonhistone protein extracted from actively growing plasmodia (gel 1) and from differentiating plasmodia just prior to sporulation (gel 2). Note the significant increase in the 46,000 dalton component (band 32) and the loss of the 34,000 mol wt protein (band 36). Observe also the many changes in the higher-mol wt proteins.

during interphase periods in *Physarum*, individual proteins are not "all-or-nothing" phosphorylated in a temporal fashion. However, temporal changes in the overall rates of protein phosphorylation and dephosphorylation might still significantly influence the quantity rather than the type of RNA transcribed [42]. For example, during maturation and heterochromatization in avian erythrocyte chromatin [76] and during starvation-induced encystment in *Physarum* [42], there is a progressive and generalized attenuation in nuclear protein phosphorylation.

C. Nonhistone Proteins During Differentiation

As pointed out in the introduction to this chapter, plasmodia of *Physarum* can be induced to undergo two morphologically and physiologically distinct forms of cellular differentiation. Starvation followed by illumination induces sporulation and meiosis in surface-grown macroplasmodia while starving microplasmodia in shake culture induces diploid encystment. Numerous studies have demonstrated altered patterns of RNA and protein synthesis [97-99] and significant changes occur in glycogen metabolism and respiration during periods of plasmodial differentiation [100,101]. Clearly discrete morphological structures (cyst walls, sporangia, spore walls, etc.) are newly formed. During differentiation, the nucleus and chromatin also undergo significant ultrastructural changes. The single centrally located nucleolus fragments into numerous dense bodies about 10 h following the initiation of starvation and the chromatin becomes progressively more heterochromatic. Nuclear size decreases to approximately one-third the interphase volume and on the establishment of quiescent states little or no RNA synthesis occurs.

The above phenomena suggest that the stimulus of starvation induces new patterns of genetic activity and evidence for this can be seen in the inhibition of differentiation by actinomycin D [8]. During sporulation and encystment, numerous changes occur in the complement of residual chromatin protein (Fig. 5). By comparing the changes which occur in *Physarum* to changes in mammalian cells induced by various stimuli, attention has been focused on the proteins of chromatin which seem to be more directly involved in regulating growth and quiescence.

More specifically, during the establishment of nonproliferative states in *Physarum* there is a loss in the amount and heterogeneity of the higher-mol wt proteins (50,000 to 250,000), and significant quantitative changes occur in several of the major lower-mol wt polypeptides as well (i.e., see bands 32 and 36, Fig. 5). In *Physarum* the changes which occur during encystment and sporulation are essentially identical, an observation which suggests that most of the proteins observed as stained bands in gels probably function in structural alterations of chromatin or in the enzymatic capabilities of chromatin rather than in the activation or inactivation of specific genes.

The changes which develop in the residual chromatin proteins on the establishment of nonproliferative cell states are reversed in a sequential temporal fashion during periods of dedifferentiation induced by refeeding encysted plasmodia (Fig. 6) [62]. Before DNA synthesis and mitosis occur, the complement of nonhistone proteins characteristic of proliferating cell states must be reestablished.

When HeLa S_3 cells are subjected to the stress of nutrient depletion, changes very similar to those which develop in *Physarum* also occur (Fig. 7) [52]. As in *Physarum*, with these changes the nucleoli fragment and the chromatin becomes more granular in appearance. In *Physarum* and in HeLa, the major protein (band 32, Fig. 7) which increases in intranuclear concentration is of identical mol wt (46,000 daltons). The major protein (band 36) which decreases more significantly in both cell types is of similar mol wt (34,000 in *Physarum*, 32,000 in HeLa). These major proteins have recently been purified and identified and their possible roles will be discussed in the next section.

The changes outlined above were induced by starvation in both HeLa and *Physarum*. Although HeLa cells do not go on to form

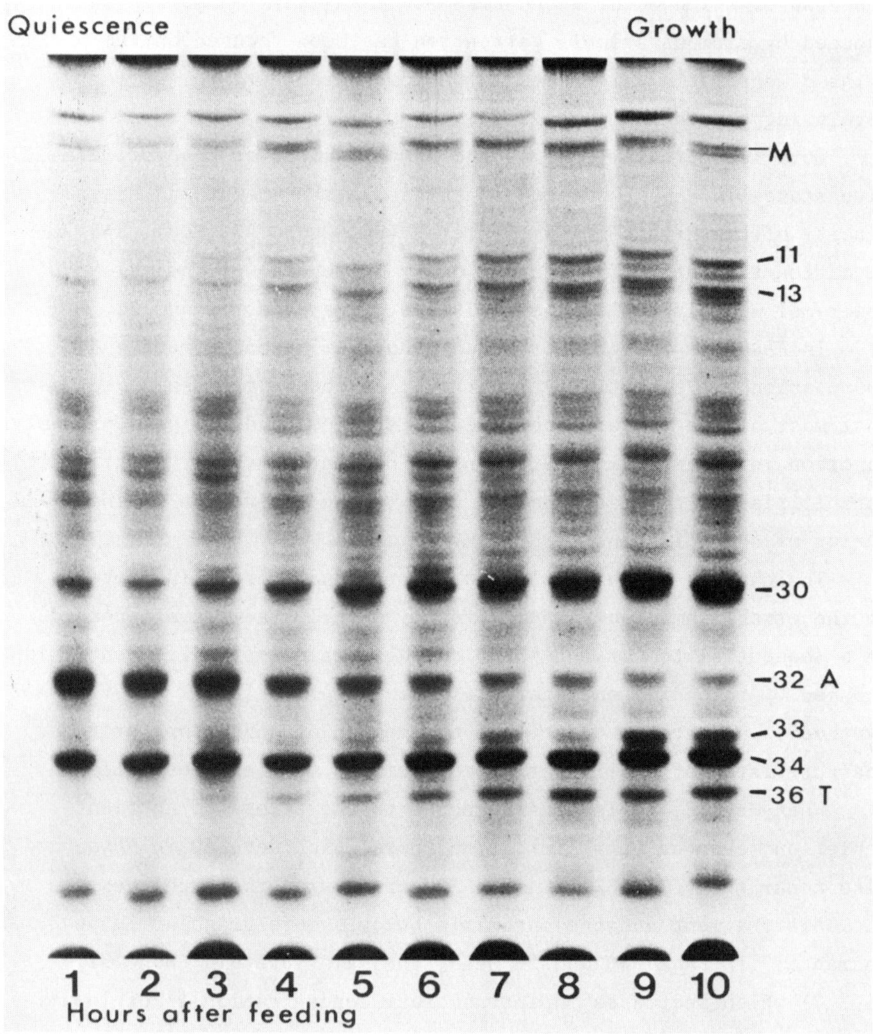

FIG. 6. Electrophoretic profiles of the residual nonhistone proteins isolated every hour during the period of dedifferentiation induced by refeeding encysting microplasmodia. Note that some proteins increase or decrease in a linear fashion while others (especially band 33) increase in intranuclear concentration maximally over a brief period of time.

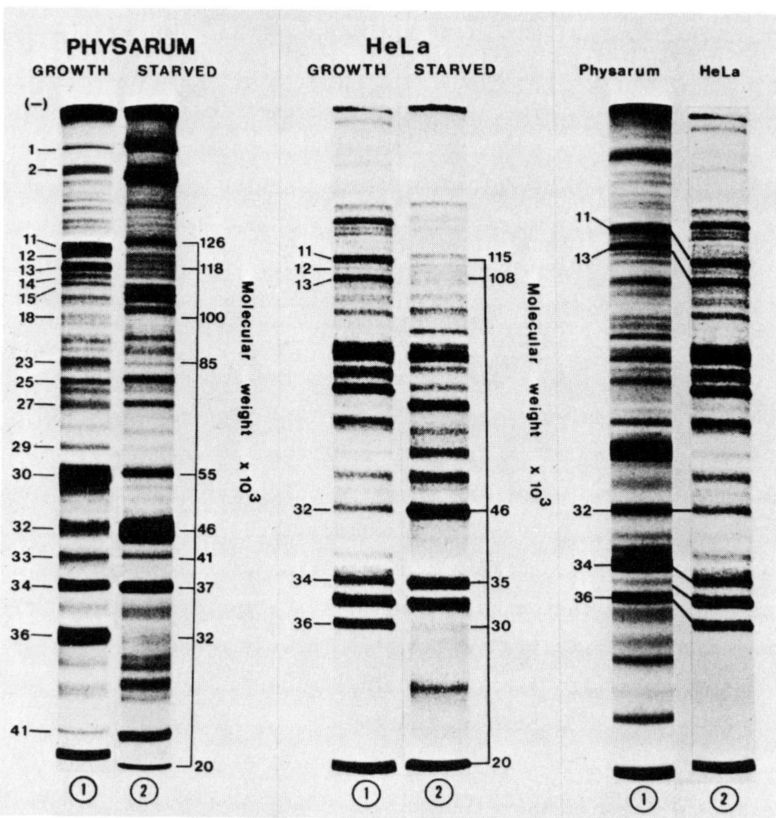

FIG. 7. Starvation-induced changes in the residual nonhistone proteins of *Physarum* and HeLa cells. Note that the changes in the major proteins (bands 32 and 36) are the same for both cell types. Note also that the major protein which increases (band 32) is of identical mol wt while the lower-mol wt component (band 36) is of slightly lower mol wt in HeLa cells.

quiescent yet viable cell states as does *Physarum*, similar changes in the normal metabolic pathways may have been induced by starvation. Starvation represents a broad-spectrum stimulus and it is not possible to determine which proteins may function in the structural interconversions of chromatin or in reorganizing the metabolic capabilities of the cells through specific gene activation. However, certain mammalian cells in tissue culture can, in the absence of starvation, differentiate into nonproliferative cell states when high cell densities are established (contact inhibition). In post-confluent inhibition of growth, the only apparent stimulus is the actual cell surface associations which develop at high cell densities. In C3H 10T1/2 mouse embryo fibroblasts [102], the only significant changes which occur in the residual chromatin proteins on the establishment of nonproliferative states are an increase in the 46,000-mol wt protein and a significant decrease in the 32,000-mol wt component (Fig. 8). Associated with these changes is an increase in chromatin granularity and a fragmentation of nucleoli [37]. These findings suggest a direct correlation between the intranuclear concentration of two specific polypeptides (bands 32 and 36) and the presence or absence of proliferative cell states in eucaryotes. The high intranuclear concentration of these proteins suggested initially that they may possess a structural role in the maintenance of an active or inactive chromatin complex or that they possess an otherwise pleiotropic influence on chromatin activity. The recent identification of the 46,000-mol wt component as actin and the 34,000-mol wt component as a probable constituent of ribonuclear protein particles (involved in the processing of nascent mRNA) is consistent with the correlations established above.

D. Characterization and Possible Developmental Role of Specific Nonhistone Proteins

As stated previously, in order to fully understand the role of specific chromosomal proteins in growth and differentiation and to gain manipulatory control over their function it will ultimately be necessary to have purified native protein for use as experimental tools. A second point previously established was the value of *Physarum* in these studies due to the large quantities of physiologically identical nuclei which can be inexpensively obtained in the laboratory. Efforts to isolate and purify the 46,000-mol wt protein

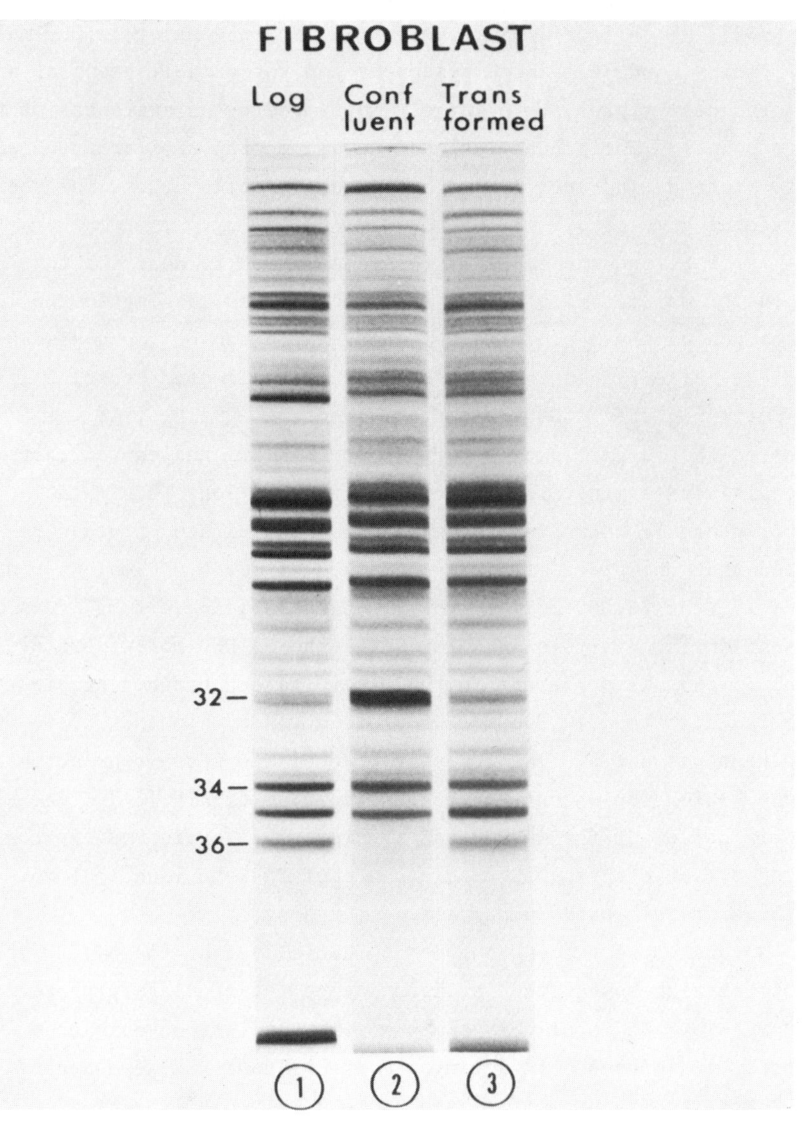

FIG. 8. Electrophoretic profiles of the residual nonhistone protein from mouse embryo fibroblasts extracted from actively growing cells (gel 1) and from cells 48 h after postconfluent inhibition of growth (gel 2). Note that in response to the stimulus of contact inhibition significant changes occur only in bands 32 and 36. Gel 3 shows the nonhistone proteins in cells oncogenically transformed with chemical carcinogens. The transformed cells, even at high cell densities, do not show the changes associated with contact inhibited cells.

(band 32), which increases in chromatin, and the major 34,000-mol wt protein (band 36), which disappears on the establishment of non-dividing cell states, were largely motivated by an awareness of the above points. The biochemical and developmental similarity between these proteins in *Physarum* and in mammalian cells argues for their functional homology.

It is beyond the scope of this discussion to describe the methods of purification or all the observations leading to the identification of these proteins. These details can be found elsewhere [37,64]. Briefly, in the course of studies directed at solubilizing and purifying the 34,000-mol wt component of *Physarum*, it was observed that a precipitate formed when 1 M KCl nuclear extracts were dialyzed against low-ionic strength solutions (0.05 M KCl). Electrophoretic characterization of the precipitated protein revealed that the precipitate was composed of a 240,000-mol wt protein (band 2), the 46,000-mol wt protein (band 32), and--if Mg^{2+} was present during dialysis into low salt--the 34,000-mol wt protein (band 36) was also present in the precipitate. Further studies revealed that the 240,000-mol wt protein was myosin and the 46,000 component was actin. The low-salt precipitate was a true actomyosin complex which would contract on the addition of ATP [37]. The identification of the 46,000-mol wt protein as actin and the conservation of actin throughout evolution explains the identical mol wt of these protein bands among all cell types.

Several observations suggested initially that the 34,000-mol wt "proliferation-associated" protein may be the regulatory protein tropomyosin. It is of very similar mol wt to tropomyosin from other sources, its presence in actomyosin precipitates is dependent on Mg^{2+}, and it disappears from the nucleus while actin increases in intranuclear concentration. These changes are associated with the condensation of chromatin. The loss of tropomyosin and other regulatory proteins from an actomyosin complex could, in vivo, lead to a cation-stabilized condensed actomyosin complex, and it was previously postulated that contractile proteins may bring about the condensation of chromatin associated with the establishment of quiescent cell states in *Physarum* [37].

However, further studies on the 34,000-mol wt protein of *Physarum* have demonstrated that it most probably is a component of ribonucleoprotein particles (informosomes) involved in the processing of newly synthesized RNA [40]. Its amino acid composition is quite unusual [37] and not similar to tropomyosin but rather is essentially identical to the major component of informosomes from mammalian cells [40]. The homologous 32,000-mol wt component (band 36) from HeLa cells has recently been found to be a major component of purified ribonucleoprotein particles [103]. The loss of this protein at times of chromatin inactivation is consistent with the generalized attenuation of RNA synthesis.

The observation that the 34,000-mol wt component of *Physarum* is not tropomyosin-like argues against an actomyosin-mediated condensation of chromatin. In addition it now seems apparent that the condensation of chromatin in avian erythrocytes is not associated with an increase in nuclear actin, since essentially no actin can be recovered from mature erythrocyte chromatin [104]. The significance of an increase in nuclear actin in quiescent cell states is as yet obscure. If, however, the finding [105] that the actin acts as a specific and potent inhibitor of deoxyribonuclease I proves to be a generalized effect, then the presence of actin in nondividing nuclei of *Physarum* could aid in stabilizing the genome over extended periods of time.

Perhaps the most likely explanation for the presence of actin in eucaryote nuclei would be if it functioned in chromosome motility during mitosis. Such a role for nuclear actin has previously been proposed (as recently reviewed, ref. 106) and some evidence exists in favor of this function. As pointed out above, actin and myosin appear to be enriched in isolated metaphase chromosomes of *Physarum* (Fig. 4). Studies conducted in other laboratories have demonstrated that newly synthesized actin specifically enters the nucleus during late G_2 in *Physarum* [22,107,108]. Furthermore, the presence of contractile proteins in the mitotic apparatus of mammalian cells has been demonstrated through immunological techniques and through the binding of heavy meromyosin to the mitotic spindle apparatus of mammalian cells [106]. If actin is required for chromosome motility and thus mitosis, then its presence may not be required in the nuclei of cells which never divide. This may explain the absence of actin in avian erythrocyte nuclei.

In recent studies myosin has been observed to be specifically concentrated in nuclear membrane fractions from rat liver [109]. Actin was not present in nuclear membrane fractions. A role for myosin might be envisioned in the nuclear pore complex and the electrophoretic characterization of isolated nuclear pore complexes, and lamina from rat liver revealed a major polypeptide which migrates similarly to myosin in gels [110]. In further studies actin and myosin have been found to bind DNA-sepharose and dissociate under different conditions of ionic strength [111]. However, this phenomenon cannot be inferred to have in vivo significance at this time, especially since contractile proteins naturally bind nucleotides.

The presence of contractile proteins in nuclei does not suggest a single metabolic role and, in preliminary studies, it is worth considering the possibility that these proteins in addition to the possibilities above may function in the transport of macromolecules across the nuclear membrane barrier.

III. FINAL COMMENTS

The information covered in the foregoing sections of this discussion does not represent a complete review of current knowledge concerning the nuclear proteins of *Physarum*. Rather, by choosing specific experimental findings (published and unpublished), an effort has been made to demonstrate the rationale for conducting fundamental genome-related research on a manageable yet highly sophisticated eucaryotic microbe. Based on the changes which occur in the residual nonhistone proteins of *Physarum* and mammalian cells during simple forms of differentiation, it seems clear that facts learned in *Physarum* will be applicable to higher eucaryotes as well.

As an experimental system *Physarum* is of value not only in genome-related studies but also in studies on reversible cytoplasmic streaming, cell motility, membrane physiology, and photochemical induced differentiation. Perhaps the most intriguing and important question concerns the mechanism which regulates and triggers millions of nuclei to divide with precise synchrony. Recently evidence has been presented which indicates that phosphorylation of H1 histone is responsible for the condensation of chromatin during prophase and the formation of chromosomes [112]. Perhaps histone H1 in the phosphorylated state causes a higher order of "beads on a string" structure. What, then, regulates H1 phosphorylation and is this event the first "trigger" for mitosis?

The phenomenon of premature chromosomal condensation in heterocaryons suggests that in fact "mitotic inductive factors" are present in the cytoplasm of cells during mitosis. For several reasons *Physarum* plasmodia represent an ideal system for studies of this nature. First, millions of nuclei in a common cytoplasm undergo mitosis with absolute synchrony. Second, heterokaryons between mitotic and interphase plasmodia can be formed by allowing simple plasmodial fusion or by physically mixing normally incompatible strains [113]. And third, methods are now available for separating nuclei in mass from artificially formed heterokaryotic plasmodia [114]. With these initial experimental capabilities and an ever increasing technology for detecting new molecules in complex mixtures, attempts to isolate specific "mitotic triggers" now seem feasible.

ACKNOWLEDGMENTS

Portions of the work described in this article were supported by grants from the Vanderbilt University Research Council and by NSF grant BMS 75-03105.

REFERENCES

1. Gray, W. D., and C. J. Alexopoulos, 1969. Biology of the Myxomycetes. The Ronald Press, New York.
2. Alexopoulos, C. J., 1962. Introductory Mycology, Second ed., John Wiley and Sons, New York.
3. Martin, G. W., and C. J. Alexopoulos, 1969. The Myxomycetes. The University of Iowa Press, Iowa City.
4. von Stosch, H. A., 1965. *In* Handbuch der Pflanzenphysiologie, Vol. 15, F. Ruhland (ed.), Springer-Verlag New York, New York, p. 641.
5. Daniel, J. W., and H. H. Baldwin, 1964. *In* Methods in Cell Physiology, Vol. 1, D. M. Prescott (ed.), Academic Press, New York, p. 9.
6. Brewer, E. N., S. Kuraishi, J. C. Garver, and F. M. Strong, 1964. Appl. Microbiol. 12: 161-164.
7. Mohberg, J., and H. P. Rusch, 1969. J. Bacteriol. 97: 1411-1418.
8. LeStourgeon, W. M., 1970. Ph.D. Dissertation, The University of Texas at Austin.
9. Sauer, H. W., K. L. Babcock, and H. P. Rusch, 1969. Exp. Cell Res. 57: 319-327.
10. LeStourgeon, W. M., C. F. Bohnstedt, and P. Thimell, 1971. Mycologia 63: 1002-1012.
11. Aldrich, H. C., 1974. Proc. Iowa Acad. Sci. 81: 1-9.
12. Aldrich, H. C., 1967. Mycologia LIX: 127-148.
13. Aldrich, H. C., and C. W. Mims, 1970. Amer. J. Bot. 57: 935-941.
14. Mohberg, J., K. L. Babcock, R. B. Haugli, and H. P. Rusch, 1973. Develop. Biol. 34: 228-245.
15. Goodman, E. M., H. W. Sauer, L. Sauer, and H. P. Rusch, 1969. Canad. J. Microbiol. 15: 1325-1331.
16. Goodman, E. M., 1972. J. Bacteriol. 111: 242-247.
17. Howard, F. L., 1932. Ann. Bot. 46: 461-477.
18. Guttes, E., S. Guttes, and H. P. Rusch, 1961. Develop. Biol. 3: 588-614.
19. Mohberg, J., 1974. *In* The Nucleus, Vol. 1, H. Busch (ed.), Academic Press, New York, p. 187.
20. Rusch, H. P., 1970. *In* Advances in Cell Physiology, Vol. 1, D. M. Prescott, L. Goldstein and E. McCarkey (eds.), New York, p. 297.

21. Cummins, J. E., 1968. *In* The Cell Cycle: Gene-Enzyme Interactions, G. Padilla, G. Whitson, and I. Cameron (eds.), Academic Press, New York, p. 141.

22. Jockusch, B. M., 1973. Ber. Deutsch. Bot. Ges. Bd. 86: 39-54.

23. Sauer, H. W., K. L. Babcock, and H. P. Rusch, 1969. Exp. Cell Res. 57: 319-327.

24. Cummins, J. E., and H. P. Rusch, 1968. Endeavour 27: 124-129.

25. Bresnick, E., 1971. *In* Methods in Cancer Research, Vol. 6, H. Busch (ed.), Academic Press, New York, p. 347.

26. LeStourgeon, W. M., C. Nations, and H. P. Rusch, 1973. Arch. Biochem. Biophys. 159: 861-872.

27. Kolata, G. B., 1975. Science 188: 1097-1099.

28. Kornberg, R. D., and J. O. Thomas, 1974. Science 184: 865-868.

29. Kornberg, R. D., 1974. Science 184: 868-871.

30. Hewisch, D. R., and L. A. Burgoyne, 1973. Biochem. Biophys. Res. Commun. 52: 504-510.

31. Noll, M., 1974. Nature 251: 249-252.

32. Noll, M., J. O. Thomas, and R. D. Kornberg, 1975. Science 187: 1203-1206.

33. Olins, A. L., and D. E. Olins, 1974. Science 183: 330-332.

34. Senior, M. B., A. L. Olins, and D. E. Olins, 1975. Science 187: 173-175.

35. Pardon, J. F., and M. H. F. Wilkins, 1972. J. Mol. Biol. 68: 115-124.

36. Kierszenbaum, A. L., and L. L. Tres, 1975. J. Cell Biol. 65: 258-270.

37. LeStourgeon, W. M., A. Forer, Y.-Z. Yang, J. S. Bertram, and H. P. Rusch, 1975. Biochim. Biophys. Acta 379: 529-552.

38. LeStourgeon, W. M., R. Totten, and A. Forer, 1974. *In* Acidic Proteins of the Nucleus, I. L. Cameron and J. Jeter (eds.), Academic Press, New York, p. 159.

39. Elgin, S. C. R., 1975. Ann. Rev. Biochem. 44: 725-774.

40. Martin, T., P. Billings, A. Levey, S. Ozarslan, T. Quinlan, H. Swift, and L. Urbas, 1973. Cold Spring Harbor Symp. Quant. Biol. 38: 921-932.

41. Magun, B. E., 1975. Arch. Biochem. Biophys. 170: 49-60.

42. Magun, B. E., 1974. *In* Acidic Proteins of the Nucleus, I. L. Cameron and J. Jeter (eds.), Academic Press, New York, p. 139.

43. Mirsky, A. E., and H. Ris, 1951. J. Gen. Physiol. 34: 475-493.

44. Dingman, C. W., and M. B. Sporn, 1964. J. Biol. Chem. 239: 3483-3492.

45. Gershey, E. L., and L. J. Kleinsmith, 1969. Biochim. Biophys. Acta 194: 519-525.

46. Vidali, G., L. C. Boffa, V. C. Littau, K. M. Allfrey, and V. G. Allfrey, 1973. J. Biol. Chem. 248: 4065-4068.

47. Frenster, J. H., V. G. Allfrey, and A. E. Mirsky, 1963. Proc. Nat. Acad. Sci. USA 50: 1026-1032.

48. Frenster, J. H., 1965. Nature 206: 680-683.

49. Dolbeare, F., and H. Koenig, 1970. Proc. Soc. Exp. Biol. Med. 135: 636-641.

50. Marushige, K., and J. Bonner, 1971. Proc. Nat. Acad. Sci. USA 68: 2941-2944.

51. Reeck, G. R., R. T. Simpson, and H. Sober, 1972. Proc. Nat. Acad. Sci. USA 69: 2317-2321.

52. LeStourgeon, W. M., W. Wray, and H. P. Rusch, 1973. Exp. Cell Res. 79: 487-492.

53. Teng, C. S., C. T. Teng, and V. G. Allfrey, 1971. J. Biol. Chem. 246: 3597-3609.

54. Loeb, J. E., and C. Cruezet, 1970. Bull. Soc. Chim. Biol. 52: 1007-1020.

55. Platz, R. D., V. M. Kish, and L. J. Kleinsmith, 1970. FEBS Letters 12: 38-40.

56. Chytil, F., and T. C. Spelsberg, 1971. Nature New Biol. 233: 215-218.

57. Spelsberg, T. C., A. W. Steggles, F. Chytil, and B. W. O'Malley, 1972. J. Biol. Chem. 247: 1368-1374.

58. Wang, T. Y., 1971. Exp. Cell Res. 69: 217-219.

59. LeStourgeon, W. M., E. M. Goodman, and H. P. Rusch, 1973. Biochim. Biophys. Acta 317: 524-528.

60. LeStourgeon, W. M., and H. P. Rusch, 1971. Science 174: 1233-1235.

61. LeStourgeon, W. M., and H. P. Rusch, 1973. Arch. Biochem. Biophys. 155: 144-158.

62. LeStourgeon, W. M., C. Nations, and H. P. Rusch, 1973. Arch. Biochem. Biophys. 159: 861-872.

63. Nations, C., W. M. LeStourgeon, B. E. Magun, and H. P. Rusch, 1974. Exp. Cell Res. 88: 207-215.

64. LeStourgeon, W. M., and W. Wray, 1974. *In* Acidic Proteins of the Nucleus, I. L. Cameron and J. Jeter (eds.), Academic Press, New York, p. 59.

65. Marushige, K., and H. Ozaki, 1967. Develop. Biol. 16: 474-488.

66. Hill, R. J., D. L. Poccia, and P. Doty, 1971. J. Mol. Biol. 61: 445-462.

67. Seale, R. L., and A. I. Aronson, 1973. J. Mol. Biol. 75: 633-645.

68. Spelsberg, T. C., W. M. Mitchell, F. Chytil, E. M. Wilson, and B. W. O'Malley, 1973. Biochim. Biophys. Acta 312: 765-778.

69. Shelton, K. R., and V. G. Allfrey, 1970. Nature 228: 132-134.

70. Teng, C. S., and T. H. Hamilton, 1970. Biochem. Biophys. Res. Commun. 40: 1231-1238.

71. Enea, V., and V. G. Allfrey, 1973. Nature 242: 265-267.

72. Stein, G., and R. Baserga, 1970. J. Biol. Chem. 245: 6097-6105.

73. Levy, R., S. Levy, S. A. Rosenberg, and R. T. Simpson, 1973. Biochemistry 12: 224-228.

74. Karn, J., E. M. Johnson, G. Vidali, and V. G. Allfrey, 1974. J. Biol. Chem. 249: 667-677.

75. Rovera, G., and R. Baserga, 1971. J. Cell Physiol. 77: 201-212.

76. Kleinsmith, L. J., 1974. *In* Acidic Proteins of the Nucleus, I. L. Cameron and J. Jeter (eds.), Academic Press, New York, p. 103.

77. Kleinsmith, L. J., V. G. Allfrey, and A. E. Mirsky, 1966. Science 154: 780-781.

78. Allfrey, V. G., C. S. Teng, and C. T. Teng, 1971. *In* Nucleic Acid-Protein Interaction--Nucleic Acid Synthesis in Viral Infection, D. W. Ribbons, J. F. Woessner, and J. Schultz (eds.), North-Holland, Amsterdam, p. 144.

79. Allfrey, V. G., E. M. Johnson, J. Karn, and G. Vidali, 1973. *In* Protein Phosphorylation in Control Mechanisms, F. Huijing and E. Y. C. Lee (eds.), Academic Press, New York, p. 217.

80. Ahmed, K., 1971. Biochim. Biophys. Acta 243: 38-48.

81. Jungmann, R. A., and J. S. Schweppe, 1972. J. Biol. Chem. 247: 5535-5542.

82. Allfrey, V. G., A. Inoue, J. Karn, E. M. Johnson, and G. Vidali, 1973. Cold Spring Harbor Symp. Quant. Biol. 38: 785-801.

83. Teng, C. T., C. S. Teng, and V. G. Allfrey, 1970. Biochem. Biophys. Res. Commun. 41: 690-696.

84. Patel, G. L., and T. L. Thomas, 1973. Proc. Nat. Acad. Sci. USA 70: 2524-2528.

85. Kleinsmith, L. J., 1973. J. Biol. Chem. 248: 5648-5653.

86. Chaudhuri, S., G. Stein, and R. Baserga, 1972. Proc. Soc. Exp. Biol. Med. 139: 1363-1366.

87. Inoue, A., V. C. Littau, and V. G. Allfrey, 1973. J. Mol. Biol.

88. van den Broek, H. W. J., L. D. Nooden, S. Sevall, and J. Bonner, 1973. Biochemistry 12: 229-236.

89. Paul, J., R. S. Gilmour, N. Affara, G. Birnie, P. Harrison, A. Hell, S. Humphries, J. Windass, and B. Young, 1973. Cold Spring Harbor Symp. Quant. Biol. 38: 885-890.

90. Allfrey, V. G., 1974. *In* Acidic Proteins of the Nucleus, I. L. Cameron and J. Jeter (eds.), Academic Press, New York, p. 2.

91. Britten, R. J., and E. H. Davidson, 1969. Science 165: 349-357.

92. Patel, G. L., 1974. *In* Acidic Proteins of the Nucleus, I. L. Cameron and J. Jeter (eds.), Academic Press, New York, p. 29.

93. Bhorjee, J. S., and T. Pederson, 1972. Proc. Nat. Acad. Sci. USA 69: 3345-3349.

94. Means, A. R., S. Woo, S. E. Harris, and B. W. O'Malley, 1975. Molecular and Cellular Biochem. 7: 33-42.

95. Spelsberg, T. C., 1974. *In* Acidic Proteins of the Nucleus, I. L. Cameron and J. Jeter (eds.), Academic Press, New York, p. 248.

96. Karn, J., E. M. Johnson, G. Vidali, and V. G. Allfrey, 1974. J. Biol. Chem. 249: 667-677.

97. Sauer, H. W., 1973. *In* Microbial Differentiation, J. E. Smith and J. M. Ashworth (eds.), 23rd Symposium of General Microbiology, Cambridge University Press, London, p. 375.

98. Grant, W. D., 1973. *In* The Cell Cycle in Development and Differentiation, M. Ball and F. S. Billett (eds.), British Soc. for Dev. Biol. Symposium, Cambridge University Press, p. 77.

99. Jockusch, B. M., H. W. Sauer, D. F. Brown, K. L. Babcock, and H. P. Rusch, 1970. J. Bacteriol. 103: 356-363.

100. Goodman, E. M., and H. P. Rusch, 1970. J. Ultrastr. Res. 30: 172-183.

101. Goodman, E. M., and T. Beck, 1974. Canad. J. Microbiol. 20: 107-111.

102. Reznikof, C. A., D. W. Brankow, and C. Heidelberger, 1973. Cancer Res. 33: 3231-3238.

103. Beyer, A. E., and W. M. LeStourgeon, unpublished results.

104. Shelton, K. E., and W. M. LeStourgeon, unpublished results.

105. Lazarides, E., and U. Lindberg, 1974. Proc. Nat. Acad. Sci. USA 71: 4742-4746.

106. Forer, A., 1974. *In* Cell Cycle Controls, G. M. Padilla, I. L. Cameron, and A. M. Zimmerman (eds.), Academic Press, New York, p. 319.

107. Jockusch, B. M., D. F. Brown, and H. P. Rusch, 1970. Biochem. Biophys. Res. Commun. 38: 279-283.

108. Jockusch, B. M., D. F. Brown, and H. P. Rusch, 1971. J. Bacteriol. 108: 704-714.

109. Wilson, E., and W. M. LeStourgeon, unpublished results.

110. Aaronson, R. P., and G. Blobel, 1974. J. Cell Biol. 63: 1a.

111. Walker, B. E., and W. M. LeStourgeon, unpublished results.

112. Bradbury, E. M., R. J. Inglis, H. R. Matthews, and T. A. Langan, 1974. Nature 249: 553-556.

113. Jeffery, W. R., and H. P. Rusch, 1974. Develop. Biol. 39: 331-335.

114. McCormick, J. J., 1974. J. Cell Biol. 62: 227-231.

UNUSUAL PHOSPHORYLATED COMPOUNDS AND TRANSCRIPTIONAL CONTROL IN *ACHLYA* AND OTHER AQUATIC MOLDS

Herb B. LéJohn, Glen R. Klassen,
David R. McNaughton, Linda E. Cameron,
Swee H. Goh, and Renate U. Meuser

Department of Microbiology
University of Manitoba
Winnipeg, Manitoba, Canada

I. INTRODUCTION

The problem of regulating transcription in eucaryotes is far from resolution at the moment. Since three major forms of DNA-dependent RNA polymerases exist in the nucleus, an understanding of their role in transcribing the various species of RNA and their mode of regulation in gene expression is a difficult undertaking. Furthermore, some experimental observations can be interpreted to mean that subspecies (isozymes?) of two of the three RNA polymerases may exist [1,2]. What seems to be certain is that RNA polymerase I is in the nucleolus and probably transcribes ribosomal RNA (rRNA) cistrons. The RNA polymerase II, which is localized in the nucleoplasm, may synthesize messenger RNA (mRNA). The role of RNA polymerase III is uncertain. It is situated in the nuclear-nucleolar interphase and it has been suggested that it may be responsible for the synthesis of transfer RNA (tRNA) and/or RNA "primers" that initiate DNA chains at replication [3,4]. If a single enzyme is responsible for transcribing the myriads of mRNAs that are used by the cell during development, then it must be under some complex and stringent control with regard to gene specificity because all mRNAs are not transcribed and used at the same time.

Bacteria and bacteriophages with their relatively simpler genomes and apparent lack of transcriptional enzyme multiplicity (not a general consensus considering the work of Travers [5]) do display complex regulatory systems. These range from the elaboration of new transcriptional enzymes and the modification of existing ones to the addition of new regulatory subunits to the basic enzyme complex [6]. Also, small molecules such as certain unusually phosphorylated guanosine polyphosphates (*ppGpp* and *pppGpp*) may modulate the activity of the bacterial RNA polymerase [7-9], although some of the data are tenuous [8,9].

Comparable definitive studies on transcriptional regulation in eucaryotes do not yet exist in the literature. With this deficiency in mind, we initiated the current studies on gross cellular transcription in *Achlya*, a simple eucaryotic microbe, and some other members of a group known as the aquatic molds. The asexual growth cycle of *Achlya* is shown in Fig. 1. Preliminary reports of some of our findings have already been published [10,11]. For a general review of fungal transcription and fungal RNA polymerases, see Griffin et al. [12].

Finamore and Warner [13] showed that the brine shrimp *Artemia salina* contains an unusually symmetrical pyrophosphate anhydride, P^1,P^4-diguanosine-5'-tetraphosphate in large amounts as acid-soluble nucleotide. A homologue, P^1,P^4-diguanosine-5'-triphosphate, was also found later, but this was present in much smaller amounts. These two high-energy polyphosphates were uncovered in 1966 in yet another aquatic organism, *Daphnia magnum* [14], but in this case the triphosphate homologue was present in much larger quantities than the tetraphosphate. Finamore and Clegg [15] have speculated on the possible importance of these nucleotides in regulating DNA synthesis during embryogenesis.

A more recent advance in eucaryotic RNA metabolism involves the detection and characterization of unusual methylated oligonucleotide fragments with the generalized structure 7mG(5')pppNmpNp at the 5' end of mRNA molecules [16-18]. These sequences are ubiquitous among mammalian viral mRNAs and have been detected in eucaryotic messages as well. What biological role they possess is obscure although the suggestion has been made that they are essential for proper translation [18]. In this regard, they are posttranscriptional modifications of the purine triphosphate 5' terminus of mRNAs. Their formation in vitro requires the presence of ribosomes [17,18].

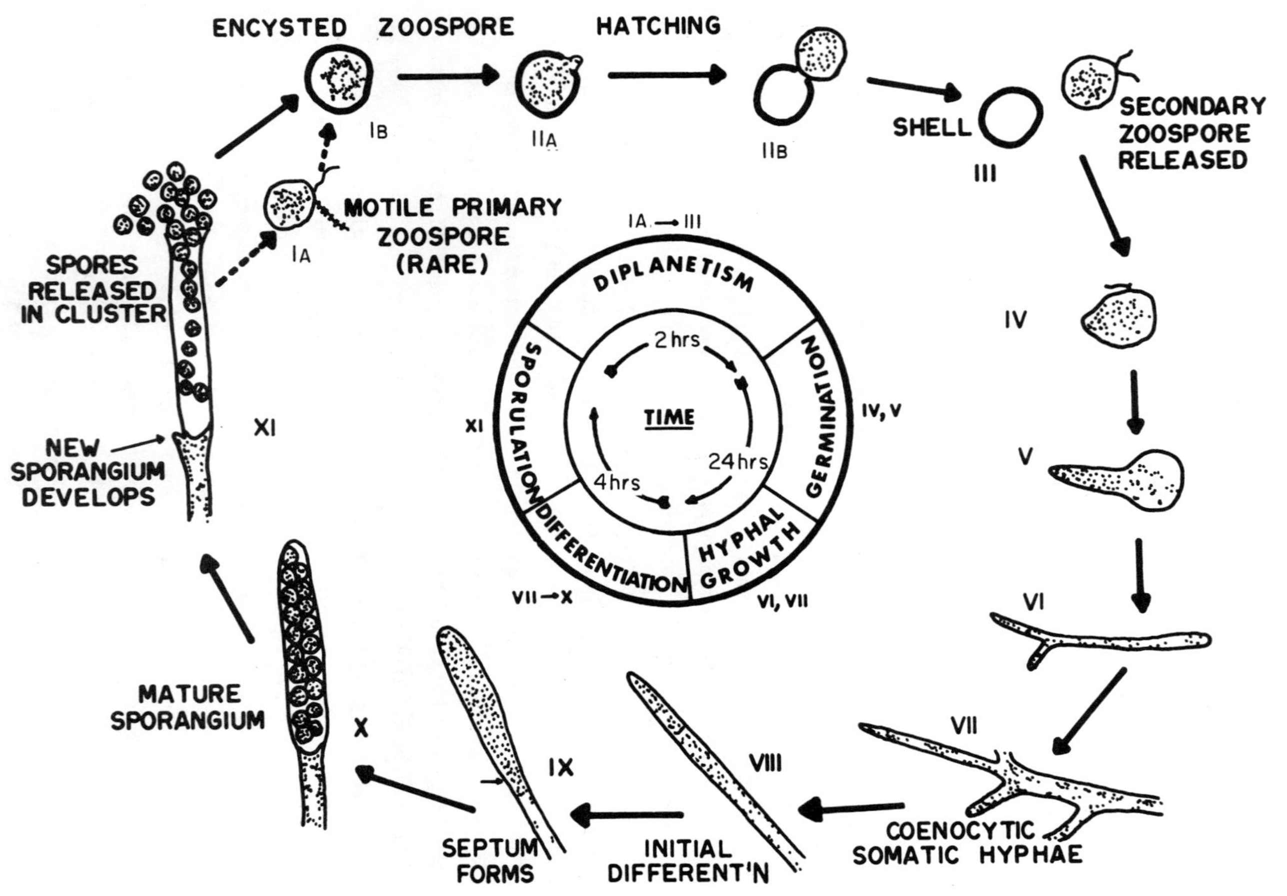

ENCYSTED ZOOSPORE
IB
HATCHING
IIA
IIB
SHELL
III
SECONDARY ZOOSPORE RELEASED
MOTILE PRIMARY ZOOSPORE (RARE)
IA
SPORES RELEASED IN CLUSTER
NEW SPORANGIUM DEVELOPS
XI
IA → III
DIPLANETISM
GERMINATION
HYPHAL GROWTH
DIFFERENTIATION
SPORULATION
TIME
2 hrs
24 hrs
4 hrs
XI
IV, V
VI, VII
VII → X
IV
V
VI
VII
COENOCYTIC SOMATIC HYPHAE
VIII
INITIAL DIFFERENT'N
IX
SEPTUM FORMS
X
MATURE SPORANGIUM

FIG. 1. A schematic summary of the life cycle path followed by the strain of *Achlya* used in these studies. The cycle is restricted to the asexual phase of development. The sexual phase of development is discussed by Horgen in this volume. Spores released from sporangia (stage I) are generally nonmotile and exist as cysts (spores). Within a few minutes after birth, the cysts "hatch" by forming a vesicle into which the cytoplasm and contents are emptied (stages IIa and IIb). When the cyst is emptied of its contents (1 to 2 min), the vesicle and contents assume the form of a kidney-shaped lateral biflagellated spore (stage III) with the flagella located at the depression. These motile spores can encyst and simultaneously produce a germination tube (hyphal protuberance) as shown in stage IV within 2 h under our growth conditions. This unique mode of two successive phases of spore production without an intervening vegetative phase is termed diplanetism. The exponential growth phase begins with elongation of the germ tube followed by hyphal outgrowth from the opposite pole of the spore (stage VI). During exponential growth, extensive dichotomy of the hyphae occurs. The cell is devoid of septa and appears as a single giant multinucleate structure. A septum forms only at the end of the proliferation stage (stage IX) to delineate cells (180 to 300 μm long) at hyphal tips. These cells mature into sporangia (stage X). During maturation, sporangial cells undergo a series of complex protoplasmic changes that terminate with the cleavage of the protoplasm into 30 to 50 segments, each containing a single nucleus. Each unit develops into a spore. The spores are released in a cluster (stage XI). Note, scale used for different stages varies.

II. CURRENT RESEARCH

A. [^{32}P]Orthophosphate Entry into HS Compounds and Nucleotides During Growth

Kinetics of [^{32}P]orthophosphate uptake into *Achlya* cells from spore to sporangial maturation stage are shown in Fig. 2. Measurements were made of the distribution of label into acid-soluble nucleotides at the same time. It was found that three highly labeled phosphorylated compounds that migrated more slowly than GTP on polyethylenimine (PEI) cellulose thin-layer sheets, when the nucleotides were separated by one-dimensional chromatography using 1.5 M KH_2PO_4 (pH 3.65) as solvent, accumulated in the cells at certain stages of the developmental cycle. These polyphosphates were referred to as "hot spots" and were arbitrarily designated HS-1, HS-2, and HS-3 in order of their migration from the point of sample application on PEI-cellulose plates.

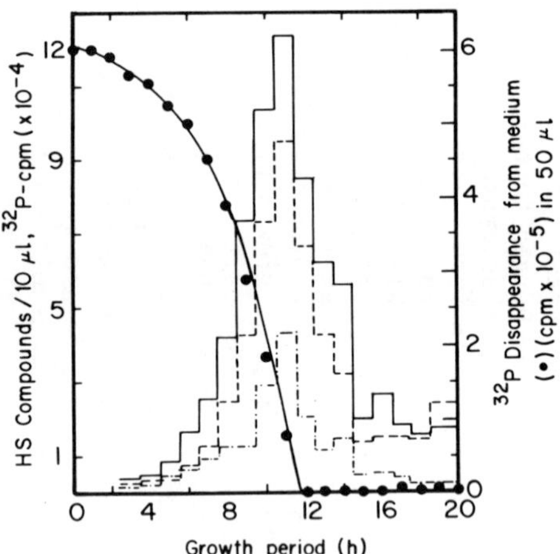

FIG. 2. Relative pool sizes of HS compounds HS-3 (———), HS-2 (-----), and HS-1 (-•-•-) in *Achlya* cells during ontogeny. A time-course profile of [^{32}P]orthophosphate uptake from the growth medium (5 g glucose with 0.5 g yeast extract per liter of distilled H_2O) is shown as (●). The volume of growth medium per culture was 13 ml and each cell sample was extracted with 0.2 ml of 1 M HCOOH. Duplicate samples were collected and the results represent an average of the duplications.

Also shown in Fig. 2 are the kinetic patterns of the relative pool sizes of the HS compounds during ontogeny. The loss of [^{32}P]-orthophosphate from the medium correlated reciprocally with appearance of label in the HS compounds. When the temperature was reduced from 28 to 22° with a corresponding increase in the developmental time from 20 to 24 h, marked fluctuations in the pool sizes of HS-3 and HS-2 were observed during the exponential growth phase (Table 1).

In the 20-h cycle study, uptake of [^{32}P]orthophosphate from the medium was completed before 12 h and simultaneously the cellular

TABLE 1

Relative HS Pool Labeling by [^{32}P]Orthophosphate During Ontogeny of *Achlya* as a Function of Cellular Protein Concentration[a]

Age culture	Protein content	$^{32}P_i$ in HS Compounds (cpm × 10^{-5})		
(h)	(mg)	HS-3	HS-2	HS-1
6	0.1	2.0	3.1	1.2
9	0.13	5.3	12.9	3.1
11	0.40	39.1	29.8	3.97
12	1.10	62.2	41.2	5.2
13	1.38	62.8	36.2	6.2
15	1.52	94.1	77.9	11.3
16	1.72	68.3	55.1	7.8
17	1.94	87.2	75.6	7.4
18	2.31	72.5	60.1	5.7
19	2.38	20.7	27.7	7.1
20	2.52	78.5	55.3	4.4
21	2.40	38.5	28.6	3.8
22	2.41	31.2	13.9	3.6
23	2.43	25.1	9.1	3.5
24	2.38	—— Sporangial maturation ——		

[a]Cultures were maintained at 22° to extend the development time to 24 h.

content of HS compounds declined to a very low level at the time of sporangial formation at 19 h.

The kinetics of [^{32}P]orthophosphate labeling of ATP (GDP) and GTP were similar to those of HS labeling in continuous feeding experiments. A different pattern was observed in pulse-labeling studies.

A more definitive picture was obtained when cultures were pulsed for 10 min with [^{32}P]orthophosphate at hourly intervals over the entire cycle of development. As seen in Fig. 3A, the bulk of HS labeling occurred after 12 h but with the same rapid increase and decrease in the amount of label detectable in HS compounds as found in cultures that were labeled continuously. An important feature is that in rate studies the synthesis of GTP continued to increase long after the synthesis of HS compounds had shown a marked decrease in rate of production (Fig. 3A). A similar observation was made for comigrating GDP and ATP but these are not shown.

B. Rates of Macromolecular Synthesis During Growth

As the cultures used in the previous study were being pulsed with [^{32}P]orthophosphate, others were pulsed simultaneously with [^{3}H]uridine (Fig. 3B), [^{3}H]thymidine (Fig. 3C), and [^{14}C]amino acids (Fig. 3D) to determine the rates of RNA, DNA, and protein synthesis, respectively. From these cultures, total cellular RNA, DNA, and protein were extracted and their specific activities and relative patterns of synthesis elucidated.

The results indicate that (1) the rates of synthesis of RNA, DNA, and protein increased dramatically prior to the appearance of ^{32}P in HS compounds; (2) there is a reciprocal relationship between the rates of RNA and HS synthesis; and (3) there are three periods in the developmental sequence when either RNA, DNA, or HS display high rates of synthesis in that order. There is a period of intense transcriptional activity that precedes and partially overlaps the period of DNA synthesis. Translational metabolism was dispersed

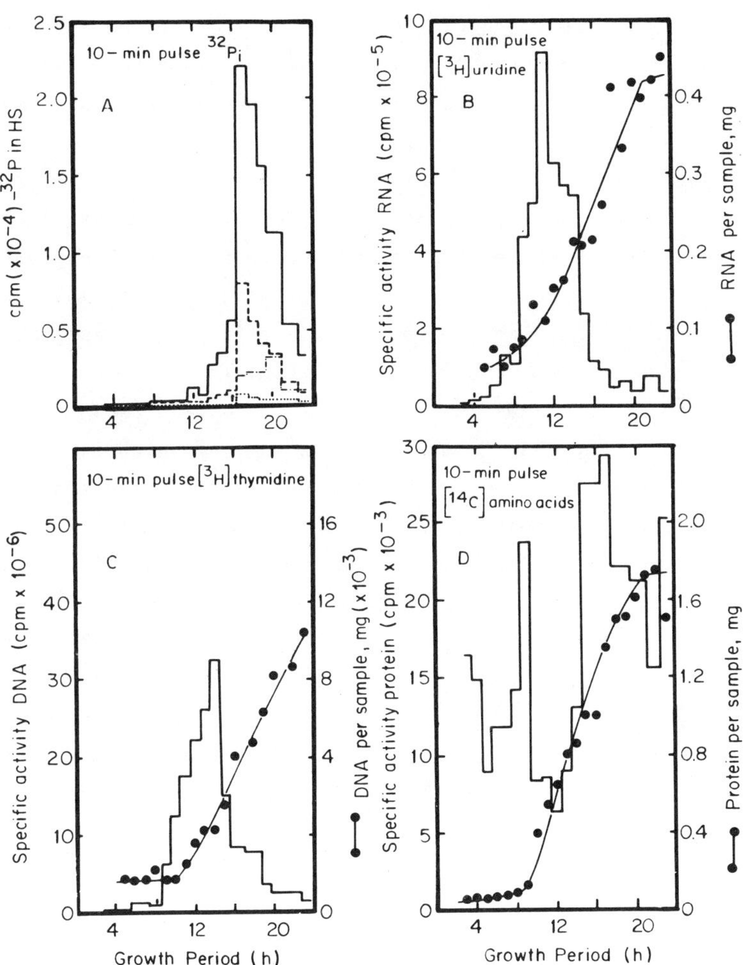

FIG. 3. Relative rates of labeling of (A) HS-3 (———), HS-2 (-----), HS-1 (•••••), and GTP (-•-•-) by [^{32}P]orthophosphate; (B) net RNA by 5,6-[^{3}H]uridine; (C) DNA by [^{3}H-methyl]thymidine; and (D) protein by ^{14}C(U)amino acid mixture (15 amino acids) during ontogeny of *Achlya*. Total RNA, DNA, and protein specified by the symbol (●) in the relevant panels. Cultures were pulsed in duplicate for 10 min at hourly intervals with the respective label before analyses.

throughout the entire life cycle with peak periods in the first 8 h. During this time, transcription rate was very low. A second period when translation rate was high (16 h) marked another low period in the rate of transcription. Data which will be presented elsewhere show that the bulk of RNA made during this intense burst of transcriptional activity is ribosomal.

The net synthesis of RNA, DNA, and protein followed the expected pattern, increasing exponentially during the period of active growth (6 to 20 h). These results are incorporated into Fig. 3B-D. Therefore, even though the rate of incorporation of macromolecular precursors declined during the second half of the developmental period, both RNA and DNA continued to accumulate. The decline in specific activities of RNA and DNA after 11 and 14 h, respectively, may be due to an increase in the cellular pool of unlabeled precursors prior to pulsing. The introduction of labeled precursors in this interval would result in an "apparent" lowering of the rate of synthesis through dilution effects and due to problems of entry of the labels into their cellular pools from which they are withdrawn for nucleic acid and protein syntheses. An observation which is at variance with this conjecture is that the high rates of labeling of HS compounds and of nucleotides such as GTP, ATP, GDP, and GMP occurred during the interval when the rates of RNA and DNA synthesis were declining (data not shown). But this important question remains unanswered at present.

C. Fate of HS Compounds During Growth

Because the cellular content of HS compounds decreased during the late stages of growth, we examined the possibility that some or all of these substances may be eliminated from the cell. Medium from cells growing synchronously was collected at hourly intervals, concentrated, and analyzed chromatographically. The results are shown in Table 2. A significant portion of each compound was excreted into the medium at these late stages of development. No HS

TABLE 2

Efflux of HS Compounds
into Growth Medium During Late Stages of Development

Age culture (h)	(cpm × 10^{-5})					
	HS Compounds in cells			HS Compounds in medium		
	HS-3	HS-2	HS-1	HS-3	HS-2	HS-1
15	92.12	46.16	26.83	2.74	0.78	2.26
16	92.03	39.01	25.99	6.12	2.67	3.98
17	58.22	34.01	12.92	7.68	4.12	6.61
18	50.42	27.98	15.73	20.14	8.89	11.49
19	33.65	29.49	7.69	---	---	---
20	31.99	25.12	4.83	33.2	20.5	17.8
21	——— Inception of sporulation ———					

compounds were detected extracellularly before 12 h. About 80% of HS-1 and 50% each of HS-2 and HS-3 were found in the medium at the peak period of excretion coincident with the stage of sporangial maturation. Thus, if these compounds were being metabolized at all, this happened during the early stages of growth. In view of their kinetics of synthesis with respect to nucleic acid biosynthesis (Fig. 3), a regulatory role is suggested. Evidence in support of this is presented in a later section.

D. Relationship Between RNA Polymerases and HS Compounds

RNA polymerases were isolated from nuclei of *Achlya* by the method of Cain and Nester [19] and from the entire cell by the procedure of Young and Whiteley [20]. The enzymes obtained by either isolation procedure showed similar chromatographic behavior on DEAE-cellulose and DEAE-sephadex A-25 columns. Typical activity profiles are shown in Fig. 4A for enzymes isolated from 7.5- and 12-h-old growing cells. The heterogeneity of the various enzyme types observed is reproducible. As shown, the same method of

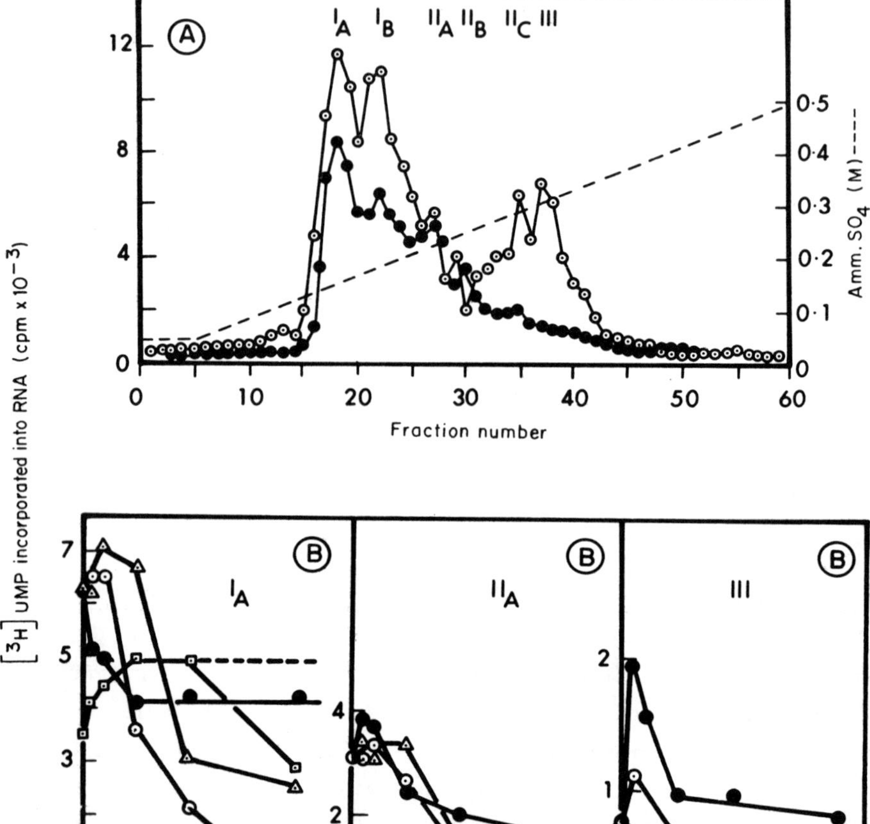

FIG. 4. Rates of synthesis of macromolecules and HS compounds. (A) Activity profiles of DEAE-sephadex A-25 column chromatography of RNA polymerases (RNA-P) from *Achlya* after 7.5 h (⊙) and 12 h (●) of growth from spores. The designations I_A through III represent enzymes that are kinetically distinguishable. (B) Differential inhibition and activation of RNA-PI, RNA-PII, and RNA-PIII by HS compounds. In the left panel, effects of HS-3 (⊙), HS-2 (◬), and HS-1 (●) on RNA-PI_A and of HS-1 (⊡) on RNA-PI_B. In the center panel, effect of HS-3 (⊙), HS-2 (◬), and HS-1 (●) on the RNA-PII_A. In the right panel, effects of HS-3 (⊙), HS-2 (◬), and HS-1 (●) on RNA-PIII. Note: Large-scale cultures of *Achlya* in 15 liters of growth medium (see Fig. 1 legend) were vigorously aerated and, at the appropriate times, samples of cells were withdrawn aseptically and fractionated for enzyme isolation. The medium was inoculated with 15 × 10^7 spores and growth temperature was 28°. Both the Cain-Nester [19] and Young-Whiteley [20] procedures were used for enzyme isolation.

isolation produced predominantly type I_A enzyme and none of type I_B at 14 h. The reverse was true for 15-h culture, suggesting that the difference may not be artefactual.

To determine whether the apparent reciprocal relation between HS production and RNA synthesis has any biochemical or physiological basis, we examined cells at different stages of growth for their cellular content of HS compounds and, at the same time, we checked to see what forms of RNA polymerases might be present. A summary of the findings is given in Table 3.

Superficially, the results indicate that there is a transition from early developmental stages up to 12 h when RNA polymerase I_A and RNA polymerase I_B predominate and to the later stages (15 to 20 h) when only RNA polymerase II and RNA polymerase III are detectable. Interestingly, this correlates with the two periods in the growth cycle when cellular HS fluctuated from high to low, irrespectively.

A complication which made the results difficult to interpret is the finding that the spectrum of enzymes detectable depends upon the material from which they were isolated. When intact cells were disrupted by the Young-Whiteley [20] procedure and enzymes isolated, no RNA polymerase II_C and RNA polymerase III were found at 12 h although

TABLE 3

Detectable Activities of *Achlya* RNA Polymerases and Relation to HS Pool Sizes During Development[a]

% of maximum HS pools	RNA Polymerases I_A	I_B	II_A	II_B	II_C	III	Age culture (h)
5-10	+	+	+	+	+	+	7.5
20-30	+	+	+	+	-	-	12.0
80-85	-	-	+	+	+	+	14.0
95-100	-	-	-	+	+	+	15.0

[a]+, enzyme detectable under optimum assay conditions; -, enzyme undetectable under optimum assay conditions.

these two enzymes were present in nuclear preparations of cells of the same age (Table 4)! No RNA polymerase I was detectable in isolated nuclei at 14 h whereas RNA polymerase I_A was present in whole cell extract of 14-h culture. This implied that the enzymes probably exist in engaged (DNA-linked) and disengaged (cytoplasmically localized) states. In the latter case, they may be in the stage of being synthesized for import into the nucleus and so would not be physiologically operative then.

Another observation which is difficult to explain is that RNA polymerase II_C and RNA polymerase III were detected in nuclear isolates but not in whole cell extracts at 15 h. The simplest explanation that can be proffered is that proteolytic enzymes may be present in the cytoplasm and are inactivating those polymerases that are most susceptible to proteolytic attack. When RNA polymerases are collected from nuclei that have been washed free of cytoplasmic material, this problem is eliminated. If this is the case, it will be indicated by mixing experiments which will be done in due course.

For two reasons--(1) an apparent reciprocal correlation between rates of RNA synthesis and HS production (Fig. 3B) and (2) the apparent absence of RNA polymerase I_A and RNA polymerase I_B in nuclei at 14 and 15 h when HS levels are optimal in the cell--we

TABLE 4

Dependence on the Method of Isolation of DNA-Dependent RNA Polymerases from *Achlya* Influencing the Species of Enzyme that Could be Detected

Age culture	RNA Polymerases detectable											
	Cain-Nester						Young-Whiteley					
(h)	I_A	I_B	II_A	II_B	II_C	III	I_A	I_B	II_A	II_B	II_C	III
7.5							+	+	+	+	+	+
12.0	+	+	+	+	+	+	+	+	+	+	-	-
14.0	-	-	+	+	+	+	+	-	+	+	+	+
15.0	-	-	-	++	+	+	-	+	+	+	-	-

decided to investigate what effects these compounds might have on the activities of the polymerases.

Individual enzymes were collected after chromatographic separation and tested for their sensitivities to HS compounds in transcription. As shown in Fig. 4B, HS-3 proved to be a potent inhibitor of all classes of RNA polymerase isolated from *Achlya*. HS-2 was also inhibitory but less so than HS-3. HS-1 showed two effects: activating RNA polymerase I_B but inhibiting RNA polymerase I_A weakly. It also stimulated RNA polymerase III threefold at low concentrations. The results of a comprehensive but incomplete study of the effects of the compounds on the various species of enzymes are given in Table 5. HS-2 showed a slight activating effect on RNA polymerase II_C only whereas HS-3 has never shown activating tendency on any enzyme.

The majority of enzyme assays were carried out with calf thymus DNA as template. We observed, however, that the HS compounds inhibited transcription better if DNA isolated from *Achlya* was used as template. Typical inhibition curves are shown in Fig. 5B for RNA polymerase I_A and RNA polymerase I_B. In this range of HS concentration, where as much as 30 and 50% of the enzymic activities were inhibited with *Achlya* DNA, no inhibitory effect was manifested

TABLE 5

Effect of Purified HS Compounds on the Activity of DNA-Dependent RNA Polymerases from 7.5- and 12-h-old Cells[a]

HS Compound/assay	RNA Polymerases					
(10-100 μg/ml)	I_A	I_B	II_A	II_B	II_C	III
HS-3	=	=	=	=	nt	=
HS-2	=	=	=	nt	++	=
HS-1	-	++	-	=	nt	+++

[a] -, weak inhibition; =, strong inhibition; +, mild activation; ++, moderate activation; +++, strong activation; nt, not tested.

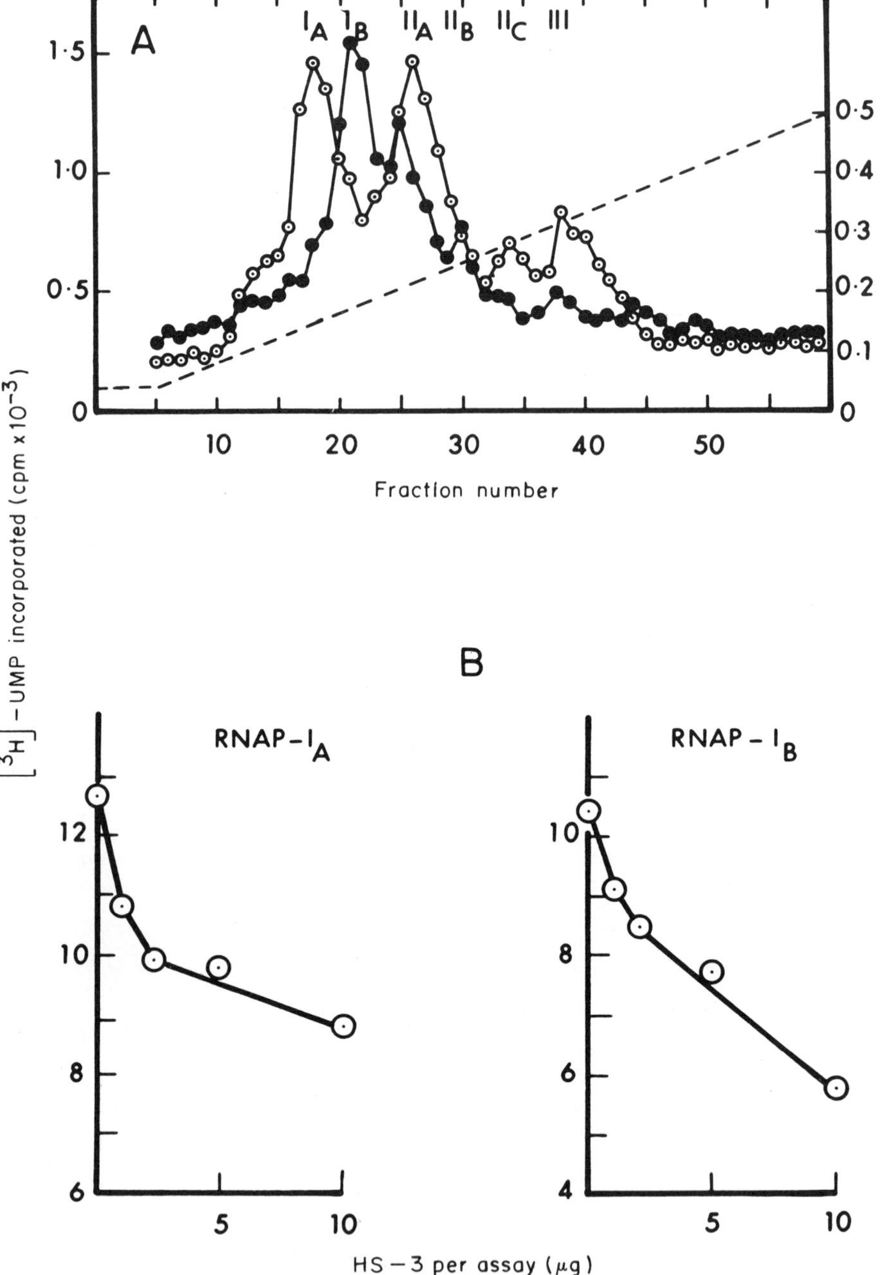

FIG. 5. (A) Activity profiles of DEAE-sephadex A-25 column chromatography of RNA polymerases (RNA-P) from *Achlya* after 14 h (⊙) and 15 h (●) of growth from spores. (B) Inhibition of RNA polymerases I_A and I_B by HS-3 using *Achlya* DNA isolated as described by Dodd et al. [29] as template. Assay conditions have been outlined [11].

with calf thymus DNA and poly[d(A-T)] as templates. To what extent this template preference is true for the other RNA polymerases is yet to be explored.

The HS compounds were without effect on the activities of the RNA polymerase holoenzyme and core enzyme of *Escherichia coli* but did inhibit RNA polymerase I_A and RNA polymerase I_B (the only enzymes tested so far) from a phylogenetically unrelated water mold, *Blastocladiella emersonii*, to the same extent as *Achlya*.

E. Isolation and Characterization of HS Compounds

The method of isolation of cellular nucleotides including HS compounds has been described [10]. The relative positions of HS-1, HS-2, and HS-3 with respect to GTP and ATP when they are separated on PEI-cellulose thin-layer sheet with pH 3.65 phosphoric acid in one dimension are shown in Fig. 6. Using the two-dimensional chromatographic system of Cashel and Kalbacher [21], but phosphate solvent at pH 3.65, the nucleotides and HS compounds were separated as shown in Fig. 7. The shaded areas depict radioactive regions of the chromatogram and all compounds identified were through the use of authentic markers.

HS compounds were isolated by the one-dimensional procedure and spectrally analyzed under acid (0.1 N HCl) and alkali (0.01 N NaOH) conditions. Estimates were made of their respective pK values and the data are summarized in Table 6. They are compared with spectral data of nucleoside triphosphates [22,23] and 7 methyl guanosine (7mG). While the pK values of the HS compounds were similar to data for 7mG, we have no evidence that the HS compounds are methylated (data not shown). The spectral data favor a relationship of HS compounds to GTP but we have not succeeded in obtaining typical guanosine UV spectral scans on the freshly isolated compounds. But when they are chromatographed with 1.6 M LiCl, pH 3.4, and heated to 50° for several days, depurination occurs and the strong UV-absorbing material collected from PEI-cellulose plates

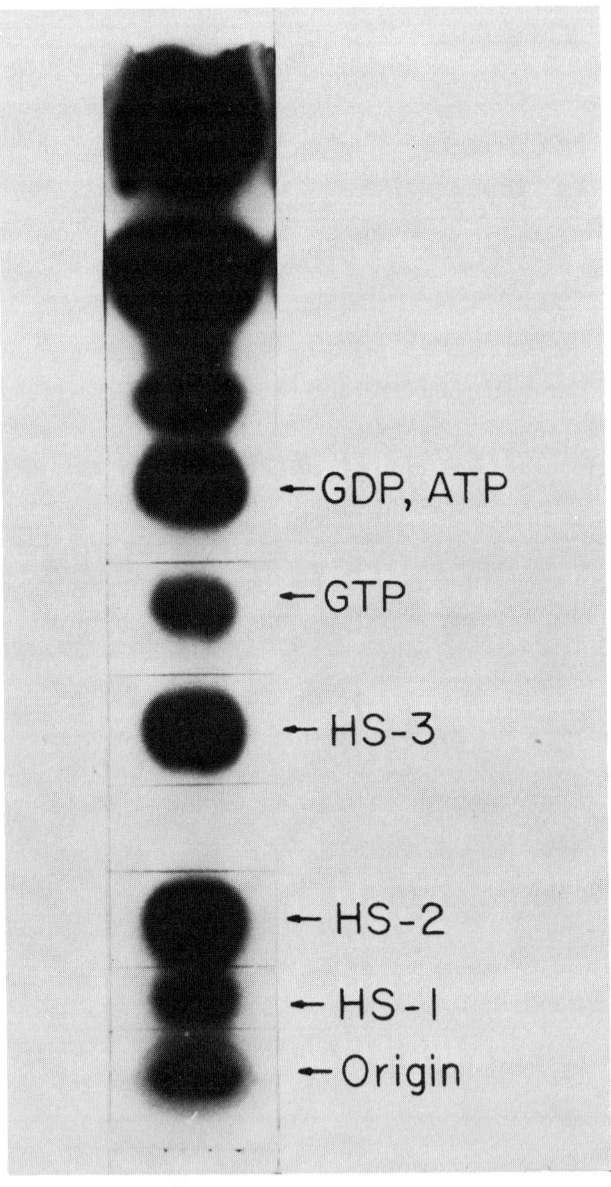

FIG. 6. Autoradiogram of a one-dimensional PEI-cellulose thin layer chromatograph of acid-soluble nucleotides and HS compounds from whole cells of *Achlya* labeled with [^{32}P]orthophosphate. The positions of readily identifiable compounds are indicated. Solvent system as defined by Cashel and Kalbacher [21] but pH 3.65 used in place of pH 3.4 phosphoric acid.

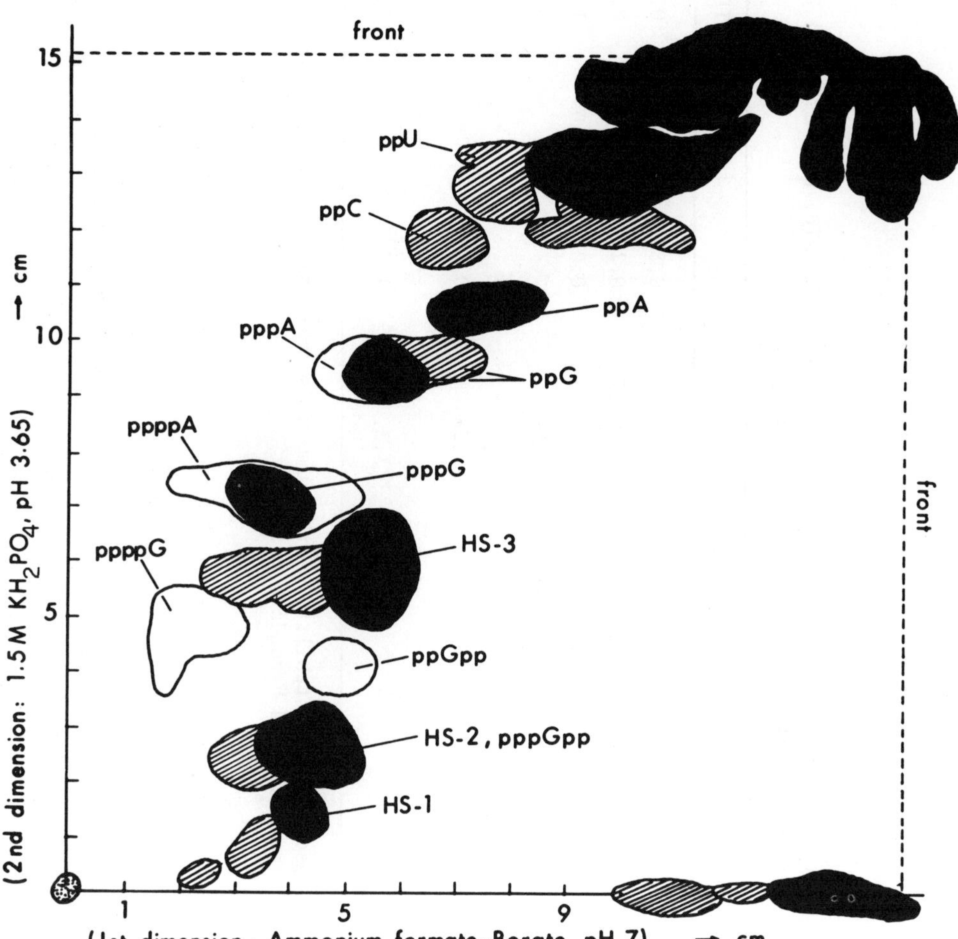

FIG. 7. Autoradiogram and uv identifiable acid-soluble nucleotides and HS compounds extracted from whole cells of *Achlya* separated by two-dimensional chromatography of PEI-cellulose thin-layer sheets. Dark areas represent strong radioactivity and cross-hatched areas weak radioactivity. Unshaded areas are nonradioactive uv absorbing markers. Other identified compounds marked are authentic samples cochromatographed with the extract. Authentic ppGpp and pppGpp kindly provided by Dr. Michael Cashel. The two-dimensional chromatographic system of Cashel and Kalbacher [21] was employed.

TABLE 6

Spectral Data of HS Compounds and Comparison with Literature Values for GTP, ATP, UTP, CTP, and 7mG[a]

Spectral ratios	HS-3	HS-2	HS-1	7mG	GTP	ATP	UTP	CTP
$A_{250-260}$	0.95	0.94	0.93	---	0.96	0.85	0.75	0.45
$A_{280/260}$	0.69	0.67	0.72	0.79	0.67	0.22	0.38	2.12
$A_{290/260}$	0.54	0.55	0.56	---	0.49	0.027	0.04	1.67
λ_{max} (pH 1)	257	258	258	256-258	256	257	262	280
λ_{max} (pH 12)	255	256	256	---	---	---	---	---
λ_{min} (pH 1)	240	242	242	230	228	230	230	242
λ_{min} (pH 12)	237	235	236	---	---	---	---	---
pK	7.2[b]	7.4[b]	---	7.1	3.3, 9.3	4.0	9.5	4.8

[a] See Refs. 22 and 23. [b] Because of the instability of the compounds, this may represent a composite of pK values for the intact molecule and breakdown products.

Note added in proof: A recent recharacterization of HS compounds has determined that the partial structures are HS3, ADP-glutamate:ribitol-phosphate:uracil-polyolpentaphosphate; HS2, UDP:ribitol-phosphate:uracil-polyoloctaphosphate; and HS1, a higher phosphorylated derivative of HS2. Preliminary notes to these characterizations can be found in references 24 and 25. A detailed report on the chemical analysis and structural determination of the HS compounds is in preparation [33].

give spectral scans identical to that of guanine in acid. In alkali, the scans for the standard guanine and the isolated substance are not identical. These results indicate that the HS compounds are likely to be nucleotides of very unusual structures.

Some further characterization of the HS compounds was achieved through limited and complete enzymic digestion. Intact, HS-3 was resistant to hydrolysis by bacterial alkaline phosphatase whereas HS-2 and HS-1 were converted to HS-3 (starting with HS-2) and to HS-3 and HS-2 (starting with HS-1) and orthophosphate. With prolonged incubation in acid or alkali, all three compounds were decomposed. HS-2 and HS-1 were decomposed in the same way as when treated with alkaline phosphatase. These results indicate that HS-2 and HS-1 are higher polyphosphates of HS-3.

The fact that HS-3 is resistant to alkaline phosphatase signifies that it does not contain free phosphates that are susceptible to enzyme attack. This would be possible if the compound is a dinucleotide with a polyphosphate chain covalently linked to nucleosides at the ends.

Attempts were made to resolve this question by the use of phosphodiesterases of spleen and viper venom on the compounds. An unexpected finding was that under special conditions (pH 7.5 to 8.0, 0.15 to 0.25 M Tris-acetate or HCl buffer, 10 mM Mg^{2+}) spleen phosphodiesterase slowly converted the parent HS compounds into a single radioactive product that migrated more slowly than the parent on PEI-cellulose under the chromatographic system described previously [10]. Such a result implies that there has been an exposure of buried negative charges (phosphate?) present in all HS compounds. Viper venom phosphodiesterase did not react with the compounds when they were freshly prepared. On aging, they became susceptible to attack. As spleen phosphodiesterase should cleave off 3'-mononucleotides from phosphate-free ends of polynucleotides, and viper venom phosphodiesterase should release 5'-mononucleotides from phosphate-free 3' terminus of a polynucleotide, the results indicate that these compounds do not have the typical polynucleotide structures. The only way spleen phosphodiesterase would react with the compounds in the pattern described is if the nucleosides are linked 5'-5' via the phosphates, thereby neutralizing some of the phosphate charges, and the enzyme were to cleave off a nucleoside at one end. Normally, such a molecule with free hydroxyls at the 3' ends should react with viper venom phosphodiesterase. But if these ends are modified or blocked, it is conceivable that the reaction would not occur. Whether blocking agent is removed upon aging so that the compounds now become susceptible to viper venom phosphodiesterase attack is a problem that needs further investigation.

The alternative is a dinucleotide structure with 3'-3' linkage. Spleen phosphodiesterase attack of this compound should normally yield a 3'-mononucleotide as well as a larger fragment. Only one product that is radioactive was found in every case, even when about 70% of the parent compound had been converted.

Nucleotide pyrophosphatase and viper venom phosphodiesterase, when they hydrolyzed aged preparations of HS compounds, yielded two products--one migrating in the area of GTP and the other in the area of mononucleotides. Sufficient quantities of material are not yet available for spectral characterization. However, since only a single nucleic acid base with spectral properties matching those of guanine (modified) was detected in depurinated HS compounds, the tentative conclusion is that these compounds are guanosine polyphosphates of extremely unusual structures.

It is worth mentioning that these HS compounds have been found in a variety of fungi [10], Chinese hamster ovary cells in culture, and SV40 transformed and untransformed 3T3 cells [24].

F. Influence of Fe^{3+} on Cellular HS Levels

During the course of nutritional studies we found that the cellular levels of HS compounds can either be elevated when Zn^{2+} is added to a defined medium culture or depressed when Fe^{3+} is added.

The same result was obtained with cells growing in glucose-yeast extract medium. The level of HS compound varied as a function of Fe^{3+} concentration (see Table 7). The tissue concentrations of HS-3 and HS-2 dropped more than tenfold and HS-1 increased about threefold when 10 to 40 μg Fe^{3+} was added as supplement. If HS-1 is a precursor for HS-2 and HS-3, it is possible that Fe^{3+} is blocking the interconversion of HS-1 to HS-2, leading to its accumulation and attendant reduction of the lower polyphosphate homologues.

A similar depression in the pool sizes of GTP, ATP, and GDP was observed, suggesting that it is a general reduction of the cellular content of purine ribonucleotides. Speculating that the HS compounds were in some way regulating the synthesis of nucleic acids, we examined the influence of these compounds on ribonucleotide reductase of *Achlya* and Chinese hamster ovary cells. The surprising result was that HS-3 and HS-2 were potent inhibitors of the enzyme [25]. It would seem then that the interconversion of ribo- and deoxyribonucleotides is a central point at which HS compounds act to mediate RNA and DNA synthesis. What effect the compounds might have on DNA polymerases of these cells is currently being examined.

TABLE 7

Influence of Fe^{3+} on the Cellular Levels of HS Compounds and Nucleotides in *Achlya*

Fe^{3+} Supplement	Optimum level of label in HS compounds and nucleotides (cpm × 10^{-4})				
(μg/ml)	HS-3	HS-2	HS-1	GTP	ATP
nil	35	25	4.5	1.5	11
20	5	4	13.5	0.5	4
30	4	1	12.0	0.25	3
40	3.5	0.5	10.0	0.22	2.3

III. FINAL COMMENTS

A. RNA Accumulation and Development

During normal growth, the fungal cell, like bacteria [26], synthesizes only that rRNA (ultimately ribosomes) that it requires for the highest efficiency in protein synthesis. A knowledge of the molecular mechanism that maintains this equilibrium between synthesis and activity is crucial to our understanding of cellular development. It is in bacteria that it was first realized that RNA and protein synthesis may be coupled [27], since depriving the cell of an essential amino acid leads to a drastic reduction in the net synthesis of RNA--a feature that was termed "stringent response." But it is not only amino acid starvation that can produce this change; alterations in the supply of carbon and/or nitrogen can cause either an "upshift" or a "downshift" in growth [28], leading to fluctuations in the rate of RNA synthesis.

A regulatory element that has been proposed for this stringent control in bacteria is the unusually phosphorylated guanosine tetraphosphate, ppGpp. But the current evidence is not entirely persuasive [5,9], since upshift and downshift conditions mediated by nutritional changes other than that imposed by amino acid supply do not differentially affect stringent and "relaxed" cells in their ability to produce and break down ppGpp [29].

Our studies on the regulatory aspect of RNA synthesis in *Achlya* have led us to the discovery of three highly phosphorylated compounds dubbed HS-3, HS-2, and HS-1. The patterns of their production during the ontogenic cycle of the mold have shown that they may relate in some unknown way with nutritional shifts, rather than amino acid starvation. But there are some features in HS synthesis that are analogous to ppGpp and pppGpp production in bacteria. For instance, there is a reciprocal relation between high rates of RNA synthesis and HS accumulation. The HS compounds, as ppGpp, inhibit the enzymic activities of native and related RNA polymerases. Both types of compounds accumulate in large quantities when growth rate declines through starvation.

To relate HS production with the path of development is even more difficult because no evidence is available to show that exogenously supplied HS can alter the established progress of cell differentiation. All evidence in this regard is indirect. The high rate of synthesis of rRNA always precedes the burst of HS accumulation and it ceases when these compounds are optimal in the cell. There is no firm direct correspondence between rates of protein DNA and HS syntheses, although this may be due to the coarse nature of the experimental design. Some insight into this may be gleaned from in vitro protein and DNA syntheses currently in progress.

We have evidence that these HS compounds are potent inhibitors of ribonucleotide reductases from *Achlya* and from Chinese hamster ovary cells. This enzyme is central in the conversion of ribonucleotides to deoxyribonucleotides. Another related finding is that Fe^{3+} supplied in microgram quantities has the remarkable effect of depressing the levels of HS-3 and HS-2 to mere trace quantities (Table 7), but of elevating the level of HS-1. As Fe^{3+} is an important component in the activity of ribonucleotide reductase in microbial systems [30], it is not unrealistic to imagine that HS compounds may play a crucial role in regulating nucleic acid biosynthesis at the level of nucleotide interconversions.

B. Inhibition and Activation of RNA Polymerases

We tentatively conclude that the in vitro inhibitory effects of HS compounds on RNA polymerase I may also be operative in vivo on the strength of the observation that the rate of rRNA synthesis declined to zero when HS pools were maximal. But what of the stimulatory effects of HS-1, particularly on RNA polymerase III? Only a conjecture can be offered here in the light of the possible involvement of HS compounds on enzymes of nucleotide biosynthesis. As synthesis or accumulation of HS-1 is not inhibited by Fe^{3+} (tendency is for Fe^{3+} to enhance its production) and HS-1 enhances the activity of RNA polymerase III, it is conceivable that RNA polymerase III may participate in some aspect of DNA biosynthesis, possibly in making RNA primers in addition to other roles.

C. Looking Ahead

It is premature to evoke models of transcriptional, translational, and replicational regulation of synthesis now. A better approach is to draw attention to some of the potentially useful areas of study. To begin with, there is scant information on the effects of HS compounds on DNA and protein syntheses in vitro. Such a study would indicate whether or not HS compounds have a global control on cellular metabolism. Some of the data hint at the idea that HS compounds are polyphosphates with energy storage roles in which case they could have a central role in the bioenergetics of the organism. How they could be involved is a potential area of study. Finally, their potential to act as extracellular signals of development, not unlike the role of cyclic AMP in slime mold development [31], may yet prove to be more grist for the experimentalist.

ACKNOWLEDGMENT

This research was supported by grant A4292 from the National Research Council of Canada.

REFERENCES

1. Chesterton, C. J., and P. H. W. Butterworth, 1971. FEBS Letters 15, 181-185.

2. Kedinger, C., P. Nuret, and P. Chambon, 1971. FEBS Letters 15, 169-174.

3. Roeder, R. G., and W. J. Rutter, 1969. Nature 224: 234-237.

4. Goldberg, M., J.-C. Perriard, G. Hager, R. B. Hallick, and W. J. Rutter, 1973. *In* Control of Transcription, B. B. Biswas, R. K. Mandal, A. Stevens, and W. E. Cohn (eds.), Plenum Press, New York, pp. 241-256.

5. Travers, A., 1973. *In* Control of Transcription, B. B. Biswas, R. K. Mandal, A. Stevens, and W. E. Cohn (eds.), Plenum Press, New York, p. 67.

6. Travers, A., 1971. Nature New Biol. 229: 69-74.

7. Aboud, M., and I. Pastan, 1975. J. Biol. Chem. 250: 2189-2195.

8. Travers, A., R. Kamen, and M. Cashel, 1970. Cold Spring Harbor Symposium Quant. Biol. 35: 415-418.

9. Haseltine, W. A., 1973. Nature 235: 329-332.

10. LéJohn, H. B., L. E. Cameron, D. R. McNaughton, and G. R. Klassen, 1975. Biochem. Biophys. Res. Commun. 66: 460-467.

11. McNaughton, D. R., G. R. Klassen, and H. B. LéJohn, 1975. Biochem. Biophys. Res. Commun. 66: 468-474.

12. Griffin, D. H., W. Timberlake, J. Cheny, and P. A. Horgen, 1975. *In* Isozymes I, C. Markert (ed.), Academic Press, New York, p. 69.

13. Finamore, F. J., and A. H. Warner, 1963. J. Biol. Chem. 238: 344-348.

14. Oikawa, T. G., and M. Smith, 1966. Biochemistry 5: 1517-1521.

15. Finamore, F. J., and J. S. Clegg, 1969. *In* The Cell Cycle: Gene-enzyme interactions, G. M. Padilla, G. L. Whitson, and I. L. Cameron (eds.), Academic Press, New York, p. 249.

16. Adams, J. M., and S. Cory, 1975. Nature 255: 28-33.

17. Muthukrishnan, S., G. W. Both, Y. Furuichi, and A. J. Shatkin, 1975. Nature 255: 33-37.

18. Abraham, B., D. P. Rhodes, and A. K. Banerjee, 1975. Nature 255: 37-40.

19. Cain, A. K., and E. W. Nester, 1973. J. Bacteriol. 115: 769-776.

20. Young, H. A., and H. R. Whiteley, 1975. Exp. Cell Res. 91: 216-222.

21. Cashel, M., and B. Kalbacher, 1970. J. Biol. Chem. 245: 2309-2318.

22. Burton, K., 1969. *In* Data for Biochemical Research, R. M. C. Dawson, D. C. Elliott, W. H. Elliott, and K. M. Jones (eds.), Second ed., Oxford University Press, London, p. 169.

23. R. H. Hall, 1971. The Modofied Nucleosides in Nucleic Acids, Columbia University Press.

24. Goh, S. H., and H. B. LeJohn, 1977. Biochem. Biophys. Res. Commun. Vol. 1, Jan. 10 (in press).

25. Lewis, W. H., D. R. McNaughton, H. B. LeJohn, and J. A. Wright, 1976. Biochem. Biophys. Res. Commun. 71: 128-135.

26. Maaloe, O., and N. O. Kjeldgaard, 1966. Control of Macromolecular Synthesis, W. A. Benjamin, New York.

27. Sands, M. K., and R. B. Roberts, 1952. J. Bacteriol. 63: 505-511.

28. Neidhardt, F. C., 1964. Prog. Nucleic Acid Res. & Molec. Biol. 3: 145-181.

29. Lazzarini, R. A., M. Cashel, and J. Gallant, 1971. J. Biol. Chem. 246: 4381-4385.

30. Coughlin, M. P., 1971. Sci. Prog. Oxf. 59: 1-23.

31. Knoijn, T. M., J. G. C. Van de Meene, J. T. Bonner, and D. S. Barkley, 1967. Proc. Nat. Acad. Sci. USA 58: 1152-1157.

32. Dodd, J. G., P. A. Horgen, and N. Straus, 1975. Cytobios 13: 31-36.

33. McNaughton, D. R., G. R. Klassen, P. C. Loewin, and H. B. LeJohn. Manuscript submitted.

CELL DIFFERENTIATION IN *NAEGLERIA*

Allan Dingle

Department of Biology
McMaster University
Hamilton, Ontario, Canada

> "If the Lord Almighty had consulted me before embarking upon the Creation, I should have recommended something simpler."
>
> Alphonso X of Castile,
> 15th Century

I. INTRODUCTION

The above prologue likely reflects the attitudes of many other contributors to this book. We are primarily interested in problems of development and especially intrigued by the phenomenon of cell differentiation. However, because of technical and conceptual difficulties involved in studying cytodifferentiation in embryonic systems, we have retreated in our studies to the relative simplicity of one or another of the "model systems" being promoted in this volume. In presenting *Naegleria gruberi* as a model for the study of cell differentiation, my goal thus becomes twofold: I wish to demonstrate that the *Naegleria* model is uniquely suited to the exploration of certain questions about cell differentiation and, secondly, I hope that my review of the methodology employed and some results already obtained may provide new perspectives to developmental biologists studying cell differentiation in embryonic systems.

Though *Naegleria* species are very widespread in nature and are readily isolated from both soil and water samples, it is not a well-known organism and the taxonomy of *Naegleria* has not been satisfactorily resolved. The usefulness of *Naegleria* as a model for cell differentiation studies becomes apparent upon inspection of what Fulton [1] has amusingly called the "life tricycle" of *Naegleria* (Fig. 1). Amoebas (10 to 12 μm in diameter), at the center of the tricycle, have three options available to them: (1) they may grow and divide indefinitely, without phenotypic change; (2) they may transform into fusiform flagellates, developing new organelles and an altered cell shape; or (3) they may cease amoeboid streaming, synthesize a different complement of molecules, and assume a

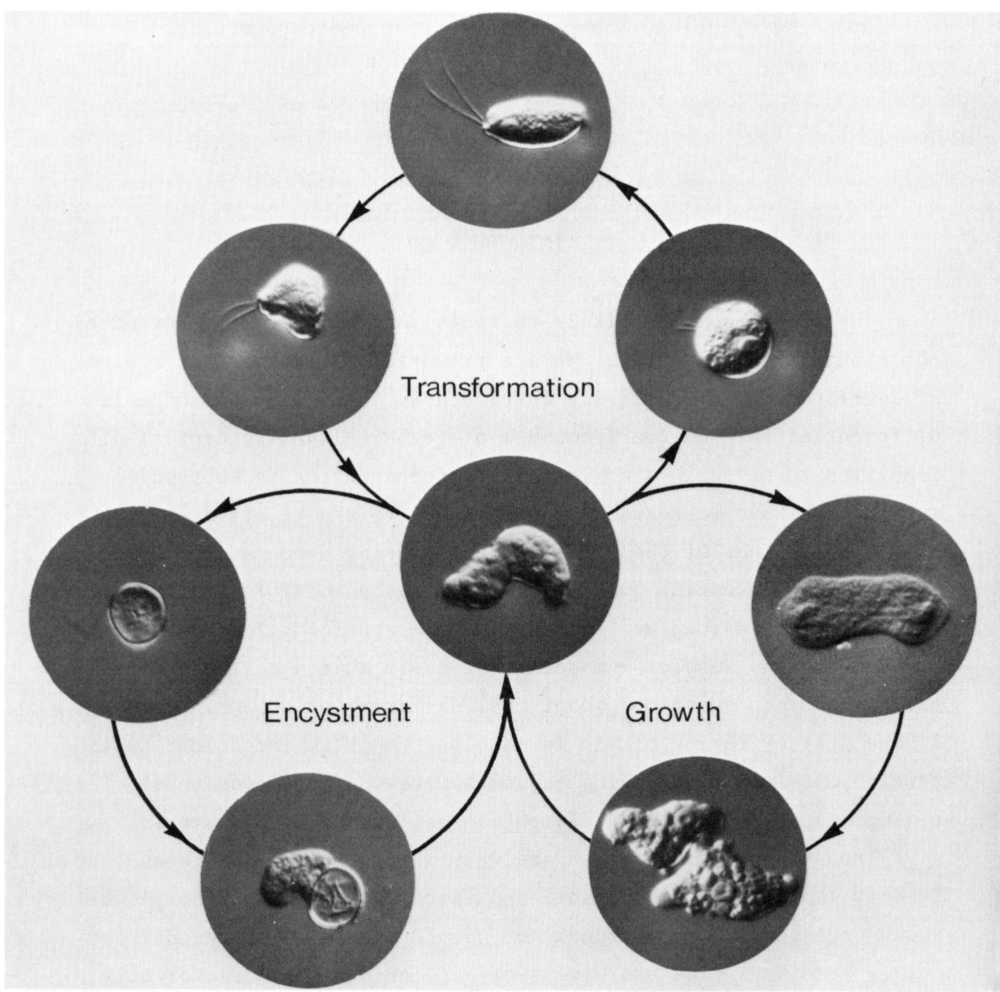

FIG. 1. The life tricycle of *Naegleria*. All cells fixed in glutaraldehyde and photographed under Leitz interference contrast optics.

double-walled spherical cyst form. Yet another cell differentiation is available for study at this point in the *Naegleria* tricycle. The dormant cyst may be induced to excyst by diverse environmental stimuli, liberating an amoeba from one of the several preformed exit pores.

With the exception of encystment, each phase of the *Naegleria* tricycle has been brought under impressive control in the laboratory, such that cells grow or differentiate both rapidly and predictably. Cell division is restricted to the amoeboid phase and methods are available for growing amoebas with doubling times less than 2 h [2] and for obtaining axenic [3], synchronized [4], or large-batch [5] cultures of *Naegleria*. The amoebo-flagellate transformation has been domesticated to a point where carefully performed experiments are "reproducible to within 3 min from day to day" [1]--a claim which surely must cause envy among developmental biologists studying cell differentiation in embryos. And, though not so impressive as the amoebo-flagellate transformation, rapid and reproducible excystment has been obtained in response to bacteria and to elevated CO_2 levels [6], or by simply immersing the cysts of some *Naegleria* strains in aqueous medium [1]. Thus, encystment is the only phase of the tricycle which remains to be brought under precise control as a model differentiation. Preliminary studies in my laboratory suggest that both plate-grown and liquid cultures of *Naegleria* amoebas can be made to encyst with approximately the same parameters as well-defined systems, such as *Acanthamoeba* encystment [7]. Aspects of encystment in *Acanthamoeba* are discussed by John Thompson in Part III of this volume.

The preceding description of *Naegleria's* life tricycle, and the laboratory control over it, illustrates the remarkable utility of the organism as a tool with which to study cell differentiation. Cells at any phase of the cycle can be stimulated by appropriate environmental cues to redirect their synthetic activities and to assume a radically altered phenotype. However, it is the amoebo-flagellate transformation cycle of *Naegleria* which provides our most appealing model for cell differentiation. I believe that no other eucaryotic differentiating system possesses the same constellation of desirable attributes: all members of a genetically homogeneous population are converted from one distinctive phenotype to another in less than 2 h from a precisely defined moment of initiation. Furthermore, the differentiation is readily measurable, highly reproducible, and relatively synchronous. The essential changes that occur when an amoeba differentiates into a flagellate include: (1) a decrease in cell volume and progressive shape change from an amoeboid shape through a sphere to an elongate spheroid; (2) an abrupt cessation of cytoplasmic streaming and relocation of the nucleus and contractile vacuole in stationary positions; and (3) synthesis of specific macromolecules and the integrated assembly of new cell organelles--basal bodies, flagella, and flagellar rootlets. Given this rapid and dramatic amoebo-flagellate transformation and the ability to measure events of differentiation with great precision, two fundamental questions about cell differentiation seem to be most appropriate for study using the *Naegleria* model:

1. What is the "inductor" of transformation, i.e., what environmental stimuli initiate differentiation along the pathway to a flagellate?
2. What essential molecular and physiological responses to these stimuli lead to the synthesis and coordinated assembly of new cell organelles?

II. CURRENT RESEARCH

A. The Stimulus for Flagellate Differentiation

Embryologists seeking insights into inductive mechanisms in cell differentiation will recognize intriguing possibilities in current studies of the initiation stimulus for the amoebo-flagellate differentiation. The essential lesson from *Naegleria* is that transformation may be initiated in response to a variety of simple physical and chemical signals, acting independently or in concert [8].

All early studies of the amoebo-flagellate transformation suffered from the inability of investigators to define the actual environmental alteration signalling start. For example, prior to the development of the so-called "transformation experiment" [2] (see below), flagellate differentiation was obtained simply by removing a loopful of amoebas from poorly defined growth cultures into a drop of water on a microscope slide [9,10] or by washing amoebas into dilute buffer or water to free them of bacteria grown with them on nutrient agar [11,12]. Even when methods became available for obtaining controlled, reproducible transformation, conclusions could not be drawn about the crucial initiating factor due to the complexity of the manipulations involved [2,13].

Critical tests of environmental transformation signals have only become feasible with the development of particulate [14] and a complex but nonparticulate [3] axenic medium for *Naegleria* growth. In a classic dissection of the environmental cues controlling flagellate differentiation [8], Fulton first showed that in the axenic medium M7 only amoebas are found, and that subculture to fresh M7 does not initiate transformation. By transferring cells to M7 media from which individual components had been deleted, it was next shown that yeast extract can affect transformation: removal of yeast extract from M7 permits flagellate differentiation and addition of yeast extract to transformation buffer inhibits differentiation. Further fractionation of the yeast extract yielded two inhibitory principles, a low-mol wt organic "factor," and electrolytes, particularly NH_4Cl. The factor, which has not been identified but is thought to be a small (mol wt $\leq$ 500) aromatic molecule, controls transformation at 0.017% (w/v)--an estimated maximum concentration of 0.001 M.

More detailed study of the action of electrolytes in controlling transformation showed that there was little specificity to their effect. A wide variety of ions prevent transformation if present at concentrations from 0.03 to 0.09 M. Moreover, nonpenetrating nonelectrolytes such as sucrose or glycerol also inhibit transformation, an effect related to elevated osmotic pressures (250 to 300 mOsm) of the medium. However, since electrolyte inhibition operates at much lower osmotic pressures (at approximately 160 mOsm), the two controls are clearly different. Electrolyte and "factor" transformation controls operate synergistically such that as little as one-fourth of the inhibitory "factor" concentration prevents differentiation if similarly subthreshold (one-eighth) amounts of electrolyte are present. Physiologically, "factor" is likely the more important of the two controls inasmuch as transformation-controlling concentrations of electrolytes (approximately 0.1 M) also inhibited protein synthesis. This was not true of "factor"-electrolyte combinations which inhibited neither RNA nor protein synthesis [15]. One highly significant methodological advance also comes from these studies of initiation-controlling factors in the *Naegleria* growth medium. This is based on the gradual adaptation of amoebas to high concentrations of "factor" in their growth medium--to a point where they ultimately become capable of transforming in medium M7 upon removal of excess "factor" [8]. The practical implication of this result is that it makes possible the initiation of flagellate differentiation either with or without nutritional stepdown: the relationship between growth and differentiation can now be studied in *Naegleria* under precise environmental control.

Fulton has recently extended this investigation [16] and can initiate flagellate differentiation of amoebas right in their growth medium, either by lowering the temperature (as little as 3°) or by increasing the mechanical agitation of the cultures. The temperature reduction and agitation stimuli also interact synergistically, and each may also reinforce the "factor"-electrolyte stimuli. Thus, to summarize the environmental cues which have been demonstrated to play a role in *Naegleria* flagellate differentiation, we may list the following: (1) reducing the yeast extract "factor;" (2) lowering the electrolyte concentration; (3) decreasing osmotic pressure; (4) lowering the growth temperature; and (5) increasing mechanical agitation.

What insights, if any, might the student of embryonic cell differentiation derive from this analysis of "inducers" of flagellate differentiation? Many embryologists still seek to account for the specificity of induction via intercellular transfer of macromolecules, particularly RNA and/or proteins [17]. Aceptance is growing, however, for an alternative concept, i.e., that cell differentiation cues may take the form of relatively minor alterations in extracellular ion concentrations [18,19]. For example, depending upon the specific ions (cations) used, the concentrations, and the length of exposure, undifferentiated amphibian gastrula cells which would normally form epithelial sheets can be redirected into nerve, pigment, or notochord differentiation. Might embryonic inductor tissues specify the differentiation of responding tissues by controlling the concentrations of electrolytes, the osmotic pressure, pH, and so on, in their microenvironment? One could even speculate that some differentiation inhibitor equivalent to yeast extract "factor" might be packaged in egg cells. A necessary prerequisite to cell differentiation would then be the dilution, or active transfer, of this "factor" from one blastomere to another: the recipient cell would then appear as a classic embryonic "inductor." The machinery for such a mechanism is well documented in the form of specific ion or molecule pumps [20]. Ample evidence exists for the crucial role of simple physiological determinants such as electrolytes, osmotic pressure, and pH in the expression of the differentiated state [18-22], and a coding mechanism utilizing such cues could be readily visualized. For example, the presence of certain ions in the microenvironment at a specific concentration over a given duration might determine a particular cell differentiation. Altering any one or combinations of the parameters could signify alternative fates, and if the stimuli interact synergistically or antagonistically, coding becomes feasible for even greater diversity. Thus, to conclude this discussion of the stimulus for *Naegleria* flagellate differentiation, I should like to emphasize that the *Naegleria* model (1) is consistent with many observations on the critical role of the composition of physiological salines in controlling differentiation and (2) supports and gives new perspective to a newly emerging concept of the embryonic inductive process.

B. Defining the Timetable of Flagellate Differentiation

When Chandler Fulton and I made our initial excursion into the *Naegleria* transformation as a potential model for cell differentiation, we were excited by the extremely precise control we were able to achieve over the process. The essence of this control lies in a protocol we developed and referred to as the "transformation experiment" [2]; the protocol has been refined and modified over the years [1], but the essential steps and results are unchanged. Cysts of a single clonal strain of *Naegleria*, either NB-1 or NEG, are plated on nutrient agar medium with a bacterial host, *Klebsiella aerogenes*, and the resulting excysted amoebas grown to early stationary phase at 33 to 34°. Amoebas are suspended in dilute iced Tris buffer and washed free of bacteria by three differential centrifugations at low speed. The final pellet is resuspended in iced buffer at a density of approximately 10^6 cells/ml and transferred to a 25° shaking water bath for transformation. During suspension and washing, cells held at 10° or lower do not begin transforming. Cells washed and resuspended in iced buffer and left in an ice bath for various times up to 100 min before being transferred to 25° thus transform with identical kinetics to those of control suspensions transferred to 25° immediately after washing. Therefore, under these conditions the moment of initiation of differentiation or time zero (T_0) is precisely defined: it is the point at which cells are transferred to 25° (or any temperature from 13 to 38°, the range over which transformation is possible).

The time course of flagellate differentiation is shown in a representative transformation experiment (Fig. 2). Amoebas begin to round up approximately 50 min after initiation, develop the first visible flagella at 55 to 60 min, and begin to elongate into the characteristic flagellate shape by 80 min. The flagellate phenotype is transient, and under these conditions cells lose their flagella and revert to the amoeboid state after about 2 h. However, within the next hour, most of the revertants transform again. In the absence of nutrients, *Naegleria* will cycle through the flagellate

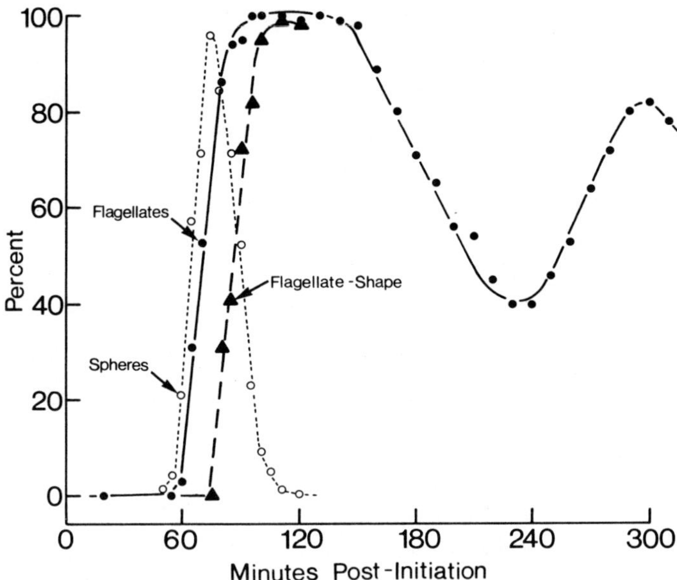

FIG. 2. Transformation, reversion, and retransformation of *Naegleria gruberi*, NB-1. Amoebas grown at 33 with *Klebsiella aerogenes* on solid medium were suspended in iced Tris buffer (2 mM, pH 7.4), washed free of bacteria and resuspended at a density of approximately 10^6/ml in Tris at 25 ± 0.1° for transformation. The T_{50} or population midpoint for cells rounding up is 64 min, followed by appearance of flagella at 70 min, and flagellate-shape at 86 min. Flagellates revert to amoebas after 2 h and transform once again after another 2 h.

phase at least four times over a 16-h period before amoebas begin encysting.

The times I have been giving are not particularly useful inasmuch as they represent unusual cells--those front-runners in the population which differentiate very rapidly. Had I expressed transformation events as the times when all cells have acquired the measured character, I would have been giving undue emphasis to the extreme laggards in the population. A more meaningful expression of any specific event is its T_{50}, the time when 50% of the cells have attained a given state. The T_{50} is routinely obtained by sampling a transforming population at appropriate short time intervals, counting the proportion of cells (out of 100) with the desired characteristic, and plotting these as percentages vs time. The time at which 50% of the cells achieve the given change, as determined from the transformation curves, is thus obtained by interpolation of a number of counts and will express the mean time of occurrence in the population. We see then from Fig. 2 that at 25° the "average cell" rounds into a sphere at 64 min ($T_{50}^{S} = 64$), develops flagella 6 min later ($T_{50}^{fl} = 70$), and elongates into the flagellate shape 16 min after that ($T_{50}^{F} = 86$).

One must be impressed by at least two aspects of this transformation curve. First, the differentiation of the first mature flagellate requires little more than 70 min, and second, virtually every cell in the population is flagellated by 90 min. The population synchrony of transformation is such that 90% of the cells undergo the measured changes (spheres, flagella, flagellate-shape) within a 25-min interval. Even more impressive, however, is the reproducibility of these measurements: the T_{50} for any carefully measured event does not vary by more than 4 min and is reproducible to ±1.5 min when scrupulous attention is paid to details of preparation, temperature, and sampling. Recognizing the unique advantages of this flagellate differentiation for defining a developmental schedule with a precision unattainable in virtually any other eucaryotic system, we set an initial goal of describing all measurable events of transformation with maximum precision.

This project was complicated considerably by changes made to the basic transformation protocol as it became clear that certain experiments could best be performed with strain NEG while the original NB-1 strain remained advantageous for other studies. Likewise, many experiments, e.g., isotope incorporation studies, require axenically grown cells whereas biochemical fractionations requiring mass cell cultures are most conveniently and economically accomplished using cells grown on nutrient agar with a bacterial host. These variations all affect transformation times, e.g., the routine 25° T_{50}^{fl} for bacteria-grown NEG is 64 min compared to 69 min for strain NB-1 [1]. Raising the transformation temperature of NEG to

28° decreases the T_{50}^{fl} to 48 min, and growing the amoebas axenically in liquid rather than with bacteria on solid medium increases the T_{50}^{fl} to 80 min [8]. Thus, in attempting to define our developmental timetable, direct comparisons cannot be made on an absolute time scale for experiments performed at different temperatures with axenic or bacteria-grown cells of one or the other *Naegleria* strain. However, it is possible to compare experiments performed under very different conditions, since the times of many events of differentiation are proportional to T_{50}^{fl}, the time of appearance of flagella. For example, for NB-1 transforming at 25°, T_{50}^{S} (spheres) = 65 min, T_{50}^{fl} (flagella) = 68 min, and T_{50}^{F} (flagellate shape) = 86 min. The two shape changes occur at times 0.96 and 1.26 relative to appearance of flagella. At 30°, T_{50}^{S} = 48 min, T_{50}^{fl} = 50 min, and T_{50}^{F} = 66 min. Although the absolute times of the shape changes are faster at the higher temperature, the events occur at virtually identical times (0.96 and 1.31) relative to appearance of flagella. In order to integrate data from diverse sources into a common developmental timetable, I have extrapolated this proportionality of events relative to the formation of visible flagella. Proportionality has been demonstrated for strain differences, shape changes, and reaction to inhibitors, but is only assumed for other events, which will be included in the timetable.

C. Early Events in Flagellate Differentiation

Upon initiation one would expect to find the earliest cellular changes in basic physiological parameters of the cell, such as in altered membrane potential, solute permeability, water uptake, cyclic AMP levels, and energy metabolism. Little information is yet available concerning such events, and those data which have been collected are not readily integrated into our differentiation timetable because they have not been measured as all-or-none quantal changes to individual cells in the population. Rather, the measurements express changes in volume or content of an entire cell population. For example, dramatic cell volume changes occur during the first few minutes of differentiation [23,24]. As measured both by packed volumes of centrifuged cells and the total sample volumes obtained by Coulter counting suspensions, we find that NB-1 amoebas (initial volume 3800 μm^3) decrease in volume within 10 min of initiation. Cells steadily shrink to as little as 54% of their initial volume by 45 min postinitiation, then expand to approximately 70% of their initial volume by 60 min as they become spheres with the earliest protruding flagella. Cell volumes remain relatively constant at 2710 μm^3 (70 to 80% of initial volume) throughout the rest of their existence in the flagellate phenotype.

What sparse information is available concerning the flow of ions into or out of transforming amoebas does not suggest a critical early function for altered concentration of either of the two ions studied (once transformation has been initiated). When cells were transformed in phosphate buffer with no added ions, internal Na^+ concentrations remained relatively constant throughout transformation, and though internal K^+ concentrations did decrease markedly--from 32.6 to 15.3 mEq/liter of cell water--this drop did not occur until after swimming flagellates appeared [23].

The diverse regulatory functions of cyclic adenosine-3',5'-monophosphate (cAMP), particularly as an intermediary between the cell surface and cytoplasm, suggests that changes in cAMP levels might be highly significant early events in flagellate differentiation. To date, however, only one preliminary study of cAMP involvement in *Naegleria* transformation has been published [25], and that with AB-2, a new isolate from Japan. Two most intriguing results come from this investigation: (1) 1 mM cAMP totally but reversibly inhibits flagellate differentiation and (2) when added at 30 min postinitiation, cAMP no longer blocked transformation. An approximate time-of-action plot (see synthesis section below for method) of the authors' data indicates that T_{50} for the "cAMP-sensitive event(s)" is 24 min in a total transformation of 50 min, i.e., 0.48 of the time to appearance of flagella. Clearly, further studies of internal cAMP levels and their control during the early minutes after initiation will constitute a major priority in the definition of critical events of flagellate differentiation.

D. Synthesis of New Molecules During Differentiation

Two approaches have been taken in studying new macromolecular synthesis during flagellate differentiation: the direct isolation and quantitation of specific molecules [26-29] and the application of selective inhibitors of synthesis as probes to determine the times at which cells complete different phases of differentiation [30-32]. Each of these approaches has its limitations, yet each contributes information about the process which would not have been evident had the alternative assay been used alone. Let us consider first the inhibitor experiments, which can be used to measure the times at which cells in the population have synthesized sufficient quantities of a given class of molecules that they complete differentiation despite blocks to further synthesis. A representative experiment, in which cells were exposed to the protein synthesis inhibitor cycloheximide at various times during transformation, is illustrated in Fig. 3. The time at which the population of cells has completed all protein synthesis essential to flagellate differentiation is measured by plotting plateau percentages of flagellates v times of inhibitor additions. Notice that the slope of the drug-insensitivity curve is identical to that of the control transformation curve; this is true of most such experiments (see Fig. 4) and reflects, once again, the heterogeneity of the population in its response to the inhibitors. We thus take the T_{50} of the derived curve (51 min) as the best measure of the time at which the population has completed "essential protein synthesis." Using the concept of time relative to appearance of flagella (64 min), this is 51/64 or 0.78 relative to the T_{50}^{fl} for this experiment.

Similar experiments using actinomycin D (an inhibitor of RNA synthesis) and mercaptoethanol (a disulfide-reducing agent) resulted in "onset of insensitivity" curves comparable to the cycloheximide

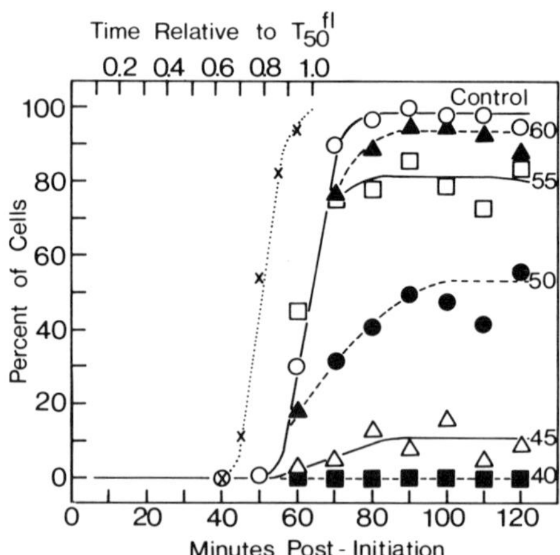

FIG. 3. Completion of "essential protein synthesis" by transforming cells. Cycloheximide, at a final concentration of 10 μg/ml, was added at 5-min intervals to suspensions of transforming *Naegleria*, and the percentage of cells capable of completing flagellum development was determined for each time of addition. For example, the populations attained plateau values of 0% flagellates when cycloheximide was added 40 min postinitiation (closed squares); 11% at 45 min (open triangles); 54% at 50 min (closed circles); 11% at 45 (open squares) and 94% at 60 min (closed triangles). The control (no addition) transformation, plotted as open circles, had a $T_{50}{}^{fl}$ of 64 min in this experiment. The dotted line is a curve of developing population insensitivity to the inhibitor derived by plotting plateau percentage transformed against time of addition.

derivative curve, though displaced to earlier relative time positions, 0.55 and 0.50, respectively. In the absence of control experiments to demonstrate that a given inhibitor acts conventionally and specifically in *Naegleria*, one can only infer from these curves that differentiation fails because of inhibition of protein or RNA synthesis, or interference with some disulfide bond-dependent conformation. These experiments are nonetheless useful in two ways: they provide clues as to the times of critical drug-sensitive phases during differentiation and they illustrate the precision with which it is possible to time particular events in the overall schedule of differentiation. Comparison of these times of drug action on bacteria-grown NB-1 with values obtained using axenically grown

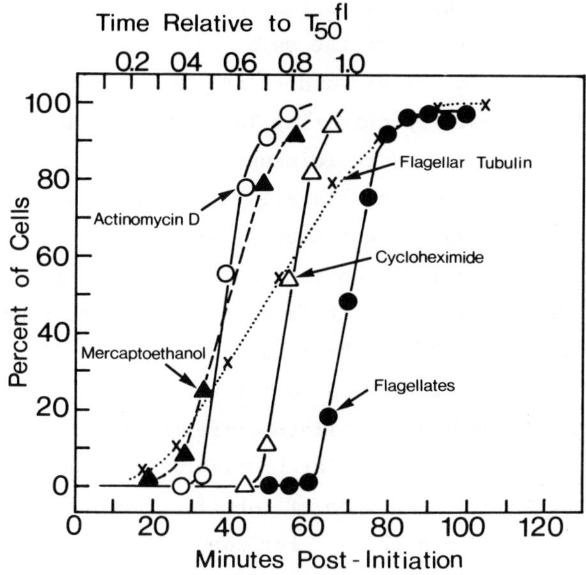

FIG. 4. Population curves of the onset of insensitivity to inhibitors of molecular synthesis and assembly. Experiments performed as in Fig. 3, using final concentrations of either 125 μg/ml actinomycin D, 10 μg/ml cycloheximide, or 30 mM mercaptoethanol. The T_{50}s for onset of insensitivity are plotted on a proportional time scale relative to appearance of flagella at 70 min, and are: mercaptoethanol (0.50), actinomycin D (0.55), cycloheximide (0.78). Plotted on the same proportional scale are data from Fulton and Kowit [33] expressing the cumulative synthesis of flagellar tubulin.

NEG demonstrates again the validity of expressing events of differentiation on a time scale relative to the appearance of flagella. Strain NEG, which transforms at 79 min, fully 10 min slower than NB-1, has T_{50}s of 40 min for actinomycin D inhibition and 60 min for cycloheximide inhibition [32]. These times represent 0.50 and 0.76 of the T_{50}^{fl}. Recall that the respective values for NB-1 were 0.55 and 0.78, truly a remarkable correspondence in view of the considerable differences between the two strains and culture conditions.

Though the inhibitor timing experiments have been informative, they are subject to reservations due to uncertainty as to the molecular action of the inhibitors. A more direct approach to measuring molecular events during differentiation is desirable, and has been achieved for at least one of the major new flagellate proteins: outer doublet tubulin. Kowit and Fulton have assayed flagellar tubulin synthesis both by measuring labeled amino acid uptake into developing flagella and by radioimmune assay of transforming cells using a specific antibody to flagellar tubulin. In a series of carefully conceived and executed experiments [27,28,33], they have established the following important points: (1) outer doublet tubulin is immunologically distinct from other tubulins found in *Naegleria*; (2) although 12% of the total cell protein is tubulin, the flagellar outer doublet tubulin comprises only 1 to 2% of the total cell tubulin; (3) during flagellate differentiation at least 70%, and perhaps all, of the flagellar protein is synthesized de novo; and (4) synthesis of outer doublet tubulin precedes assembly into flagella by about 30 min (25°). For comparison with our inhibitor sensitivity curves, I have added to Fig. 4 some data of Fulton and Kowit [33] showing the cumulative synthesis of new flagellar tubulin as measured by the direct isolation of molecules labeled during differentiation. The uptake of labeled amino acid into outer doublet protein during pulse labeling of transforming cells has been plotted as percentage of the total amino acid uptake into flagella, and times have been converted to our time scale relative to T_{50}^{fl} = 70 min. As might be expected, tubulin synthesis starts before and finishes after the critical times of cycloheximide sensitivity, i.e., completion of essential protein synthesis. One can estimate a "T_{50}" for accumulation of tubulin molecules at 50 min, or 0.71 on our relative time scale. Although this corresponds closely to the T_{50} for cycloheximide sensitivity (0.76), one cannot compare the times directly for various reasons: the slopes of the two curves are quite dissimilar, the one curve measures quantal events and the other measures continuous accumulation, and we can only infer what contribution population heterogeneity might make to the cumulative synthesis curve.

Patterns of synthesis during *Naegleria* flagellate differentiation have been studied by direct isolation of only two other molecular species: RNA and RNA polymerases. Walsh and Fulton [29] showed that RNA synthesis shifts within 5 min after initiation of transformation to a pattern of preferential messenger RNA and proportionally limited ribosomal RNA synthesis. During the RNA synthesis-dependent transformation period, there was no qualitative or quantitative change detectable in any of three RNA polymerases isolated from *Naegleria* [26].

E. Assembly of the Flagellar Apparatus

The definitive event in flagellate differentiation is the appearance of a flagellar apparatus--an assemblage of cell organelles which includes basal bodies, flagella, associated cytoplasmic microtubules, and a flagellar rootlet. And the striking aspect of this event is the dramatic speed with which the separate pieces fall into place, creating a highly integrated functional unit within approximately 10 min. We have previously published detailed descriptions of the fine structure of the flagellar apparatus [34] whose essential features may be seen in Fig. 5. Basal bodies looking like pinwheels composed of nine triplet microtubules are situated just beneath the cell surface (Fig. 5d). They are continuous distally with flagella having the characteristic nine peripheral doublets plus two central single microtubules (Fig. 5c). Proximally, the basal bodies are attached to a banded flagellar rootlet or rhizoplast (Fig. 5e) by a palisade of very short microtubules and occasionally by a triangular array of banded filaments. So firm is this attachment that when the isolated flagellar apparatus is subjected to severe shear forces, flagella break away from the basal bodies before the basal body-rootlet connection is severed. The flagellar apparatus may be isolated intact (Fig. 5b), and by a combination of mechanical disruption and differential solubilization, its components have been dissected from one another (Fig. 5c-e).

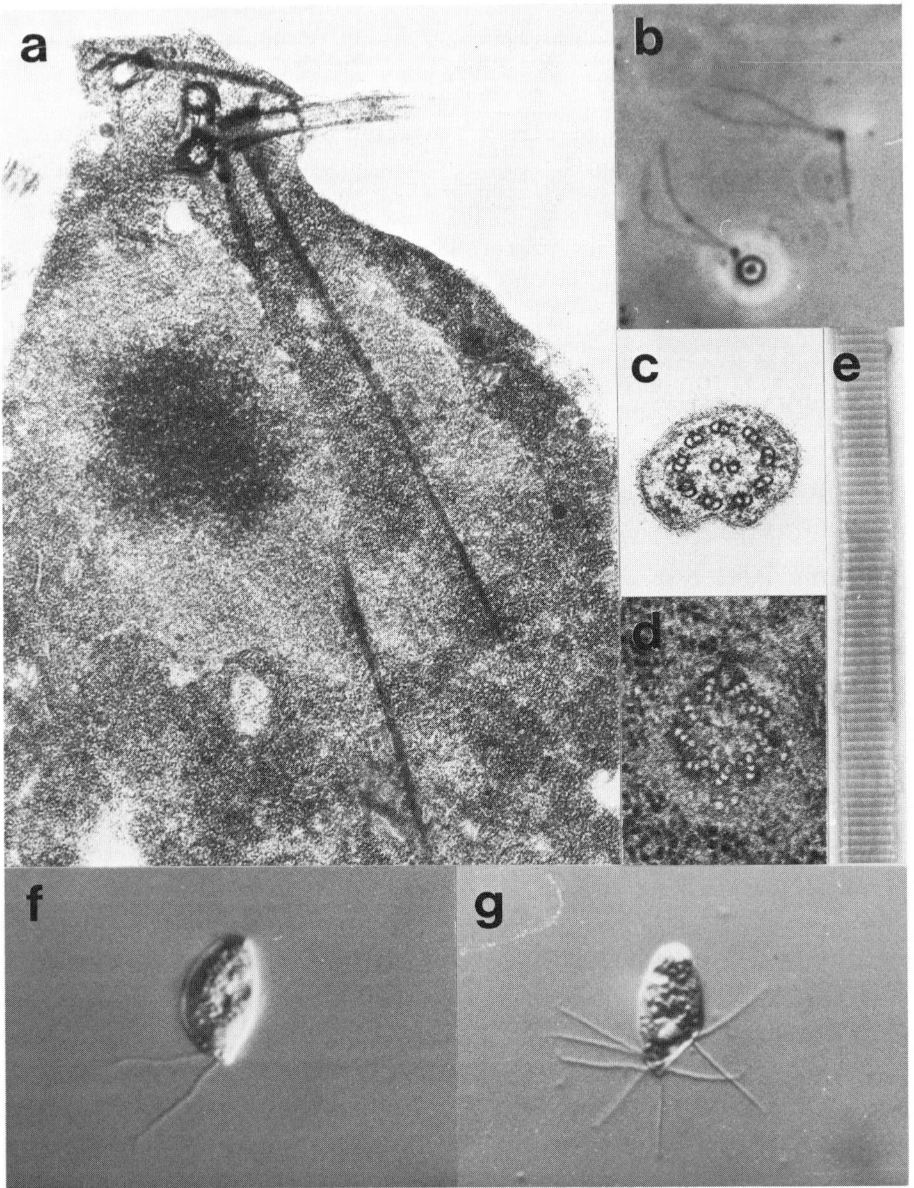

FIG. 5. *Naegleria* flagellate morphology and components of the flagellar apparatus. (a) Electron micrograph of the anterior end of a temperature-shocked "hairy" flagellate, showing three transversely sectioned basal bodies and one longitudinally sectioned basal body and flagellum. Also present are three banded flagellar rootlets, one running below the anterior cell surface and two extending alongside the nucleus. Approximate magnification × 25,000. (b) Phase contrast photomicrograph of two isolated flagellar apparatuses, one consisting of rootlet, basal bodies, and two flagella, and the other with a highly refractile nucleus still attached. Approximate magnification × 2,500. (c) Transverse section through a flagellum showing the typical 9 + 2

Flagella

Assembly of *Naegleria* flagella has been described previously [34]. The process appears to be essentially one of microtubule elongation by addition to the tips of basal bodies lying near the cell surface. Developing axonemes extend into an outpocketing cell membrane which expands to accommodate the growing flagella. In the earliest stages of flagellum development, axonemes may appear as a ring of nine mixed single and doublet microtubules or sometimes as an incomplete ring of fewer than nine doublets. By the time flagella attain a length of about 2 μm they become readily visible in the light microscope, and they then increase in length at an average rate of approximately 0.5 μm/min until they reach their mature length of 19 μm.

The time when flagella become visible by light microscopy is of course the standard by which we time all other transformation events: for strain NB-1 prepared by method C and transformed at 25°, the T_{50}^{fl} is 67 to 71 min (arbitrarily defined as 70 min for convenience of calculations; Figs. 4 and 6). Flagella do not attain their maximum length until much later in transformation. The population T_{50} for development of full-length flagella has been measured for NEG at a time of 1.6 relative to flagellum appearance [16].

Basal bodies

One of the most challenging and hopefully rewarding tasks undertaken in depicting the events of flagellate differentiation has been that of describing the origin and development of the basal bodies. The task is challenging because normally only two basal bodies develop in each cell and they are particularly small organelles--approximately 0.2 × 1.0 μm. Second, they are assembled very rapidly. For a long while we were unable to identify assembly

---

(Continued from page 114.)
pattern and "arms" on some doublets (× 75,000). (d) Section through the proximal end of a basal body showing a single central filament with "spokes" radiating towards a cylinder of nine triplets of microtubules (× 75,000). (e) An isolated flagellar rootlet negatively stained with uranyl acetate, which reveals a 22 nm repeating light and dark period (× 75,000). (f) Interference contrast photomicrograph of a normal glutaraldehyde-fixed biflagellate cell. (g) Same of a "hairy" cell subjected to a 38.2° temperature shock 35 to 70 min postinitiation and bearing nine basal bodies and flagella. Approximate magnification × 2,500.

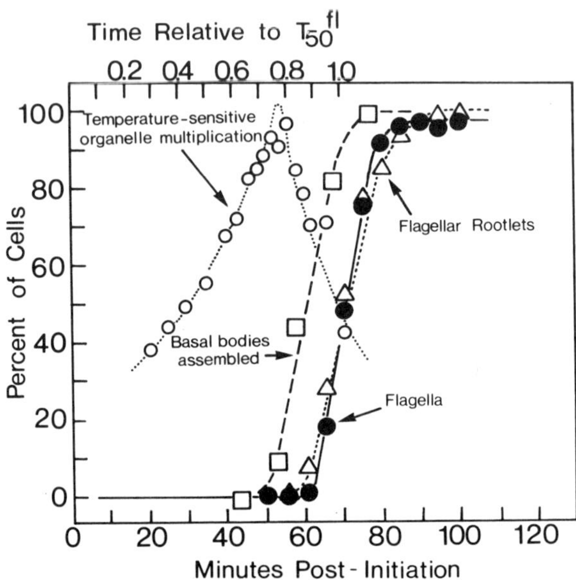

FIG. 6. Population curves of the appearance of components of the flagellar apparatus. The curve of basal body appearance was obtained by scanning 250 mid-cell sections by electron microscopy and expressing each count as a percentage relative to the number of profiles containing basal bodies in fully formed flagellates (see Ref. 35 for details). The appearance of flagellated cells was evaluated by counting samples of 200 Lugol's iodine cells using phase optics. Rootlet appearance was likewise assayed in samples of 200 cells by fixing 4 drops cells:1 drop one-third strength Lugol's iodine, and then lysing the lightly fixed cells by applying a drop of the detergent Sarkosyl to the edge of a wet mount preparation of the cells. As the cells slowly lysed, their flagellar rootlets came free of the cytoplasm and were briefly visible for counting under phase optics. The time of maximum susceptibility to a temperature-induced multiplication of flagellar apparatus organelles is indicated by replotting data from Fig. 8: all data in this figure are plotted relative to a T_{50}^{fl} of 70 min.

stages, even in sections of cells fixed during the critical 15-min interval when essentially all basal bodies appeared in the cells [35]. Third, no basal body precursor has yet been identified. There is no preexisting centriole-like structure as in spermatogenesis [36], no specialized precursor organelles such as the blepharoplasts found in ferns and cycads [37], nor are there any identifiable electron-dense amorphous condensation forms such as are found in mammalian ciliogenesis [38]. Looking for the basal body precursor in *Naegleria* is like looking for a pinwheel in a haystack and we have not, as yet, been able to identify any likely candidate.

Nevertheless, it has been possible to quantitatively measure the appearance of basal bodies during transformation [35]. By selecting randomly oriented midcell profiles for intensive scanning by electron microscopy, we found that basal bodies first become visible 55 min after initiation, and the maximal number has developed by 80 min. A T_{50} for basal body appearance was determined by the following rationale: in mature flagellate samples in which each cell has two basal bodies, the organelles are seen in only 32 of every 250 cell profiles scanned, or approximately 13%. We can thus plot a transformation curve for basal bodies by normalizing counts of earlier samples to 32/250 = 100%. Because the scanning involved in such an experiment is extremely tedious and demanding, we have attempted only one such T_{50}^{Bb} determination. But the results were internally consistent and compatible with all other aspects of the developmental schedule. Basal bodies are constructed just 10 min before flagella become visible by light microscopy; the T_{50}^{Bb} was 62 min in an experiment whose T_{50}^{fl} = 72 min, i.e., the time of basal body appearance is 62/72 = 0.86 on the relative time scale (data replotted in Fig. 6).

There are no structural precursors in *Naegleria* amoebas for the basal bodies which develop during flagellate differentiation. During mitosis no centrioles are seen at the poles of the mitotic spindle, and centriole-like structures are never seen until 50 to 55 min after transformation is initiated. I cannot state unequivocally that centrioles do not exist in amoebas solely on the basis of my inability to ever see them. However, based on electron microscope counts of 63 centriole-like structures in 500 cell sections of flagellates v 0/500 for amoebas, the probability that a centriole-like organelle of equivalent visibility exists in amoebas can be calculated. The result is negligible--less than 10^{-30} [35]. *Naegleria* transformation thus becomes a particularly useful system in which to study basal body assembly. The resolution of events, both synthetic and morphogenetic, must be much greater in *Naegleria*, where centrioles increase from 0 to 2, than in cells which have a conventional centriole cycle of 2 to 4 [39]; and the control, reproducibility, and precise timing of organelle development make *Naegleria* superior in some respects to systems such as ciliate protozoa [40] or ciliated epithelia [38], in which literally hundreds of basal bodies develop. We are, nevertheless, at a considerable disadvantage using *Naegleria* to study basal body development inasmuch as we must search for two very small organelles amidst a relatively huge volume of cytoplasm.

The remedy for this problem has been found once again in our ability to manipulate transformation in a controlled and reproducible way. Initially it was observed that if amoebas of strain NB-1 were exposed to a 45-min sublethal (38°) temperature shock from time 0 to 45 min during transformation, they became multiflagellate or "hairy" (Fig. 5G), producing anywhere from 1 to 27 supernumerary basal bodies and flagella [41]. An intensive study of cellular and environmental parameters affecting the hairiness phenomenon [42] has resulted in a method which gives a population mean of nearly 6 basal bodies and flagella per flagellate, compared to the normal mean of 2. The major variables affecting the number of organelles assembled are temperature, length of exposure, and the time post-initiation during which the temperature shock is applied. A sharp temperature optimum has been demonstrated at 38.2°. The effect virtually disappears at temperatures beyond the narrow limits of 37.5 to 38.5°.

The hairiness phenomenon depends upon both the duration of the temperature shock and the interval during transformation when it is given (Fig. 7). The term $R_{f/F}$ in that figure is a measure of the increase in the mean flagellum number of the population relative to a control hairiness protocol (defined in the figure legend): an $R_{f/F}$ of 1.0 corresponds to about 4.5 flagella per flagellate. Some increase in organelle number results from as little as 5 min of exposure to 38.2°, and the maximum increase follows a 35-min

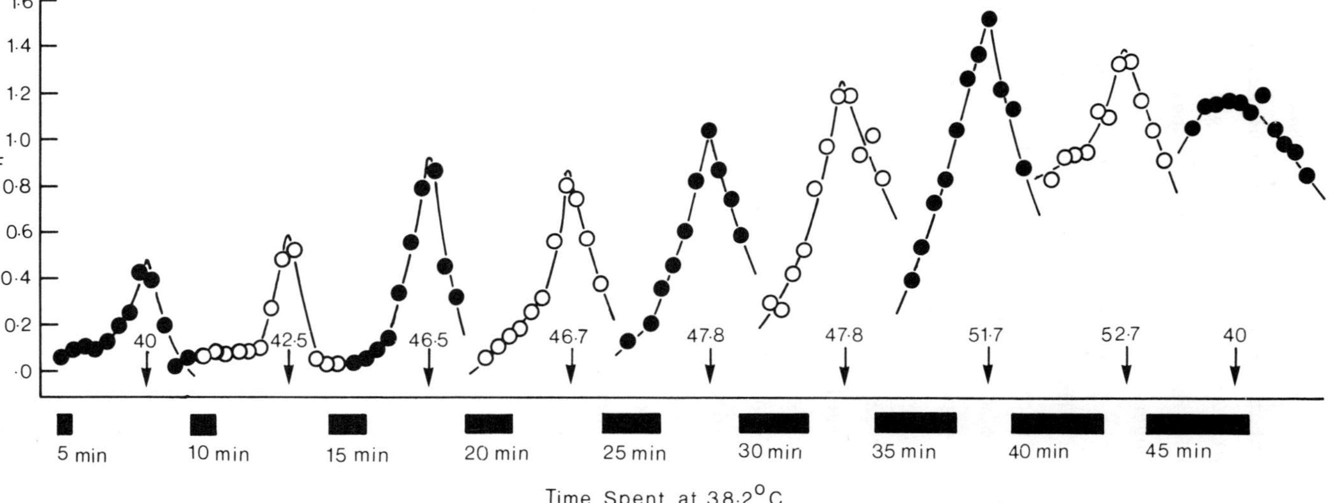

FIG. 7. Increase in flagellum numbers of cells subjected to high-temperature shocks during transformation. Amoebas suspended from 33° growth plates, washed free of bacteria and resuspended in iced Tris buffer, were given 38.2° temperature shocks of increasing duration (5 to 45 min) immediately upon initiating transformation, or were taken to 38.2° at successive 5-min intervals from the standard 25° transformation temperature.

In order to evaluate the relative increase in flagellum number using data from several separate experiments, the measure $R_{f/F}$ has been used [42]. Included in each experiment was a "standard hairiness control," a 38.2° temperature shock applied from t_0-45 min postinitiation. $R_{f/F}$ is then given by

$$\frac{\text{\# flagella/300 cells (experimental) - \# flagella/300 cells (25° control)}}{\text{\# flagella/300 cells (standard hairiness) - \# flagella/300 cells (25° control)}}$$

Each $R_{f/F}$ is then plotted vs the midpoint of the particular temperature interval.

temperature shock applied 35 to 70 min postinitiation. These observations, taken together, suggest that the controls which normally ensure that *Naegleria* develops only 2 basal bodies and flagella can be disrupted by a 5- to 10-min sojourn at the elevated temperature. We know that the population heterogeneity--time from the first to the last flagellate appearing in the population--is on the order of 25 to 30 min in these experiments. The 35-min optimal duration may therefore represent the 5 to 10 min exposure sufficient to induce hairiness in an individual cell, to which must be added 25 to 30 min in order for all cells in the population to pass through the temperature-sensitive stage.

Another form of heterogeneity in this system lies in the number of basal bodies and flagella developing on individual cells: typically 65 to 70% of a temperature-shocked population will have between 5 and 7 flagella, but individuals bearing any number from 1 to 13 flagella are not uncommon. Ideally, a system for studying the control of organelle assembly would be uniform, with individual cells producing some exact multiple of the normal number. Possibly *Naegleria* cells do not respond uniformly to temperature shocking because the population is again heterogeneous with respect to the temperature-sensitive phase when subjected to the 38.2° shock. Attempts have been made to obtain uniform organelle multiplication by temperature-shocking populations of growth-synchronized cells, but since growth synchronization is based on a 38° inhibition of karyokinesis [4], and since long exposures to such temperatures delay and prevent transformation, these attempts have been unsuccessful. We are hopeful of repeating these experiments using amoebas which have been synchronized by hydroxyurea inhibition and by physical (size) selection of uniform cell populations.

Note that the midpoints of the most effective treatments are indicated beneath the arrows in Fig. 7. For the optimal 35-min exposure, this was 51.7 min postinitiation. This midpoint has been precisely defined by displacing the 35-min 38.2° treatment by a series of 2-min steps later into transformation (Fig. 8). It was again 51.5 min, or 51.5/68 = 0.76 on a time scale relative to

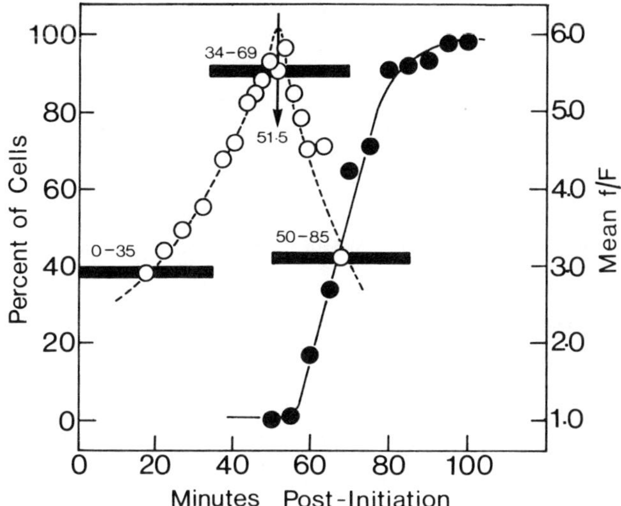

FIG. 8. Precise definition of the 35-min high-temperature interval, which induces the maximal number of supernumerary flagella in populations of differentiating flagellates. Experimental conditions as described for Fig. 7. Each point (open circles) represents the mean flagellum number for 300 flagellated cells counted in samples subjected to the 35-min 38.2° sojourn, and is plotted at the midpoint of the high-temperature interval (for example, see 0-35, 34-69, and 50-85 min treatments). A 25° control transformation was included in this experiment (closed circles): the T_{50}^{fl} was 68 min.

T_{50}^{fl} = 68 min for this experiment. This corresponds exactly--whether fortuitously or in some related way we cannot say--to the time of completion of essential protein synthesis as measured by insensitivity to cycloheximide inhibition.

This excursion into the phenomenon of hairiness has provided two worthwhile additions to the *Naegleria* differentiation study. One more event in the program has been identified and precisely timed. But much more significantly, the phenomenon has been manipulated to provide material for the otherwise unapproachable study of basal body assembly in *Naegleria*. Although this investigation is not yet complete, the outline is clear: the first putative basal body microtubules have been identified in cells fixed 95 min post-initiation (temperature-shocked cells are delayed in transformation:

the T_{50} for these cells was 111 min). Seen in transverse section, the microtubules formed irregular cylinders of seven, eight, or nine single microtubules: no "hub and spokes" or "cartwheel" is associated with these earliest stages. Sections of later samples had recognizable though incomplete pinwheels, with either nine sets of singlet and doublet microtubules or doublet and triplet combinations. Longitudinally sectioned developing basal bodies were approximately 240 nm long when first recognized and appear simply as shortened copies of the mature organelle, elongating to about 800 nm when fully formed. All developing basal bodies have been found in the peripheral cytoplasm near the cell surface and, though basal bodies in multiflagellate cells tend to be found in groups of four or five organelles, each basal body appears to condense in the cytoplasm independently of others around it. There is no evidence of a "mother-daughter" relationship such as is seen in centriole [36] and kinetosome [40] replication, nor of an intermediate source of generative material such as the electron-dense masses implicated in producing the hundreds of basal bodies of ciliated epithelia [38].

Flagellar rootlet

Anchoring the basal bodies and flagella in the cell is a periodically banded flagellar rootlet or rhizoplast, a long (approximately 12 μm) tapering fiber which sometimes runs free in the cytoplasm but more usually attaches tightly to the nuclear envelope [34]. Studies of rhizoplast structure and composition raise intriguing questions about its function. For example, we have found variable periodicity in organelles fixed in situ: the 18.5-nm overall band width varies from 13.8 to 25.2 nm [34,43]. Isolated rootlets, on the other hand, have virtually uniform 21.7-nm band widths (Fig. 5E), ranging from 20.0 to 23.6 nm [43]. On the basis of mol wt of the principal protein (240,000 by SDS acrylamide gel electrophoresis), solubility properties, and enzymatic digestion of the rootlets [44], it has been suggested that the flagellar rootlets of *Naegleria* may be composed of a primitive form of collagen, and function therefore as subcellular "tendons" anchoring the basal bodies and flagella in the cytoplasm.

Because the rootlet is long but very narrow (approximately 0.2 μm), quantitative electron microscope studies of its development are precluded. In early stages of differentiation, when rootlets are present but very short, the probability of seeing one in a cell section would be very much lower than in sections of mature flagellates. We do not see rootlets in electron micrographs of transforming cells until about 65 min after initiation. However, it is possible to construct a curve of rootlet appearance analogous to the flagellum development curve by light microscopy [45]. Cells are lightly fixed in dilute Lugol's iodine, then lysed on the microscope slide by infusing a small drop of the detergent Sarkosyl under the coverglass. As the cells slowly expand and lyse, the rootlets (just barely visible under phase optics) may be briefly seen and recorded before disintegrating. The first flagellar rootlets were seen 55 min after initiation, and rootlet assembly appears to parallel flagellum development: the T_{50} for rootlets was 69 min in an experiment in which T_{50}^{fl} was 67.5 min. Thus, within the accuracy of the methods for scoring both organelles, it is obvious that rootlet assembly proceeds concurrently with flagellum assembly (time 1.02 on the proportional time scale). These data are replotted relative to T_{50}^{fl} = 70 min in Fig. 6.

III. FINAL COMMENTS

The general conclusion I wish to draw from our work on *Naegleria* is that the system is remarkably well suited as a model with which to study cell differentiation. The cells, which can be grown rapidly and inexpensively under well defined environmental control, respond to appropriate stimuli by assuming one of three highly distinctive phenotypes. In each case the phenotypic conversion occurs relatively rapidly and synchronously, and transforms virtually every individual in the population. Furthermore, the moment of initiation of differentiation can be defined and experimentally manipulated. And because *Naegleria* is unicellular at all stages of its life cycle, it is also possible to quantitatively evaluate cell-to-cell variability, i.e., the population heterogeneity for any phenotypic change.

Transformation of amoebas to flagellates provides the most appealing model for studying cell differentiation in *Naegleria*. In the course of this transformation, amoebas undergo major changes in size, shape, and metabolic activity, synthesizing new molecules and assembling new organelles. The *Naegleria* transformation model lends itself particularly well to the study of two fundamental problems of cell differentiation: (1) what is the stimulus to cellular change? and (2) what is the program of events leading to the flagellate differentiation? The initial question has been reviewed and conclusions have been drawn earlier in this paper. To summarize, flagellate differentiation is controlled by diverse environmental factors, including osmotic pressure, electrolyte concentration, a differentiation-inhibiting factor in the growth medium, temperature reduction, and mechanical agitation. Since embryonic cell differentiation takes place against a physiological background which could include any of these parameters, it is proposed that specificity in embryonic induction could be coded for by specific combinations of these physiological cues rather than through intercellular transfers of instructional molecules.

The emphasis in this present study has been on defining the timetable of events followed by *Naegleria* amoebas as they transform into flagellates. The current status of this enterprise is illustrated in Fig. 9, in which has been plotted the T_{50} of every event for which we have been able to collect adequate data. By plotting the data on a time scale relative to the appearance of flagella, I have been able to collect measurements from diverse sources--from different laboratories using different strains and culture conditions--and to order the sequence of known cell changes, assigning to each of them a precise time in the differentiation schedule. Rather than dealing in proportional times, however, I have

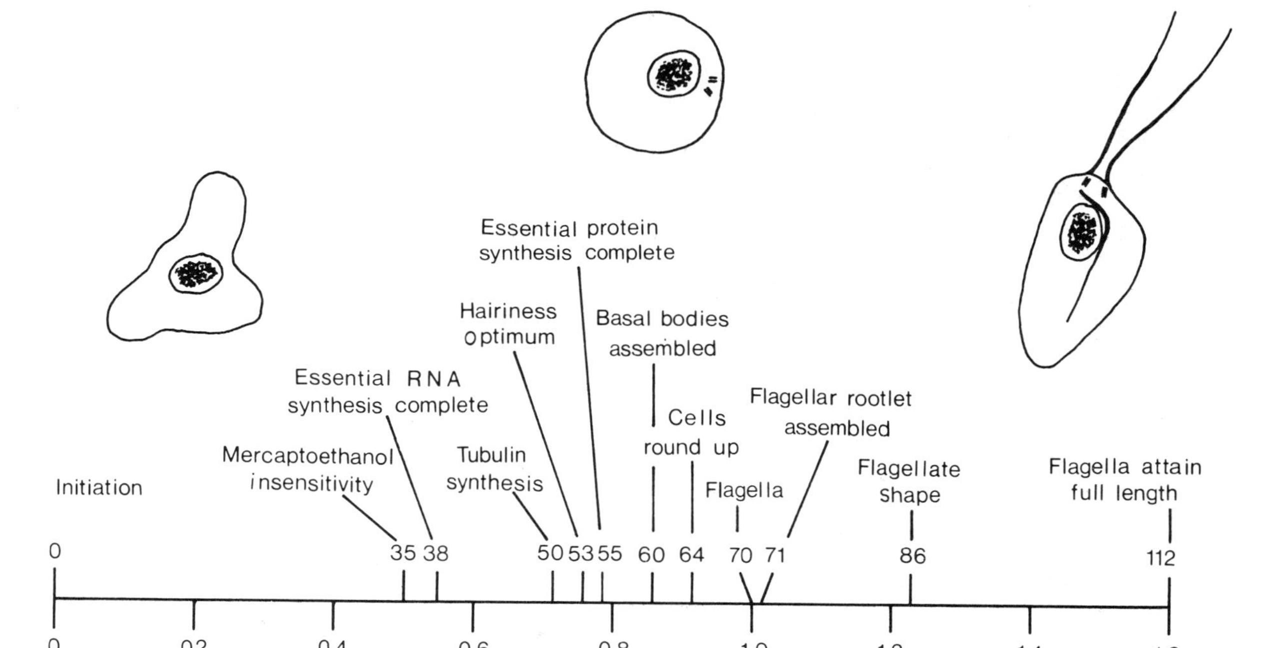

FIG. 9. Timetable of *Naegleria* flagellate differentiation. The T_{50}s for acquisition of a number of flagellate properties were measured relative to the appearance of flagella (T_{50}^{fl}) in a series of independent experiments, and plotted as "proportional times" below the abscissa. Although the T_{50}^{fl} varied from 64 to 71 min in these experiments, it most often fell between 69 and 71 min. In this figure T_{50}^{fl} has been arbitrarily defined as 70 min, and events timed in minutes postinitiation above the abscissa.

arbitrarily defined T_{50}^{fl} = 70 min and converted all measured events to that time scale. The schedule is straightforward and, in large measure, predictable; no significant event has been recorded until about midway through the schedule when differentiation becomes insensitive to the disulfide reducing agent mercaptoethanol at 35 min, and then insensitive to actinomycin D (indicating completion of "essential RNA synthesis") at 38 min. We know of three important changes earlier in the schedule--the very early (beginning within 5 min of initiation) shift to preferential messenger RNA synthesis, the dramatic reduction in cell volume occurring at approximately 20 min, and the cyclic AMP-sensitive event which was calculated very roughly at 34 min. These are not included in the summary timetable because accurate T_{50}s have not been obtained; nonetheless they represent significant events in flagellate differentiation. Synthesis of the tubulin for flagellar microtubules occurs next at 50 min post-initiation. This corresponds closely to the completion of "essential protein synthesis" at 55 min, as indicated by the onset of insensitivity to cycloheximide inhibition, and also to the time (53 min) of maximum sensitivity to the high-temperature induction of multiple basal bodies, flagella, and rootlets. Next on the agenda is the assembly of basal bodies at 60 min, followed quickly by the change to a spherical shape at 64 min, and outgrowth of visible flagella at 70 min. Coincident with flagellum development is the assembly of flagellar rootlets at 71 min. Completing the measured schedule is the change to flagellate shape which occurs at 86 min and the attainment of full-length flagella 112 min after the amoeboflagellate transformation was initiated.

This summary of the measured events of *Naegleria* flagellate differentiation is nowhere near as comprehensive as those which could be constructed for more extensively investigated developmental systems such as the cellular slime molds or water molds. It is nevertheless impressive in that it represents one of the most precisely timed descriptions of cellular development to be found in the literature. If there is one major contribution of *Naegleria* as a model developmental system, it lies in the methodology of the "transformation experiment," a set of procedures in which (1) the moment of initiation of a particular differentiation can be defined and controlled, (2) single characteristics are selected for measurement from the complex continuum of changes constituting differentiation, and (3) these events are measured as all or none, quantal changes in individual cells to define the contribution of cell heterogeneity to the measured times and to select a midpoint most representative of the entire population. Though difficult to achieve with differentiating multicellular tissues, this methodology should be applicable to most differentiating cell systems. The exceptional reproducibility and precision attainable by embryologists adopting this methodology should prove well worth the effort required to look at cell differentiation from this perspective.

ACKNOWLEDGMENTS

The unpublished research described in this paper has been supported by Grant A2815 from the National Research Council of Canada. I wish to thank my long-term collaborator, Chandler Fulton, for many valuable contributions to this study, and to acknowledge the help of Ray Stephens. Peter Simpson, Fred Green, and Lynda Ewers all contributed significantly to this enterprise. The technical assistance of several undergraduate assistants--Joseph Tai, Don Anthony, Sandi Farrell, and P. B. Au Yeung--has been a major contribution to the study.

REFERENCES

1. Fulton, C., 1970. *In* Methods in Cell Physiol., Vol. 4, D. M. Prescott (ed.), Academic Press, New York, p. 341.
2. Fulton, C., and A. D. Dingle, 1967. Develop. Biol. 15: 165-191.
3. Fulton, C., 1974. Exp. Cell Res. 88: 365-370.
4. Fulton, C., and A. M. Guerrini, 1969. Exp. Cell Res. 56: 194-200.

5. Lastovica, A. J., and A. D. Dingle, 1971. Exp. Cell Res. 66: 337-345.

6. Averner, M., and C. Fulton, 1966. J. Gen. Microbiol. 42: 245-255.

7. Neff, R. J., S. A. Ray, W. F. Benton, and M. Wilborn, 1964. *In* Methods in Cell Physiol., Vol. 1, D. M. Prescott (ed.), Academic Press, New York, p. 55.

8. Fulton, C., 1972. Develop. Biol. 28: 603-619.

9. Schardinger, F., 1899. Sitzber. Akad. Wiss. (Wien) Math. Natur. Abt. I. 108: 713-734.

10. Wilson, C. W., 1916. Univ. Calif. Publ. Zool. 16: 241-292.

11. Willmer, E. N., 1956. J. Exp. Biol. 33: 583-603.

12. Schuster, F., 1963. J. Protozool. 10: 297-313.

13. Yuyama, S., 1971. *In* Developmental Aspects of the Cell Cycle, I. L. Cameron, G. M. Padilla, and A. M. Zimmerman (eds.), Academic Press, New York, p. 41.

14. Balamuth, W., 1964. J. Protozool. 11: Suppl., 19-20.

15. Fulton, C., 1972, unpublished results.

16. Fulton, C., 1974. Personal communication.

17. Hamburgh, M., 1971. Theories of Differentiation, Edward Arnold, London.

18. Barth, L. G., and L. J. Barth, 1974. Develop. Biol. 39: 1-22.

19. Barth, L. J., and L. G. Barth, 1974. Biol. Bull. 146: 313-325.

20. Bittar, E. E. (ed.), 1970. Membranes and Ion Transport, Vols. 1-3, Wiley-Interscience, London.

21. Kostellow, A. B., and G. A. Morrill, 1968. Exp. Cell Res. 50: 639-694.

22. Lash, J. W., K. Rosene, R. R. Minor, J. C. Daniel, and R. A. Kosher, 1973. Develop. Biol. 35: 370-375.

23. Kahn, A. J., R. Boroff, and R. W. Cutler, 1968. J. Protozool. 15: Suppl., 29.

24. Au Yeung, P. B., and A. Dingle, 1975. Unpublished results.

25. Okubo, S., and S. Inoki, 1973. Biken J. 16: 181-184.

26. Soll, D. R., and C. Fulton, 1974. Develop. Biol. 36: 236-244.

27. Kowit, J. D., and C. Fulton, 1974. J. Biol. Chem. 249: 3638-3646.

28. Kowit, J. D., and C. Fulton, 1974. Proc. Nat. Acad. Sci. USA 71: 2877-2881.

29. Walsh, C., and C. Fulton, 1973. Biochim. Biophys. Acta 312: 52-71.

30. Wade, H. A., and P. Satir, 1968. Exp. Cell Res. 50: 81-92.

31. Yuyama, S., 1971. J. Protozool. 18: 337-343.

32. Fulton, C., and C. Walsh, 1977. Manuscript submitted.

33. Fulton, C., and J. D. Kowit, 1975. N. Y. Acad. Sci. Ann. 253: 318-332.

34. Dingle, A. D., and C. Fulton, 1966. J. Cell Biol. 31: 43-54.

35. Fulton, C., and A. D. Dingle, 1971. J. Cell Biol. 51: 826-836.

36. Gall, J. G., 1961. J. Biophys. Biochem. Cytol. 10: 163-193.

37. Mizumaki, I., and J. Gall, 1966. J. Cell Biol. 29: 97-111.

38. Dirksen, E. R., and T. T. Crocker, 1966. J. Microsc. 5: 629-644.

39. Robbins, E., G. Jentzch, and A. Micali, 1968. J. Cell Biol. 36: 329-339.

40. Dippell, R. V., 1968. Proc. Nat. Acad. Sci. USA 61: 461-468.

41. Dingle, A. D., 1970. J. Cell Sci. 7: 463-482.

42. Dingle, A. D., 1977. Manuscript submitted.

43. Simpson, P. A., and A. D. Dingle, 1971. J. Cell Biol. 51: 323-328.

44. Dingle, A. D., and A. F. Green, 1974. Proc. Canad. Fed. Biol. Soc. 17: 235.

45. Green, A. F., 1974. MSc Thesis. McMaster University, Hamilton, Ontario, Canada.

CONTROLS OF SPORULATION IN *SACCHAROMYCES CEREVISIAE*

James E. Haber, Peter J. Wejksnora,
Deborah D. Wygal, and Elaine Y. Lai

Department of Biology
Rosenstiel Basic Medical Sciences Research Center
Brandeis University
Waltham, Massachusetts

I. INTRODUCTION

A. Aspects of Sporulation

Sporulation of *Saccharomyces cerevisiae* provides an excellent opportunity to investigate the control of some essential features of gametogenesis. One of the main advantages of studying this process of intracellular differentiation in yeast is that sporulation takes place in a single cell, free of the complexities of interaction with other cells not undergoing the same process of specialization. It is therefore possible to study not only the sporulation process itself but to relate these events to those of the predifferentiation vegetative cell cycle under nearly equivalent conditions. In addition to the relative simplicity of studying unicellular differentiation, *S. cerevisiae* is amenable to extensive genetic analysis, so that mutations affecting growth and/or differentiation can be isolated and characterized in considerable detail.

Sporulation of *Saccharomyces cerevisiae* involves the formation of four haploid ascospores within an ascus from a diploid vegetative cell. Sporulation occurs in diploid cells heterozygous (a/α) for mating type, while diploids which are a/a or α/α fail to undergo premeiotic DNA synthesis or spore formation [1,2]. The conditions for sporulation are substantially different from those which support vegetative growth. Exponentially growing yeast divide approximately every 2 to 3 h while sporogenesis requires 12 to 24 h, depending on culture conditions. Sporulation occurs under semistarvation conditions, in the absence of a nitrogen source and usually in 1% potassium acetate, pH 7 [3,4]. While most strains of *Saccharomyces cerevisiae* are able to grow at 37°, there is little or no sporulation above 35°. During sporulation the pH of the medium rises to pH 8 to 9. This pH rise is apparently obligate, as sporulating cells buffered at pH 6 or 7 fail to complete the process [3,5]. Sporulation is also subject to both glucose and nitrogen repression [5-8] so that the process requires that the diploid be preadapted to respiration of acetate [4]. Cells are generally either transferred from stationary phase after growth on glucose [5] or from exponential growth in an acetate growth medium [9,10]. When cells are preadapted to growth on acetate, there is no actual requirement for the expression of the mitochondrial genome during sporulation [11].

The efficiency of sporulation is strain-dependent. In both homothallic and heterothallic lines, the quantity of asci ranges from approximately 50 to nearly 100%.

The process of sporulation generally takes 24 h for completion. Premeiotic DNA replication is observed at approximately 4 to 8 h after initiation [2,4,6]. Although chromosomes cannot be visualized sufficiently to permit characterization of the stages of meiosis in great detail [12], the two meiotic nuclear divisions to form four haploid nuclei can be determined either by staining with a DNA indicator (Giemsa stain) to observe mono-, bi-, and tetranucleate forms with the light microscope [12], or by electron microscopic examination of change in nuclear structure [13,14].

B. Control of Sporogenesis

The control of the transition from mitosis and vegetative growth to meiosis and sporulation is complex, involving an interaction of genetic and physiological signals to begin a new developmental pathway. One primary aspect of the regulation of this event is the shift of cells to a medium lacking a nitrogen source. Diploid cells, whether they are genetically capable of sporulation or not, and even haploid cells, all show similar changes in the synthesis of proteins [15], carbohydrates [15,16], and lipids [17,18]. Thus, some of the events during sporulation reflect the response of cells to a new, restrictive environment and are not specific for cells actually able to undergo meiosis.

The most certain genetic determinant of the ability of diploid yeast to sporulate is the mating-type locus. Diploid cells heterozygous (a/α) for mating type sporulate, whereas homozygotes (a/a and α/α) do not [1]. Tetraploid cells [($a/a/a/\alpha$) or ($a/\alpha/\alpha/\alpha$)] are able to sporulate as well as ($a/a/\alpha/\alpha$) [19]. A number of biochemical events appear to be specifically controlled by the presence of both a and α alleles in the cell:

1. Premeiotic DNA synthesis occurs only in a/α diploids [2].
2. Ribosomal RNA synthesis is altered with the accumulation of an unmethylated 20-S RNA [20-22], as well as normally methylated 26-S and 18-S ribosomal RNA. Under sporulation conditions, uptake of phosphate is delayed in a/α cells [23].
3. During sporulation, glycogen utilization is observed only in a/α diploids, whereas carbohydrate biosynthesis is found in a/a and α/α cells, as well [14,15].
4. Neither mating-type-specific hormones, α factor [24], nor a factor [25] appears to be synthesized in a/α cells, which are also nonmating.
5. Mitotic recombination, induced by UV light, is significantly greater in a/α diploids than in a/a or α/α cells [26].

Although the mating-type locus clearly plays a central role in the control of meiosis, there are yet few data on the mode of its action in establishing and maintaining the prerequisites for meiosis.

There is some evidence that meiotic events under the control of an a/α hybrid locus are separable from mitotic events, as mutant strains of α/α or a/a phenotype have been isolated which are able to sporulate, although less efficiently than a/α strains. Reports from two laboratories have shown that the ability of a/a or α/α strains to sporulate is controlled by either a dominant [27] or recessive gene [28].

C. Extent of Sporulation-specific Genetic Information

Beyond the complex role of the mating-type locus, a number of recessive mutants affecting sporulation and meiosis without affecting vegetative growth have been isolated [29-31]. A few dominant, temperature-sensitive mutations, in which cells sporulate at 25° and do not sporulate at 34°, have also been found [32]. Although several of these mutants have been characterized biochemically and cytologically [32-36], in no case has a clear biochemical role been assigned. On the other hand, it is clear that many of the genes involved in the mitotic cell cycle are also required for meiosis and sporulation. Simchen [37] has shown that a number of cell division cycle (cdc) mutants isolated by Hartwell [38], when homozygous, also control temperature-sensitive events in sporulation.

Thus it is not yet clear to what extent the transition from vegetative growth to sporulation in yeast requires the expression of new genetic information. A genetic study of temperature-sensitive mutations which appear to affect only sporulation and not vegetative growth suggests that approximately 50 sporulation-specific functions are required for the completion of sporulation [32]. An analysis of proteins synthesized during sporulation shows that there are a number of labeled bands on polyacrylamide gels which are not found in vegetative cells [15]. Evidence for a significant change in the control of RNA transcription during sporulation is at present uncertain.

D. RNA Synthesis During Sporulation

The synthesis of ribosomal RNA (rRNA) during sporulation is significantly different from that observed during vegetative growth. First, while rRNA synthesis is continuous during the vegetative cell cycle [39], it is limited to the period between approximately T_4 to T_{10} during sporulation [5,21,23], which is also the period of premeiotic DNA synthesis [2]. Second, there is extensive turnover of rRNA during sporulation, so that by the end of the process approximately 50% of the ribosomes are newly synthesized, without an increase in total RNA content [15,22]. The rate of rRNA synthesis is much slower than that observed during growth [22,39]. Finally, a new sporulation-specific 20-S RNA accumulates during sporulation [20]. This species is most unusual: whereas it is largely homologous with 18-S RNA and is the same size as a transient 20-S RNA found during the processing of vegetative rRNA from a 35-S precursor [40], 20-S RNA found in sporulating cells is totally unmethylated [21,22]. The appearance of this new species of RNA and the differences in the control of rRNA synthesis may reflect significant changes in the translational apparatus of the cell or may simply reflect changes in the metabolism of sporulating cells. One crucial question is whether new ribosome synthesis during sporulation is essential for the completion of differentiation.

Information about the metabolism of messenger RNA (mRNA) during sporulation is distinctly limited. Perhaps because of a dramatic increase in the level of ribonucleases during sporulation and greater difficulty in extraction of polyribosomes from cells, it is considerably more difficult to isolate polyribosomes as large as those found in vegetative cells [41,42]. Nevertheless, large polyribosomes can be isolated to obtain intact mRNA. An analysis of the fraction of ribosomes involved in protein synthesis (i.e., in polysomes) at different times during sporulation found no significant changes during the process [43].

III. CURRENT RESEARCH

A. Conditional Mutants

As indicated in the introduction, the mating-type locus plays a pivotal role in controlling the differentiation of *Saccharomyces cerevisiae*. One of our major efforts has been to isolate conditional mutants affecting the mating-type locus of yeast, in order to examine the role of the mating-type genes in the control of sporulation. Because diploids heterozygous (a/α) for mating type are able to undergo meiosis and sporulate whereas cells homozygous for mating type (a/a or α/α) cannot, we have sought mutants which are temperature-sensitive for mating type. Specifically, we were interested in finding a haploid strain which exhibited an α mating phenotype at 25° and a mating type at 34°. One might then construct a strain which was heterozygous for mating type at one temperature and homozygous for mating type at the other. Such a strain would be invaluable in assessing which functions were directly under mating type control and whether some or all of these genes could be ordered in a temporal sequence.

It has been possible to isolate many conversions of mating type by mutation, although none have proven to be temperature-sensitive. In order to select for the rare conversion of an α mating type to an a mating type, we have used a forced mating technique, illustrated in Fig. 1. One haploid strain is treated with a mutagen and then mixed with a second haploid strain of the same mating type. The two α strains are allowed to grow together to a high density, either in liquid or on agar plates, and samples are then spread or replica-plated to agar plates containing minimal medium plus leucine, so that only diploids which were wild type by complementation for all nutritional markers except leucine could grow. The diploids obtained in this way were tested for both sporulation and mating type. Nonmating diploids able to sporulate were then tested for meiotic recombination by replica-plating sporulated colonies to leucine dropout medium but including canavanine. In this way only haploid spores which were canavanine-resistant and prototrophic for leucine will be able to grow. The heteroallelic markers *leu*2-1 and *leu*2-27 do not complement; however, colonies prototrophic for leucine can be obtained by intragenic recombination or gene conversion [33].

Over 3,000 diploid colonies which grew under forced mating conditions have been tested for mating and sporulation at both 34° and 25°. No strains have exhibited a temperature-sensitive mating

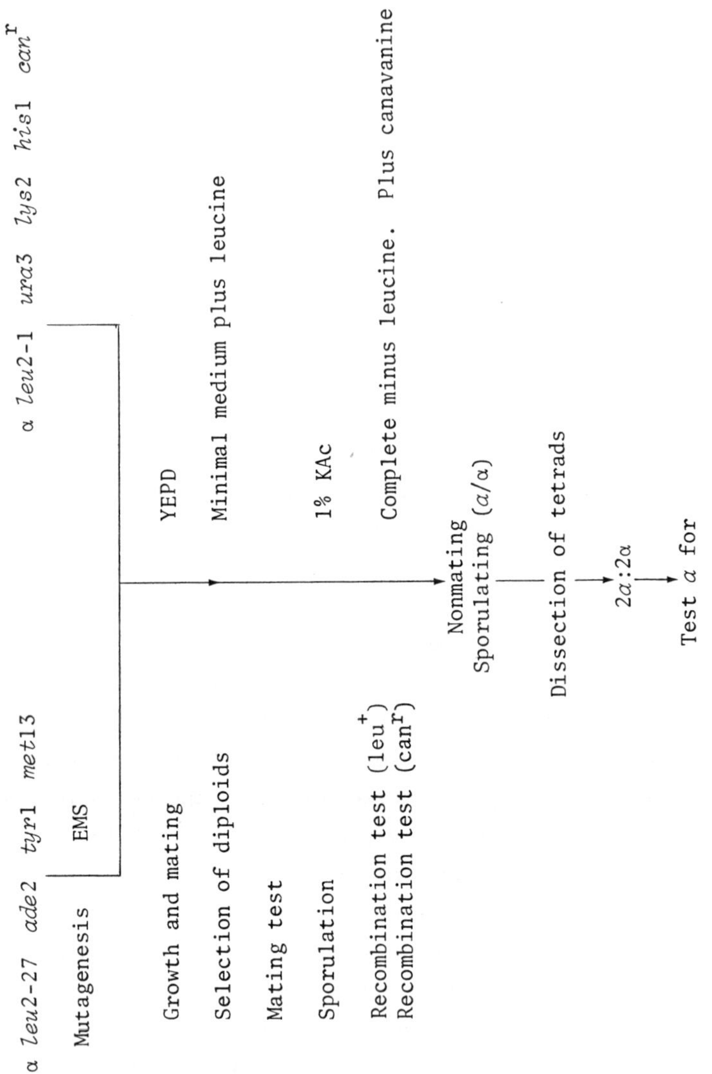

FIG. 1. Procedure to select mutants of the mating type.

phenotype; however, several different other mutants affecting sporulation have been found. One such class are colonies which have α mating characteristics but which are nevertheless able to sporulate. Strains of this sort appear to be quite similar to those described in detail by Hopper and Hall [27] and Gerlach [28], in which the normal inability of α/α strains to sporulate has been bypassed by mutation.

In the absence of any temperature-sensitive mating-type strains, we have elected an alternative approach to obtain strains which are temperature-sensitive for continued gene expression of the mating-type locus. As previously suggested by Hawthorne [44], we reasoned that *a* mating-type cells which have derived from the original α mating type might contain a suppressible amber mutation. Such a suppressible *a* allele could be combined with a dominant, temperature-sensitive amber suppressor tRNA (*SUP* 4-3) isolated by Rasse-Messenguy and Fink [45]. According to the scheme shown in Fig. 2, at 34°,

I. Isolation of *a* phenotype by mutation of α strain. Assume some $a = \alpha_{amber}$.

II. Cross new *a* strains with α carrying a dominant, temperature-sensitive amber suppressor (*SUP*4-3). The suppressor is active at 25° but inactive at 34°. *SUP*4-3 *met*8-*a* *trp*1-*a* *ade*2 requires only adenine for growth at 25° but requires adenine, methionine, and tryptophan for growth at 34°.

III. Test diploids for sporulation and mating at 25 and 34°.

Genotype	Predicted behavior 25°	Predicted behavior 34°
$\frac{\text{"}a\text{"}}{\alpha}$ $\frac{+}{SUP4\text{-}3}$	$\frac{\text{"}\alpha\text{"}}{\alpha}$ mate with *a* no sporulation	$\frac{\text{"}a\text{"}}{\alpha}$ nonmating sporulation

FIG. 2. Predicted behavior of suppressible mutant of the mating type in diploids.

the diploids formed from such crosses should be nonmating *a*/α diploids able to sporulate (as the suppressor is inactive). At 25°, however, the amber suppressor is active, and diploids in which the *a* allele is suppressible should then be unable to sporulate but able to mate as an α phenotype cell.

We obtained several *a* mating-type segregants from each of 700 of the diploid colonies selected as described above. Each *a* segregant was then crossed with an α strain carrying the dominant temperature-sensitive suppressor and tested for temperature sensitivity of sporulation and mating. Of the 700 colonies, we have found at least four strains which appear to be suppressible for sporulation in the manner described. The results for these four *a* segregants, when crossed both with an α strain which carries the temperature-sensitive suppressor and another α strain which does not carry a suppressor, are shown in Table 1. When these *a* segregants are mated to form diploids without the suppressor, sporulation of the resulting diploids is essentially identical at either 25 or 34°. On the other hand, when these diploids are crossed to α strains carrying the *SUP*4-3, there is little or no sporulation in cells grown and sporulated at 25° but substantial sporulation in cells grown and sporulated at 34°. The temperature-sensitive sporulation observed here is a function of both the particular *a* mutant selected and the temperature-sensitive suppressor. The suppressor exerts no temperature-sensitive effect in diploids derived from other *a* strains in our laboratory collection or in other *a* mutants obtained from the same selection procedure (Table 1B).

In these diploid strains, it is presumed that the presence of the temperature-sensitive amber suppressor permits the conditional regulation of continued gene expression but does not affect the thermolability of gene products already synthesized. We therefore also investigated whether diploids which have been grown under preconditions which apparently permit sporulation (i.e., 34°) could sporulate even if the cells were transferred to sporulation medium at 25° and, conversely, whether cells pregrown at 25° would still be incapable of sporulation if incubated at 34°. The results of

TABLE 1[a]

Genetic Analysis of Four *a* Mating Type Segregants

	Cross	Pregrowth sporulation	Percent sporulation 25° 25°	34° 34°	34° 25°	25° 34°
A.	10E2-10 × I1a ($SUP4^{ts}$)		3	29	2	38
	10E2-10 × A28 ($SUP4^{+}$)		51	49	31	52
	D1-28 × I1a		3	32	11	43
	D1-28 × A28		32	52	26	58
	G14-5 × I1a		5	53	7	52
	G14-5 × A28		59	60	51	80
	H18-15 × I1a		0	38	4	37
	H18-15 × A28		43	73	47	64
B.	A27 × I1a		65	59	51	62
	10E1-7 × I1a		27	21	47	43

[a]Four *a* segregants (10E2-10, D1-28, G14-5, and H18-15) exhibit temperature-sensitive suppression of sporulation in diploids containing $SUP4^{ts}$ but temperature-independent sporulation if crossed with another (SUP^{+}) α strain. The ts suppressor does not affect sporulation of other *a* strains (A27 and 10E1-7).

this test are also shown in Table 1. There is apparently a continued requirement for the expression of new (suppressible) gene product during sporulation itself, as cells pregrown at 34° were unable to sporulate if then incubated at 25°, whereas cells grown at 25° sporulated well when incubated at the permissive temperature of 34°. By shifting cells from permissive to nonpermissive temperature at intervals during sporulation, it appears that temperature-dependent suppression of sporulation continued until approximately one-third of the time necessary to complete sporulation (Fig. 3). In one case, the release from temperature sensitivity occurred within the first 6 h of sporulation. We do not yet know the basis

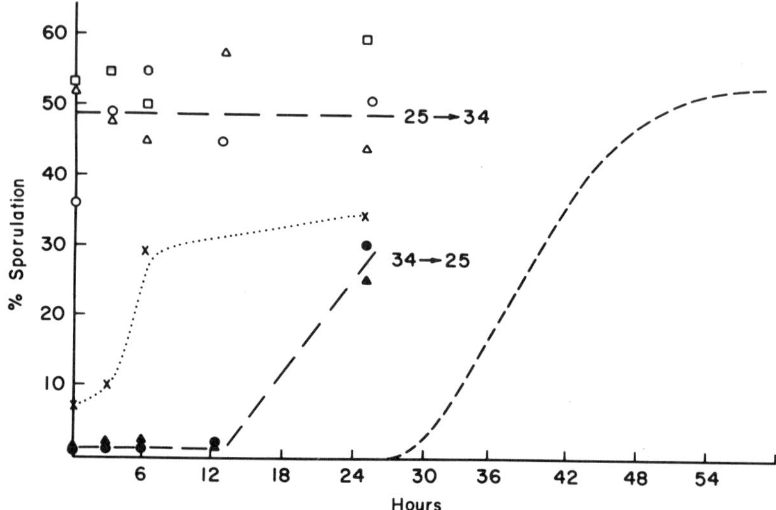

FIG. 3. Period of temperature-sensitive suppression of sporulation. Three diploids constructed to carry a mutant *a* allele and the dominant ts suppression were grown at either 25 or 34° and the replica plated to sporulation medium. The plates were shifted from one temperature to the other at the time indicated. Sporulation was determined after 72 h. The dotted line indicates the time of appearance of spores on agar plates maintained at the permissive temperature, 34°. All three diploids are able to sporulate at 34°, even after incubation at 25° for 24 h. The need for continued gene expression of the mating-type locus is seen from the lack of sporulation if cells are transferred from 34 to 25° during the first 12 h of sporulation. One strain, G14-15 (X), requires less time at 34° during sporulation than do the other two strains, 10E2-10 (●) and D1-28 (▲).

for the difference between these strains, but it can be seen that this last diploid was not as completely suppressed at 25° as the other two diploids which are shown.

These experiments provide an indication of the kind of investigation which we hope to be able to carry out using such diploids temperature-sensitive for sporulation in this way. Of course, it is necessary to demonstrate more completely that the suppressible locus which is affecting sporulation is indeed the particular *a* allele of the mating-type locus. To confirm that mating-type locus is indeed the site of suppression, we have examined the segregation of the suppressible locus by the dissection of tetrads in sporulated cultures. For this test, the diploids formed by crossing mutant *a* allele with an α strain which did not contain the temperature-sensitive suppressor were used. The four spores dissected from one ascus were allowed to germinate and grow and then were tested for mating type and the segregation of several nutritional markers. Each segregant was then crossed with either an *a* or an α strain carrying the dominant temperature-sensitive suppressor. The four new diploids were then isolated and tested for temperature sensitivity of sporulation by growing and sporulating the cells at 25 and 34°. The results of three such tetrads are shown in Table 2. Not all *a* segregants are temperature-sensitive when tested in this fashion; however, most *a* segregants are temperature-sensitive whereas no α segregants show temperature-sensitive suppression. In this preliminary study, 17 *a* segregants exhibited temperature-sensitive sporulation when crossed with an α strain carrying the *SUP*4-3. Only a single α segregant exhibited temperature-sensitive sporulation when crossed with an *a* strain carrying the temperature-sensitive suppressor. These results indicate that the suppressible locus is either the *a* mating-type allele itself or very closely linked to this locus. These results do, however, indicate one serious problem which we have encountered in working with these strains; that is, when the mutant *a* alleles are cultivated as *a*/α diploids for many generations, there appears to be a loss of the suppressibility. When diploids which originally showed significant temperature sensitivity to sporulation and then lost this suppressibility are then studied by tetrad analysis, we find that the temperature-sensitive suppressor is still present and that it is the *a* locus itself which has "reverted." Such instability of mutants of the mating-type locus may be a general phenomenon as a number of other mating-type mutants also show a high likelihood of reverting when cultivated as a heterozygous diploid [46].

A word should also be offered about the mating behavior of these strains. The scheme presented in Fig. 2 indicates that at 25° diploids in which the *a* allele was suppressed by an amber suppressor

TABLE 2

A. Tetrad Analysis of $\frac{''a''}{\alpha}$ $\frac{+}{SUP4^{ts}}$

	Sporulation	
	25°	34°
10E2-10 × I1a (*SUP4*^{ts})	3%	29%
10E2-10 × A28 (*SUP4*$^{+}$)	51%	49%

B. Dissection of 10E2-10 A28

Tetrad		Spore mating type	Cross with	Sporulation	
				25°	34°
1	A	α	FM11 (*a* *SUP4*^{ts})	24	39
	B	*a*	I1a (α *SUP4*^{ts})	9	41
	C	*a*	I1a	2	38
	D	α	FM11	60	62
2	A	α	FM11	34	42
	B	*a*	I1a	9	41
	C	*a*	I1a	18	39
	D	α	FM11	53	54
3	A	*a*	I1a	2	55
	B	α	FM11	76	81
	C	*a*	I1a	14	46
	D	α	FM11	64	62

should have an α mating phenotype whereas at 34° these cells would have a nonmating phenotype. We have not found mating phenotypes to conform to this prediction. These strains are usually nonmating at both temperatures.

The results presented here give a strong preliminary indication that an α allele of the mating-type locus of yeast can be converted by an amber mutation to an *a* mating phenotype cell. This mutant *a* strain, when crossed with an α strain carrying a dominant temperature-sensitive amber suppressor, then exhibits conditional temperature sensitivity for sporulation. We are hopeful that such strains will make it possible to understand the role of the mating-type locus in the control of genetic recombination, meiosis, and the differentiation of the yeast cell to an ascus containing four haploid spores.

B. Control of RNA Synthesis

There are a number of differences in the control of RNA synthesis during sporulation when compared to cells in vegetative growth. The most well characterized of these events to date involves the synthesis of rRNA. As indicated above, rRNA synthesis in sporulating cells is restricted to a limited period of the sporulation cycle and after most protein synthesis has already occurred. There is also extensive breakdown and resynthesis of rRNA which continues throughout sporulation, so that new ribosomal synthesis accounts for approximately half of the rRNA which is found at the end of sporulation [22]. We have been examining several aspects of the control of ribosomal biosynthesis during sporulation.

The apparent rate of synthesis of rRNA during sporulation is considerably slower than that observed in vegetative cells. After pulse labeling, the first appearance of 26-S and 18-S RNA (which are derived from a common 35-S precursor RNA) requires nearly 30 min during sporulation [22], while the same event takes only several minutes during vegetative growth [40]. The measurement of the kinetics of rRNA synthesis during sporulation is complicated by the fact that there is a substantial pH dependence on the incorporation of radioactive precursors such as [^{3}H]adenine, so that when the pH of sporulation medium rises from pH 7 to nearly pH 9, the incorporation of label is substantially decreased [5]. We have found that the effect of changing the pH of a medium of sporulating cells is not restricted to the uptake of radioactive precursors but can also be seen in the actual rate of processing of rRNA within the cell. If cells are labeled for 20 min at pH 8.8 and pH 7.0, one finds a significant difference in the pattern of labeled RNA (Fig. 4). Those which are incubated at the normal high pH of sporulation medium accumulate labeled RNA virtually only into the region of 35-S precursor RNA, while cells labeled in sporulation medium adjusted to pH 7.0 contain labeled 26-S and 18-S RNA as well as their slightly larger 27-S and 20-S precursors. This difference can be attributed to a change in the rate of processing of the precursor, as shown in Fig. 5. In this experiment, cells which had been labeled for 20 min at the high pH (cf. Fig. 4A) were then followed for an additional 20 min in the presence of excess of unlabeled adenine to prevent further incorporation of label into newly synthesized RNA. Half of the culture was permitted to continue at pH 8.8 while the other half was adjusted to pH 7.0. Again, it is evident that cells maintained at a high pH still contain a large fraction of the RNA in the form of precursors, whereas virtually all of the RNA has been processed to 26-S and 18-S species when the pH is adjusted to 7.0. These results suggest that some of the reduction in the rate of rRNA synthesis during sporulation is a direct response of the cells to a change in the pH of the medium. Because of the tight coupling between continued protein synthesis and the accumulation of new ribosomes, we cannot yet rule out that the primary effect of changing pH is a general effect on protein synthesis. This explanation seems unlikely, however, in view of the fact that Mills [5] did not observe any significant difference in the accumulation of labeled amino acids into protein under different pH conditions during sporulation.

In view of the fact that yeast sporulate under pseudostarvation conditions, one wonders why there is such an extensive breakdown of rRNA and resynthesis. It is possible that this process is an expensive method of generating sufficient nucleotides for another round of (premeiotic) DNA synthesis; on the other hand, there may be some significant difference between the rRNA or ribosomes which are

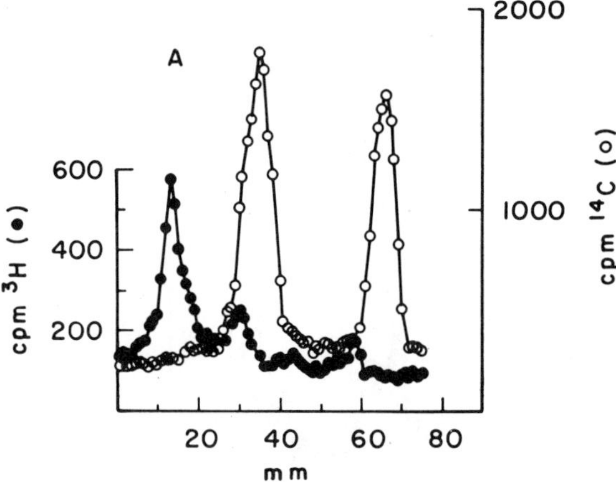

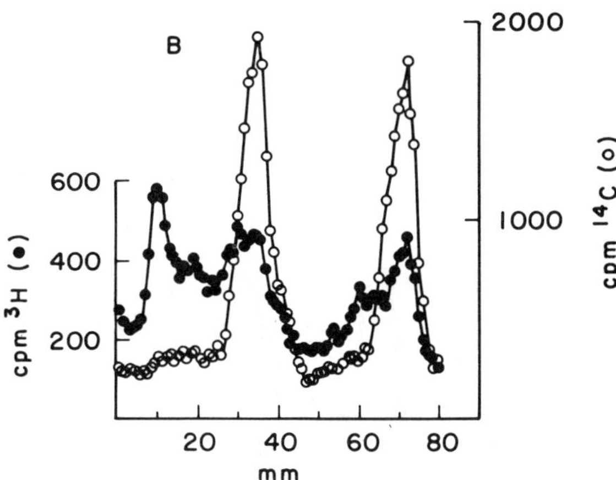

FIG. 4. Synthesis of rRNA labeled for 20 min in sporulation media at pH 7.0 and pH 8.8. Cells of strain J1 [22] were grown to stationary phase in the presence of 0.2 μCi of [^{14}C]adenine/ml, washed and resuspended in 1% potassium acetate, pH 7, containing 0.33 mM L-leucine and 0.13 mM L-methionine. After 6.5 h the cells were removed by centrifugation and resuspended in sporulation media either at pH 8.8 (A) containing 35 μCi [^{3}H]adenine/ml, or at pH 7.0 (B) containing 3 μCi [^{3}H]adenine/ml. Cells were harvested 20 min later. The RNA was separated on 3% polyacrylamide gels. RNA synthesized during 20 min at pH 8.8 appears as labeled material only in the 35-S, 27-S, and 20-S rRNA precursor species as well as 18-S and 26-S rRNA.

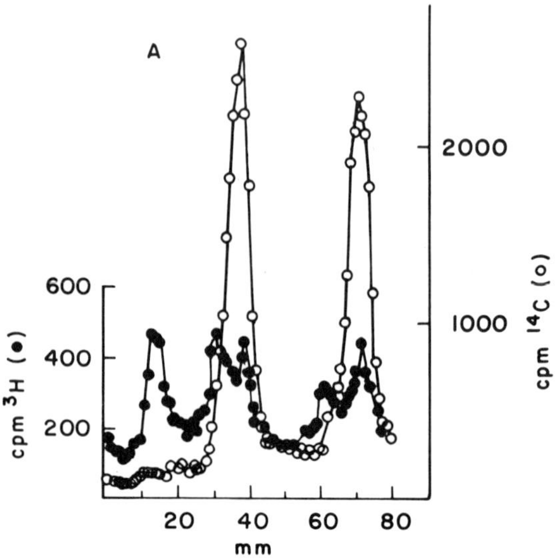

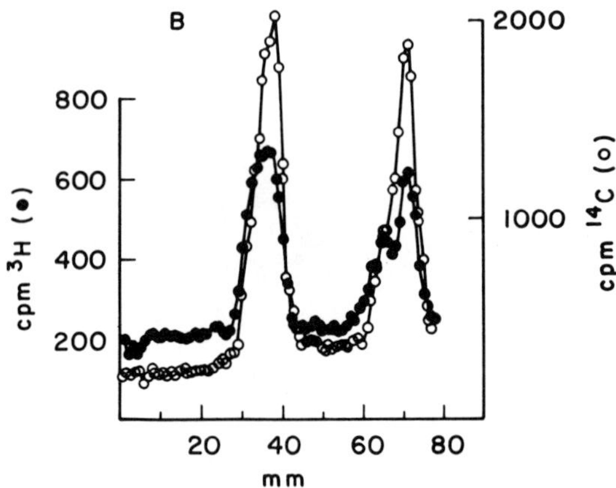

FIG. 5. The processing of prelabeled RNA in 20 min at pH 8.8 and 7.0. Sporulating cells which had been labeled for 20 min in sporulation medium containing 35 μc [^{3}H]adenine/ml at pH 8.8 (see Fig. 4A) were harvested and resuspended in sporulation media at either 8.8 or 7.0 containing 200 μg adenine/ml. After 20 min incubation the cells were harvested and the purified RNA was separated on 3% polyacrylamide gels. For cells remaining at pH 8.8 (Fig. 4A) a considerable amount of labeled RNA is found in the 35-S, 27-S, and 20-S precursor regions of the gel. In contrast, the tritium-labeled RNA found in cells transferred to pH 7.0 is nearly all mature 26-S and 18-S rRNA species (Fig. 4B).

made during sporulation from those which existed during vegetative growth. In order to determine whether new rRNA synthesis during sporulation is an essential aspect of the process or whether it is merely gratuitous, we have constructed diploids homozygous for temperature-sensitive mutations known to block rRNA synthesis during vegetative growth. These mutants, drawn from those described by Hartwell et al. [47], block new rRNA accumulation without significantly inhibiting mRNA synthesis and protein synthesis. Two diploid strains were used--one homozygous for the temperature-sensitive gene *rna*2 and the other homozygous for *rna*6. The behavior of these diploids in vegetative conditions was identical to that previously described for haploid strains carrying these same mutations [47]. The cells grew normally at 25° but grew only for approximately one generation after transfer to 34°. When cells were pulse-labeled with [^{3}H]adenine, the ratio of radioactivity incorporated into acid precipitable RNA at 34 vs 25° was approximately 0.2 in both cases, a value quite similar to that obtained for haploid strains carrying these same mutations. When the same diploids were transferred to sporulation medium, a similar temperature sensitivity was observed. Approximately 50% of the cells sporulated at 25° whereas no sporulation was observed at 34°. The lack of sporulation appears to be controlled by the same genes which prevent vegetative growth at 34°, as revertants selected for growth at 34° also become able to sporulate at the same, previously nonpermissive, temperature (Table 3).

There is no significant inhibition of either premeiotic DNA synthesis or protein synthesis during sporulation in the cells homozygous for the *rna* mutations at 34°. Both strains exhibited a release from temperature sensitivity of sporulation between 6 and 9 h of incubation, as shown in Fig. 6. This is the same interval during which most rRNA synthesis is completed.

In view of these results it was quite surprising to discover that rRNA synthesis itself was not temperature-sensitive during sporulation. The incorporation of either [^{3}H]adenine or ^{32}P into RNA was nearly identical at 25 and 34° during sporulation, even though there was a fivefold difference in accumulation of labeled

TABLE 3

Comparison of Diploid Strains Temperature-sensitive for Growth with Temperature-independent Revertants

Strain	Growth[c] 25°	Growth[c] 34°	Percentage sporulation[d] 25°	Percentage sporulation[d] 34°
J200[a]	+	-	51	0
J200R1	+	+	53	34
J200R2	+	+	60	38
J205[b]	+	-	64	0
J205R1	+	+	48	18

[a]J200 genotype: $\frac{a}{\alpha}\ \frac{rna6}{rna6}\ \frac{ade2}{ade2}\ \frac{his1}{his1}\ \frac{his7}{+}\ \frac{+}{ade2}\ \frac{+}{lys2}\ \frac{+}{tyr1}\ \frac{+}{ura1}$

[b]J205 genotype: $\frac{a}{\alpha}\ \frac{rna2}{rna2}\ \frac{ura1}{+}\ \frac{+}{lys2}\ \frac{+}{tyr1}$

[c]Growth was monitored on YEPD agar plates at 25 or 34°.

[d]Sporulation was measured after 48 h in sporulation after transfer from late logarithmic growth in AC medium.

RNA species during vegetative growth at the same temperatures. We have found that this lack of temperature sensitivity in the incorporation of label into RNA can be observed both during short pulse labeling and long-term incubation. In addition, the labeled RNA appears to be fully functional, as judged by the appearance of virtually all of the radioactivity in polyribosomes (Fig. 7).

These results suggest that there is a significant difference in the regulation of rRNA synthesis during sporulation when compared with vegetative growth. Processes which when thermolabile lead to an almost complete inhibition of rRNA synthesis during vegetative growth do not exhibit this same control during sporulation. On the other hand, these strains are still unable to sporulate even though rRNA synthesis does not seem to be significantly impaired. One possible explanation for the failure of these strains to sporulate is that there is indeed another element in the genetic

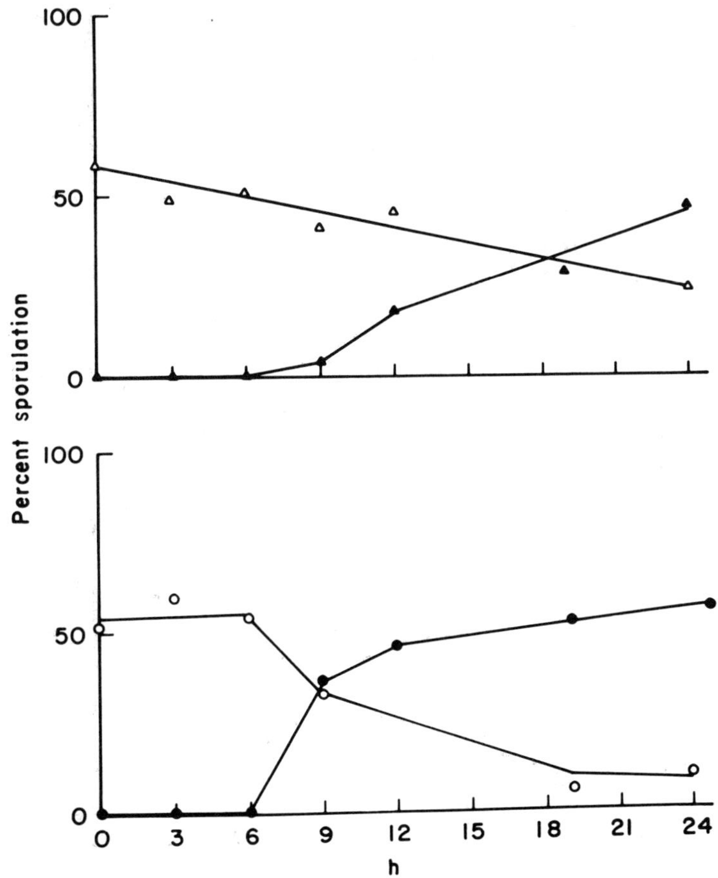

FIG. 6. Temperature dependence of sporulation of strain J205 (top) and J200 (bottom). Cells grown at 25° in acetate growth medium were transferred to sporulation medium. Cells sporulating at 25° were shifted at the time indicated to 34° (▲, ●). Similarly, cells incubated at 34° were shifted to 25° at different times (△, ○). The percentage of sporulation was determined after 48 h.

background of these strains which prevents sporulation at 34°. This explanation seems unlikely in view of the fact that several revertants for growth at 34° isolated to date are all able to sporulate, although at a somewhat reduced frequency (Table 3). It is more likely that the functions of the gene product of *rna*2 and *rna*6 may

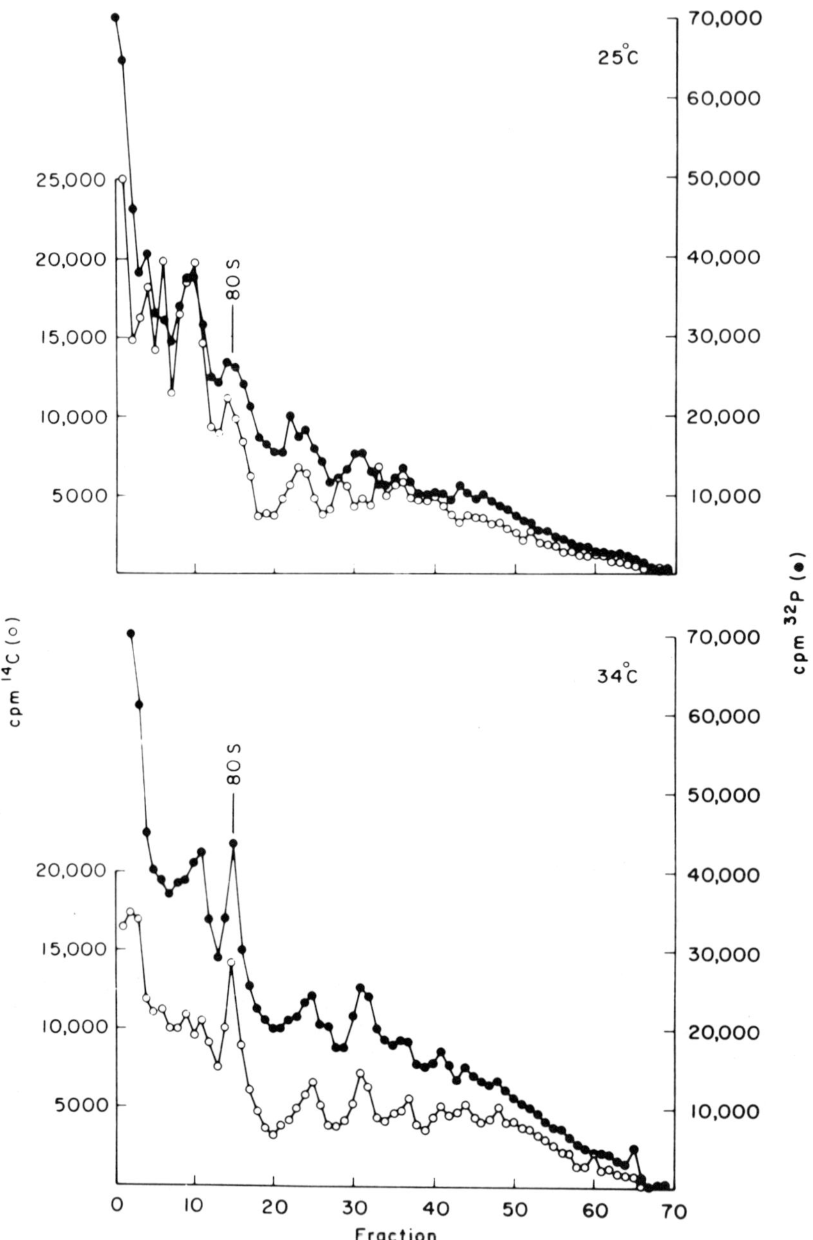

FIG. 7. Incorporation of ^{32}P into rRNA in polyribosomes during sporulation. Cells of strain J200 were grown in the presence of 0.2 μc [^{14}C] adenine during vegetative growth at 14 and 25 , then transferred to sporulation medium at 25°. After 6 h the culture was divided in half, one half remaining at 25°, the other half at 34°. After 10 min, 0.4 μc ^{32}P was added. The cells were harvested after a total of 10 h in sporulation medium and then broken in a Bronwill homogenizer with glass beads. The homogenate was then centrifuged to remove debris and layered on a 10-40% sucrose gradient. The polyribosomal fractions contain both ^{14}C (O) and ^{32}P (●). Virtually all of the label is in stable rRNA.

affect rRNA synthesis only indirectly, so that sporulation is blocked in another fashion but still caused by the same lesion. These explanations, nevertheless, acknowledge the fact that rRNA synthesis itself does not seem to be temperature-sensitive under the control of two different genes, both of which have been identified as affecting primarily rRNA as during vegetative growth. A more thorough investigation of these strains and of other mutants which apparently block rRNA synthesis may reveal the extent of differences in the control of rRNA synthesis during different phases of cellular differentiation.

Finally, we have begun to study changes in the expression of mRNA during differentiation. One striking observation is that the extent of polyadenylation of mRNA is significantly less during

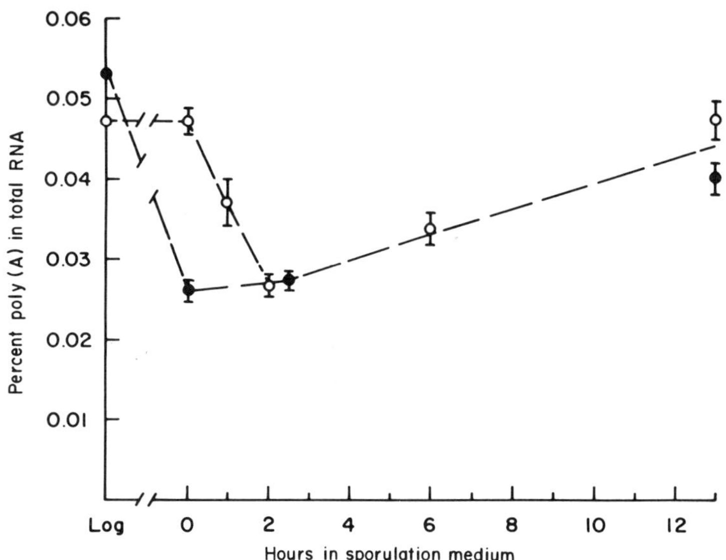

FIG. 8. Changes in the total poly(A) content of RNA during different stages of differentiation. Glucose-grown cells (●) were grown to stationary phase and transferred to sporulation medium. Acetate-grown cells (○) were transferred directly from logarithmic growth. The total amount of poly(A) in RNA extracted from cells at each stage indicated determined by hybridization with [^{3}H]poly(A), as described by Rosbash and Ford [49].

sporulation (and at stationary phase) than in vegetative growth. This 30% reduction is observed in the total amount of poly(A) in whole-cell RNA (Fig. 8) as well as in RNA extracts from polyribosomes.

Our preliminary efforts to measure the average size of poly(A) on mRNA isolated at different stages of growth and differentiation indicate that the difference in polyadenylation is actually a difference in the length of poly(A). These observations are supported by a report by Saunders et al. [48] that activity of a cytoplasmic poly(A) synthetase is greatly reduced during stationary phase and sporulation. It remains to be seen what role this difference in poly(A) may play in the control of translation during different stages of differentiation.

III. FINAL COMMENTS

Saccharomyces cerevisiae is a very simple organism which undergoes a process of meiosis and haploidization similar in form to virtually all higher cell systems. The ease with which mutants specific to growth or differentiation can be obtained and the increasing resolution of molecular events provide strong reason to expect that yeast will provide continuing evidence of the kinds of controls and changes associated with many differentiating systems.

ACKNOWLEDGMENTS

Daniel Klein provided excellent technical assistance for much of this work. This research was supported by NIH grant GM20056.

REFERENCES

1. Roman, H., and S. M. Sands, 1953. Proc. Nat. Acad. Sci. USA 39: 171-179.
2. Roth, R., and K. Lusnak, 1970. Science 168: 493-494.
3. Fowell, R. R., 1969. *In* The Yeasts, Vol. 1, A. H. Rose and J. S. Harrison (eds.), Academic Press, New York, p. 303.
4. Croes, A. F., 1967. Planta 76: 209-226.
5. Mills, D., 1972. J. Bacteriol. 112: 519-526.
6. Esposito, M. S., R. E. Esposito, M. Arnaud, and H. O. Halvorson, 1969. J. Bacteriol. 100: 180-186.
7. Miller, J. J., 1963. Can. J. Microbiol. 9: 259-277.
8. Miller, J. J., and O. Hofman-Ostenhof, 1964. Allg. Microbiol. 4: 273-289.
9. Roth, R., and H. O. Halvorson, 1969. J. Bacteriol. 98: 831-832.
10. Fast, D., 1973. J. Bacteriol. 116: 925-930.
11. Kuenzi, M. T., M. A. Tingle, and H. O. Halvorson, 1974. J. Bacteriol. 117: 80-88.
12. Pontrefact, R. D., and J. J. Miller, 1962. Can. J. Microbiol. 8: 573-584.
13. Moens, P. B., and E. Rapport, 1971. J. Cell Biol. 50: 344-361.
14. Guth, E., J. Hashimoto, and S. F. Conti, 1972. J. Bacteriol. 109: 869-880.
15. Hopper, A. K., P. T. Magee, S. K. Welch, M. Friedman, and B. D. Hall, 1974. J. Bacteriol. 119: 619-628.
16. Kane, S. M., and R. Roth, 1974. J. Bacteriol. 118: 8-14.
17. Henry, S. A., and H. O. Halvorson, 1973. J. Bacteriol. 114: 1158-1163.
18. Illingworth, R. F., A. H. Rose, and A. J. Beckett, 1973. J. Bacteriol. 113: 373-386.
19. Gunge, N., and Y. Nakatomi, 1970. Genetics 70: 41-47.
20. Kadowaki, K., and H. O. Halvorson, 1971. J. Bacteriol. 105: 826-830.
21. Sogin, S. J., J. E. Haber, and H. O. Halvorson, 1972. J. Bacteriol. 112: 806-814.
22. Wejksnora, P. J., and J. E. Haber, 1974. J. Bacteriol. 120: 1344-1356.

23. Chaffin, W. L., S. J. Sogin, and H. O. Halvorson, 1974. J. Bacteriol. 120: 872-879.

24. Duntze, W., V. MacKay, and T. R. Manney, 1970. Science 168: 1472-1473.

25. Wilkinson, L. E., and J. R. Pringle, 1974. Exp. Cell Res. 89: 175-187.

26. Friis, J., and H. Roman, 1968. Genetics 59: 33-36.

27. Hopper, A. K., and B. D. Hall, 1975. Genetics 80: 41-59.

28. Gerlach, W. L., 1974. Heredity 32: 241-249.

29. Esposito, M. S., and R. E. Esposito, 1969. Genetics 61: 79-89.

30. Roth, R., and S. Fogel, 1971. Mol. Gen. Genet. 112: 295-305.

31. Roth, R., and S. Fogel, 1974. Mol. Gen. Genet. 130: 189-201.

32. Esposito, M. S., R. E. Esposito, M. Arnaud, and H. O. Halvorson, 1970. J. Bacteriol. 104: 202-210.

33. Roth, R., 1973. Proc. Nat. Acad. Sci. USA 70: 3087-3091.

34. Moens, P. B., R. E. Esposito, and M. S. Esposito, 1974. Exp. Cell Res. 83: 166-174.

35. Esposito, R. E., and M. S. Esposito, 1974. Genetics 78: 215-225.

36. Pinon, R., Y. Salts, and G. Simchen, 1974. Exp. Cell Res. 83: 231-238.

37. Simchen, G., 1974. Genetics 76: 745-753.

38. Hartwell, L. H., 1974. Bacteriol. Rev. 38: 164-198.

39. Sogin, S. J., B. L. A. Carter, and H. O. Halvorson, 1974. Exp. Cell Res. 89: 127-138.

40. Udem, S. A., and J. R. Warner, 1972. J. Mol. Biol. 65: 226-242.

41. Mills, D., 1974. Appl. Microbiol. 27: 944-948.

42. Peterson, N., and C. S. McLaughlin, 1973. J. Mol. Biol. 81: 33-45.

43. Mills, D., and K. Frank, 1973. Genetics 74: 5182.

44. Hawthorne, D. Personal communication.

45. Rasse-Messenguy, G., and G. Fink, 1973. Genetics 75: 459-464.

46. McKay, V. Personal communication.

47. Hartwell, L. H., C. S. McLaughlin, and J. R. Warner, 1970. Mol. Gen. Genet. 109: 42-56.

48. Saunders, C. A., S. J. Sogin, and H. O. Halvorson, 1975. Fed. Proc. 34: 1736.

49. Rosbash, M., and P. J. Ford, 1974. J. Mol. Biol. 85: 87-101.

GENETICS AND DEVELOPMENT IN *DICTYOSTELIUM* AND *POLYSPHONDYLIUM*

David W. Francis, Robert M. Eisenberg, and Mark MacInnes

Department of Biological Sciences
University of Delaware
Newark, Delaware

I. INTRODUCTION

Figure 1 shows the life history of a cellular slime mold, *Polysphondylium pallidum*. Of the three cycles which produce resting stages, only one, the macrocyst cycle, normally involves a sexual event. In addition, genetic exchange occurs very rarely during the aggregation phase by a parasexual process. Most probably, parasexuality is of little significance in nature.

We will examine the history of discoveries concerning both sexual processes in the cellular slime molds, then turn more specifically to the peculiarities of meiosis in macrocysts. Finally, applications of genetics to the study of development in the cellular slime molds will be discussed.

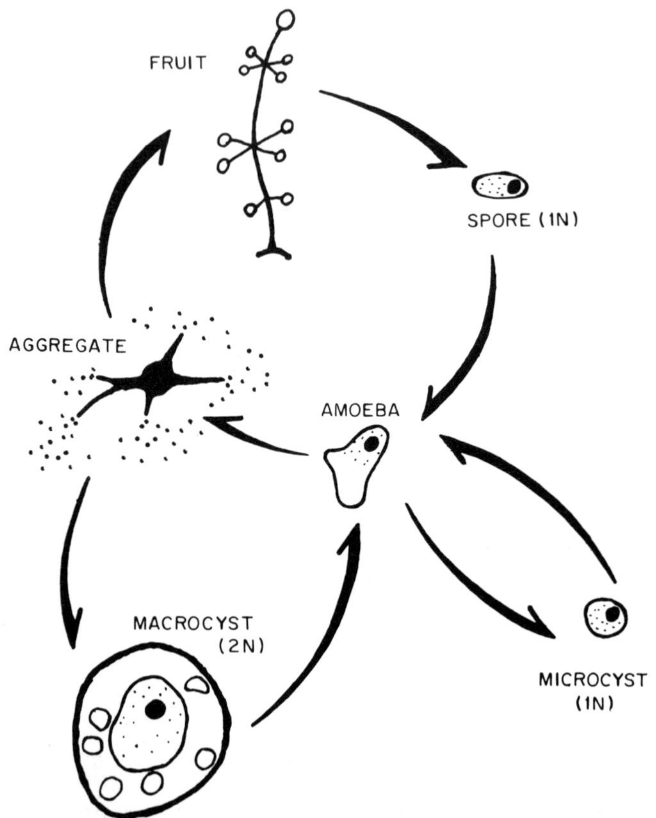

FIG. 1. The life cycles of *Polysphondylium pallidum*. Which pathway is followed depends on environmental conditions. Darkness, osmotic concentration higher than 0.1 M, and crowded conditions favor microcyst formation. Light, low osmotic pressure and less dense populations favor aggregation and spore and stalk formation. Darkness, crowding, and the presence of cells of opposite mating type result in macrocyst formation.

A. Early Apocrypha

The first to propose that sex occurred in a cellular slime mold was Skupienski [1], working with *D. mucoroides*. He claimed to have observed amoebas coupling in pairs at the end of the feeding stage. The coupling was by fusion of pseudopods and not by engulfment, which is important when considering later work. Skupienski was using hanging-drop cultures. He noted that cell fusion occurred best in darkness. Darkness and wetness are two conditions later shown to favor macrocyst formation in this homothallic species, and it seems very possible that he observed the first stage of macrocyst development, which is formation of the zygote.

His account of what happened after cell fusion now seems unlikely. He described the amoeboid zygotes aggregating and forming fruiting structures with spores. Stained slides showed the occurrence of meiosis in some cells during spore formation.

Although it is possible that Skupienski [1] actually observed zygote formation in a homothallic strain of *D. mucoroides*, this was not realized for many years. In the meantime, Wilson [2,3] and Ross [5] made a strong (but in the end, abortive) effort to repeat his observations. Unfortunately these workers used *D. discoideum*, in which most strains are heterothallic. They used relatively dry plates apparently kept under room lighting. These conditions are not favorable for macrocyst development. What Wilson [2,3] saw in stained slides were concentric figures, which he interpreted as gametes fusing by means of one cell engulfing another. He described meiosis as occurring in the aggregate just after this stage.

This work gave rise to two lines of exploration. The first, the study of cell engulfment, was finally pursued to an unrewarding end by Huffman [4]. He used a variety of different isolates of *D. discoideum* and *D. mucoroides* and mixed them in both intra- and interspecific combinations (clearly motivated by the notion that mating types might be involved). Although cell anastomoses and engulfments were observed frequently even between cells of the two different species, nuclear fusions were never seen. No phenotypes having recombinant characteristics were recovered from mixtures of two differently appearing strains. The conclusion was that engulfment of one amoeba by another, while it does occur rather frequently, is a kind of cannibalism rather than a kind of sexual union.

The second lead from Wilson's work began with Ross's [5] observation that strain NC-4 of *D. discoideum* can occur in both haploid and diploid phases. At one period all of the cells in a culture would be haploid (as shown by aceto-orcein stains of cells in mitotic metaphase); on another day all might be diploid. Clearly, some mechanism of doubling and halving the number of chromosomes existed. The work which followed along this line led to the discovery of the parasexual cycle, the details of which have been pieced together by the efforts of several laboratories.

B. Parasexuality

Sussman and Sussman [6,7], after confirming Ross's [5] observation, were able to isolate a diploid line from a mixture of two mutant substrains of NC-4 which produced two characteristic pigments. The diploid was isolated by visual selection as it differed from both parents in spore color. Upon cloning this strain it was found that a small proportion of the cells were haploid and had the pigment characters of one or the other parent. These results showed for the first time that diploid nuclei were not due to the failure of mitosis within a single cell but rather a product of fusion of nuclei from two different cells. Furthermore, these diploids were unstable and returned to the haploid state. It seemed likely that these daughter haploids might have different combinations of genes from their parents, although this was not demonstrated.

What was needed before the cycle became a routine tool for studying genetic exchange was a method of selecting the diploid cells which normally appeared only very rarely. This method was provided by Loomis [8], who isolated two different mutants of haploid NC-4 *D. discoideum*, each of which would grow normally at 22 but not at 27°. When these were mixed, rare temperature-resistant strains appeared which could be easily selected for by incubating the mixture at 27°. These resistant strains were apparently heterozygous diploids and indicated that the two temperature-sensitive mutations were at different loci and were both recessive. They appeared at a rate too high (more than 1 in 10^5 cells of the mixture) to be revertents, which occurred at a rate of less than 1 in 10^6 cells. As the Sussmans [6,7] had found earlier, the diploids were unstable. Besides this Loomis [8] found that parents which contained a second marker mutation generated progeny with new combinations of properties, indicating that segregation and assortment of genes occurred during the process of haploidization.

The other necessary element was a way for selecting the haploids which arise from the heterozygous diploids. These normally appear at a rate of less than 1 in 10^3 cells [8a]. A selection technique was provided by Katz and Sussman [9], who loaded the parental temperature-resistant strains with a cycloheximide-resistant mutation as well as with morphological markers. Heterozygous diploids were first selected as Loomis [8] had done, and were then plated on cycloheximide-containing agar. Cycloheximide resistance is recessive, so that the only clones which appear are haploids for this marker, with rare exceptions.

These two selection techniques have made the parasexual cycle into a practical and simple way of doing genetics with the cellular slime molds, in spite of its low frequency of occurrence. In the 5 years since Katz and Sussman's [9] publication of the method, many papers have appeared in which the technique has been used. Most of this work has been with *D. discoideum*, for which a genetic map with five linkage groups and thirteen markers is present [10]. The cycle operates in *P. violaceum* as well, however [11], and it is probably universal in the group. An important finding is that the mechanism of ploidy reduction is by random loss of individual chromosomes over a period of several mitoses until the haploid number of seven is reached [12,13]. No meiosis occurs in this parasexual cycle, and it resembles in this respect the parasexual cycles operating in the fungus *Aspergillus* and in cultured mammalian cells. Of extreme importance for application to certain problems is the fact that recombination within a linkage group can occur, apparently by crossing over during mitoses in the diploid phase. Such mitotic crossing over occurs rarely (about 1 in 10^5 diploid cells in the case of one marker), but can be detected by selection for antibiotic-resistant diploid lines arising from the standard parasexual parents described above. Haploid segregants showed that these diploids became homozygous by crossing over rather than by mutation or other means [10,14-16].

The possibility for detecting recombination within a linkage group means that the parasexual system is useful for a great variety of genetic studies. The only obvious limitation it has is that mapping within a linkage group depends on the location of a selectable marker distal to the loci being mapped, a somewhat stringent condition.

C. The Macrocyst Cycle

Blaskovics and Raper [17] first called attention to the existence of macrocysts in 1957. They described them as walled, multicellular, resting stages which appeared in cultures of some isolates. Under appropriate conditions, a small percentage germinated and released normal appearing vegetative amoebas. Their significance was unknown, but one speculation among others put forth was that they might represent a sexual phase. Further support for this idea came only much later when the fine structure of the macrocyst was investigated by Erdos et al. [18] using *P. violaceum* and by Filosa and Dengler [19] using *D. mucoroides*. Both groups saw that a single large cell with large nucleus came to fill the entire macrocyst. This cell was seen in life by Erdos et al. [18] in young unwalled macrocysts, but its probable origin by fusion of two amoebas has not been observed even now, except perhaps by Skupienski [1]. Most important was the observation of a synaptonemal complex in a pregerminating macrocyst, implying the presence of meiosis [18]. This was a key discovery, and it has been followed by findings elucidating the details of the cycle.

II. CURRENT RESEARCH

A. Detection of Mating Types

First, it has been shown that macrocysts occur infrequently in cultures of single strains because most species are heterothallic; cells of two opposite mating types must be present for a zygote to be formed. This is not universally true; most strains of *D. mucoroides* are homothallic [20], and occasional homothallic strains are found in other species of *Dictyostelium* and *Polysphondylium* (Table 1). In both species of *Polysphondylium* the situation is made more complex by the apparent existence of syngens, or breeding groups of subspecific level. Isolates belonging to different syngens of the same species are morphologically very similar and fall into the same taxonomic species, but will not form macrocysts when mixed. Isolates belonging to different mating types of the same syngen do form macrocysts. It is an assumption that members of a syngen share a potentially common gene pool; to prove this, macrocysts of a large number of intrasyngenic crosses would have to be shown to yield viable progeny. *D. giganteum* is unusual in that four mating types have been found in the single syngen so far identified [23]. An isolate of one mating type will mate only with members of any of the three other mating types. Two other heterothallic species so far examined have displayed only two mating types [20,22].

TABLE 1

Breeding Systems in Three Species of Cellular Slime Molds

Species	Total isolates	Hetero-thallic isolates	Homo-thallic isolates	Non-mating isolates	Syngens	Mating types per syngen	Ref.
P. violaceum	49	39	0	10	2	2	21
P. pallidum	161	128	4	29	2	2	22
D. giganteum	40	24	0	16	1	4	23

In *P. violaceum*, two syngens have been found, each having two mating types [21]. Isolates of both syngens have been found in Texas and in Delaware. The situation is somewhat different in *P. pallidum*. Here also two syngens have been described of which syngen Provost appears to be cosmopolitan in distribution; isolates having been collected in Georgia, Ontario, Delaware, Nova Scotia, and Switzerland [22]. Syngen two, on the other hand, has been isolated only from Florida.

The fact that *D. discoideum* has mating types partly explains why Wilson [2,3] had difficulty in observing the fusion of gametes, which Skupienski [1] had reported for *D. mucoroides*. Cell and nuclear fusion occur only extremely rarely between cells belonging to the same mating type, as abundantly proved by the work on parasexuality.

The mating type situation raises a group of questions of relevance to the population biology of these organisms. For example: how frequently does mating occur in nature, and when does it occur? The first condition for mating--that compatible cells meet each other--must be frequently realized, as it is common to find two mating types in a single 5-g soil sample, at least with *P. pallidum* [22]. There are clear environmental conditions which favor either formation or germination of macrocysts [24,25], which suggest the occurrence of either one or two cycles/year with germination during the cool periods of either fall or spring.

Another emphasis of current work has been to characterize the mechanism of ploidy reduction which occurs in the macrocyst before germination, which will be considered in detail.

B. Meiosis in Macrocysts

The first clear genetic demonstration of meiosis in macrocysts was given by the work of MacInnes and Francis with the homothallic *D. mucoroides* [26]. Three marker mutations were used: *cactus*, with thicker stalks than wild type; *tag*, which makes unusually small aggregations that give normal fruiting bodies; and *plaque*, which gives clones with a halo of spreading amoebas when grown on a bacterial lawn. The first two of these are morphogenetic mutants, meaning that their action is only visible during a developmental process. Macrocysts were produced using equal number cell mixes of a pair of mutants, and the progeny of single macrocysts were examined.

Roughly half of the macrocysts (Table 2) yielded clones of only one or the other parental type. Such cases are not informative, as these macrocysts might have (1) involved no sexual union or (2) contained zygotes resulting from fusion of two cells of the same genotype or (3) involved meiosis without reassortment. The remaining macrocysts gave rise to either clones of two nonparental phenotypes (called nonparental ditype or NPD macrocysts), to clones of two parental phenotypes (parental ditype macrocysts), or to clones of four different phenotypes all in high frequency (tetratype macrocysts). These results are all expected if meiosis occurs. They are completely different from what is found during the parasexual cycle in a similar cross where sexual fusion is very rare

TABLE 2

Types of Macrocysts in the Cross *Plaque* × *Cactus* in *D. mucoroides*[a]

Number of clones of phenotype					
Plaque	*Cactus*	Wild Type	*Plaque*, *Cactus*	Macrocyst type	Number of macrocysts
145	0	0	0	PT	2
0	398	0	0	PT	5
34	37	0	0	PD	2
0	0	16	19	NPD	1
49	72	36	36	TT	6
					16 Total

[a]Macrocyst types: PT, parental type; PD, parental ditype; NPD, nonparental ditype; TT, tetratype. Data from Ref. 26.

and where the wild-type diploid phase decays slowly to give haploid progeny of mutant phenotypes. The result which argues most strongly against the occurrence of parasexual reduction, however, is the absence of mixed ditype macrocysts. These would be expected to be more frequent than either parental or nonparental ditypes if chromosome reduction were occurring by a parasexual mechanism rather than by meiosis (Fig. 2). This is true regardless of whether the diploid cell divides mitotically before chromosome loss begins. In addition, tritype macrocysts, giving rise to three phenotypic classes of progeny, would be expected by this mechanism of chromosome reduction, and these were never observed. The data imply further that only a single zygote is present in each macrocyst. If two zygotes were present in some macrocysts, the number of apparent uniparental

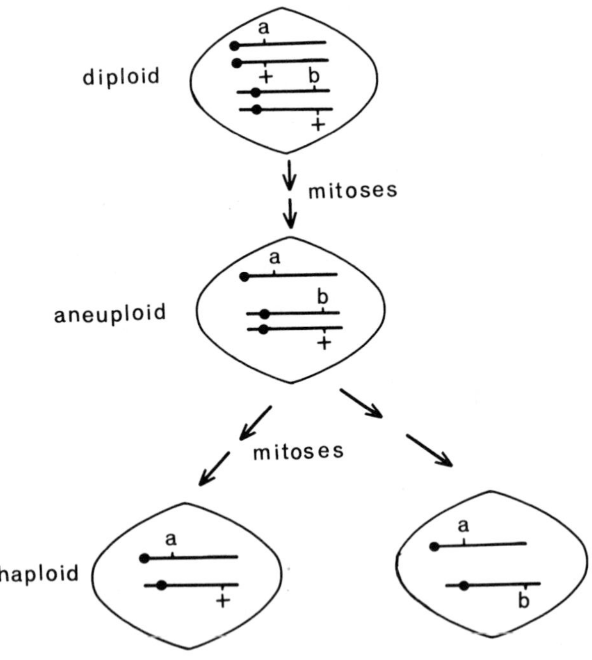

FIG. 2. A possible way of generating mixed ditype progeny from a single diploid cell. This is postulated to occur in the parasexual cycle but does not happen in the macrocyst cycle.

macrocysts would be very low, and tritypes would be expected to occur as well. Neither of these conditions was found. In *D. mucoroides* then, a single zygote undergoes meiosis. Haploid progeny of four types may emerge from a single dihybrid macrocyst.

A somewhat different situation has been reported for *D. giganteum* [23]. This species is heterothallic with four mating types. Several dihybrid crosses were made and single macrocysts germinated by Erdos. The progeny of a single macrocyst were all identical, although a population of macrocysts gave four progeny types in a 1:1:1:1 ratio, as expected from a dihybrid cross. This result implies that three of the four products of meiosis die in each macrocyst in a random fashion and that each zygote is the product of two cells of different mating types, as otherwise the parental classes would be overweighted. The results indicate in addition that mating type is controlled by four alleles at a single locus.

C. Developmental Genetics with Macrocysts

It is generally expected that one of the most rewarding areas to apply genetics in the cellular slime molds will be in the study of developmental mutants. Three studies have already been performed in which parasexual diploids have been used to estimate the number of complementation groups controlling aggregation [11,27,28]. Recent results show that the macrocyst sexual cycle will also be of value.

The first requirement for using macrocyst genetics is that mutants of interest be able to form viable macrocysts. The initial expectation might be that mutants unable to aggregate would not be able to form macrocysts, since the building of a macrocyst occurs after aggregation. Experiments with morphogenetic mutants of *P. pallidum* show that this limitation is a partial one. A number of mutants falling into several morphogenetically deficient classes were each mixed with wild-type cells of the appropriate mating type. Of 10 aggregateless strains tested, only one failed to make macrocysts. Five morphogenetic mutants which are abnormal later in development were all able to form macrocysts. Similar results have been obtained for *P. violaceum*, as well [29]. The proportion of failures might be expected to be higher when two aggregateless mutants are mated, and available data support this notion [29].

Other results indicate that this situation may arise because cells of only one partner of the mating-type pair are actually needed to build the macrocyst. Experiments of Machac [30] indicate that only a very few cells of *P. pallidum* strain 5 of mating-type I are needed to initiate macrocyst formation in a population of cells of strain 2.1 of mating-type II. It would appear that in *P. pallidum* a morphogenetically deficient mutant need only be able to enter into zygote formation to become incorporated into a macrocyst. It need not be capable of building the macrocyst structure itself. Even more startling is the case of *D. discoideum*, where O'Day and Lewis [31] have shown that cells of strain V-12 form macrocysts when stimulated by the extracellular medium of the opposite mating type (NC-4). This work also reveals the presence of sexual hormones in cellular slime molds.

Of course the macrocysts must also germinate. In species like *D. giganteum*, where germination is due to the activity of a number of identical haploid progeny, this may be a real problem. Certain combinations of mutant genes could well result in an inability to digest the macrocyst wall, for example. In species like *D. mucoroides*, where mixed haploid progeny are present in each germination macrocyst, this problem would be more rarely encountered, since at least one group of offspring in a macrocyst may have genes required for germination.

The conclusion to be drawn from the above is that there are developmental mutants which can be used in the macrocyst sexual cycle. The inheritance of these mutations promises to yield valuable information about the formal structure of the developmental program.

Several years ago [33] it was postulated that the developmental events which occur during aggregation and fruiting body formation must have underlying them a special kind of program or "timing sequence" which causes developmental events to be arranged in a linear order in time. Available evidence indicates that the timing sequence can be partly described as a series of events, each of which causes the next event in time.

The work of Loomis [32] and others on changing enzyme activities in developmental mutants has been most instructive in this regard. It has long been known that the specific activities of many enzymes undergo rapid changes correlated with particular morphogenetic events of development. In general, mutants which are blocked early in development lack all later-appearing enzymes [32], and the earlier the block, the more enzyme changes are missing. Clearly these mutants imply that there is a sequence of gene-controlled events during development, each event being essential for the occurrence of the next essential event and for the occurrence of such detectable processes as changes in enzyme specific activity and appearance of morphological organization. Unfortunately, in no case to date has the protein of a morphogenetic mutation been identified. In particular, there is no reason for believing a mutation to be in the gene for an enzyme which fails to appear during development. The mutation could just as well be in a regulator gene controlling the structural gene of the enzyme. In fact, Free and Loomis [34] have shown that mutations in the structural gene of a developmental enzyme may block the activity of the enzyme but nevertheless not interrupt development. Similarly, the changes in specific activities of certain developmentally correlated enzymes can be altered by unusual environmental conditions, and yet later events of development proceed unhindered [35]. These results indicate that many of the easily assayed developmental enzymes do not belong to the timing sequence of essential events.

The idea of a timing sequence is supported by the occurrence of morphogenetic mutants which stop development. Such mutants are interpreted as being in genes controlling events which are nearly or directly on the backbone of the timing sequence, since these are the only genes in which mutation would stop development. The number of such genes for the aggregative stage of development is given as 15 to 60 by the three studies of complementation mentioned earlier [11,27,28]. This number of essential developmental events is probably much lower than the total number of genes which are transcribed differentially during development. In fact, in a later paper in this volume, Loomis suggests that as many as 115 genes may be involved in aggregation. In his paper, Loomis reviews in detail the aforementioned material on enzyme and morphological mutants and expresses his own views on linked and nonlinked developmental events. In his article, the term "dependent sequences" is essentially equivalent to our "timing sequences," in that subsequent developmental events are dependent on previous events, while "independent sequences" are equivalent to events that are not part of the timing sequence of essential events.

An important corollary to the postulate of a linear timing sequence--and a testable one--is that an essential developmental event directly affects only the step immediately following it, and influences stages after that only via a chain of intermediate events. This suggests that in some cases it might be possible to fix a developmental defect by supplying some developmental event past the block, and so reinitiating the sequence of events. The developmental events which might easily be supplied are those consisting of a diffusible chemical which might normally be produced by the entire mass of codeveloping cells.

The same notion had earlier been suggested by the work of Yamada et al. [38]. They examined the asexual development cooperation of pairs of developmental mutants blocked at different stages of development, using *D. purpureum*. Certain pairs showed synergism; that is, the mixture developed wild-type fruiting bodies even though neither mutant could do so by itself. The only pairs in which synergism occurred were those in which development was blocked at stages widely separated in time; two mutants both blocked at aggregation, for example, did not cooperate. Such synergism most probably operates by means of extracellular chemicals; an early-blocked mutant cannot be induced to continue development by a second mutant blocked at the same stage because it too lacks the required chemical; only a mutant blocked at a stage after the production of the deficient molecule will reinitiate development.

To test this idea, cells of early-blocked mutants of *P. pallidum* were each mixed in known proportions with wild-type cells. Spores were harvested from the fruiting structures which formed and, after examination for the absence of amoebas, cloned. Those spores which gave rise to clones of mutant phenotype must have developed from mutant amoebas. The results with several developmental mutants are shown in Table 3 [39]. Some of the mutants which never form spores by themselves can be induced to do so by mixing with wild-type amoebas. Clearly the developmental block in these cases affected only the immediately succeeding stages, and not much later ones. This result fits better with the idea of a linear timing sequence (Fig. 3A) than with a multiply branched and interconnected one (Fig. 3B), in which loss of an essential element at any stage would have effects at many later stages. Other mutants cannot be "fixed" in this way. The significance of these is unclear, as they may result from either a failure of the assay method or from a netted portion of the timing sequence.

These mutants yield information of another sort about properties of the timing sequence, when tested for their ability to develop along alternative pathways of development. The argument is that if each of the three developmental cycles shown in Fig. 1 is caused by an independent timing sequence, then mutants which interrupt one cycle should have no effect on development along the other cycles. If, however, there are essential events common to two cycles, then a mutant blocked in fruiting, for example, may also fail to make cysts.

To examine these alternatives, the developmental mutants of *P. pallidum* which had been selected as being unable to complete the sporulation cycle were tested for their ability to form microcysts

TABLE 3

Developmental Capabilities
of Morphogenetic Mutants of *Polysphondylium pallidum*[a]

Mutant	Phenotype	Spores	Microcysts	Macrocysts
Wild type	Wild type	+	+	+ or --
6M1	Aggless	0	+	+
5M3	Aggless	+	0	0
5M5	Aggless	0	+	+
5M14	Aggless	0	+	+
5M18	Aggless	0	+	+
NG-2	Aggless	+	0	--
NG-4	Aggless	0	+	--
NG-6	Aggless	0	0	--
NG-3	Aggless	+	+	--
5M-2	Fruity	+	+	+
5M7	Fruity	+	+	+
5M13	Fruity	+	+	+
5M15	Fruity	+	+	+
MM8	Fruity	+	+	--

[a]Explanation: aggless, mutants which fail to complete aggregation; fruity, mutants which aggregate but produce abnormal fruiting bodies. (+), >30% of cells differentiate terminally; (0), <5% of cells differentiate terminally; (--), these mutants were derived from WS320, a wild-type isolate incapable of forming macrocysts.

when exposed to high salt concentrations, and where possible for their ability to form macrocysts when mixed with a strain of opposite mating types. The results, shown in Table 3, are of surprising complexity. All combinations of ability or inability for microcyst formation with ability or inability to make spores occur, although most of the aggregateless mutants can make microcysts. The single mutant which could not be induced to form macrocysts failed to make microcysts as well, but was able to make spores.

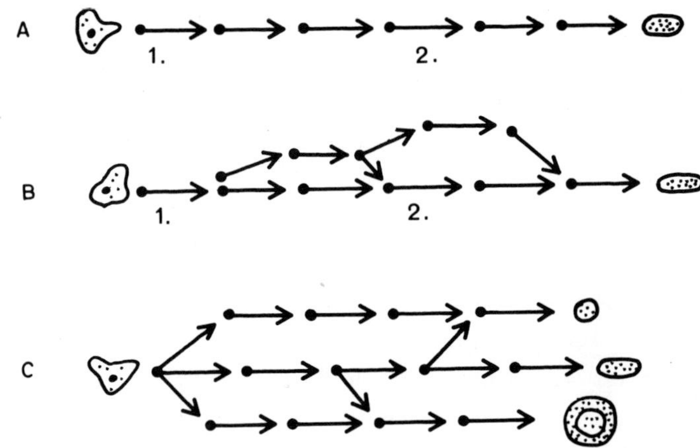

FIG. 3. Three types of timing sequences: (A) linear; (B) network; and (C) branched with rare connections. In type (A), damage caused by a mutation at step 1 can be fixed by artificially supplying step 2. In (B), supplying step 2 will not restore normal development because one of the events affected by step 1 takes place after step 2. Diagram (C) shows how mutations which affect early events in the timing sequence for sporulation might also block development of macrocysts or microcysts.

TABLE 4

Phenotypes of Progeny of the Cross Mutant A × Mutant B in *P. pallidum*[a]

Phenotype	Number of clones	Mating type			
		I	II	Asexual	Selfers
Wild type	34	21	7	0	0
A	122	15	22	3	0
B	265	49	1	0	1

[a]Not all of the clones were tested for mating type. Asexual clones are ones which will not mate with cells of either mating type. Selfer clones are ones which form macrocysts in the absence of cells of either mating type. Data from Ref. 37.

A tentative conclusion is that the timing sequences for the pathways to spore, microcyst, and macrocyst are largely independent, arguing from the fact that most aggregateless mutants and all fruity mutants are able to make macrocysts and microcysts. Nevertheless, there must be some events necessary for aggregation which are also necessary for microcyst and macrocyst development. The complete timing sequence might look like Fig. 3C.

In the case of another kind of cell differentiation, sporulation of *Bacillus*, genetic manipulations have been used to determine whether a particular morphogenetic mutation is in the linear backbone of the timing sequence or on a side branch [37]. The technique was to examine phenotypes of constructed double mutants. Two examples with cellular slime molds will show that the same approach can be used here, although very little has been done as yet. Two morphogenetic mutants of *P. pallidum*, both deficient near the aggregative stage of development, are depicted in Fig. 4. In mutant A, aggregation is triggered by light as it is in wild-type, but few streams form and the soft, low aggregation mounds disperse again into scattered amoebas after a few hours instead of continuing to form a fruiting body. Aggregation of mutant B is more nearly normal, with streams and compact aggregation centers. The centers have difficulty in forming the tip when stalk formation begins, but under ideal conditions delicate unbranched stalks form which carry a sorus containing normal spores.

What we would like to do is determine whether mutation A, which apparently acts at an earlier stage in development than mutation B, is on the timing sequence backbone or on a side branch by examining the phenotype of a constructed double mutant. When the two mutants were crossed using the macrocyst cycle, no progeny had an obvious combined phenotype although reassortment for mating-type did occur (Table 4) [37]. By back-crossing a series of progeny of A phenotype against wild-type, it was demonstrated that some of them contained both A and B mutations and were, in fact, double mutants. The result suggests that the strain containing both mutants is

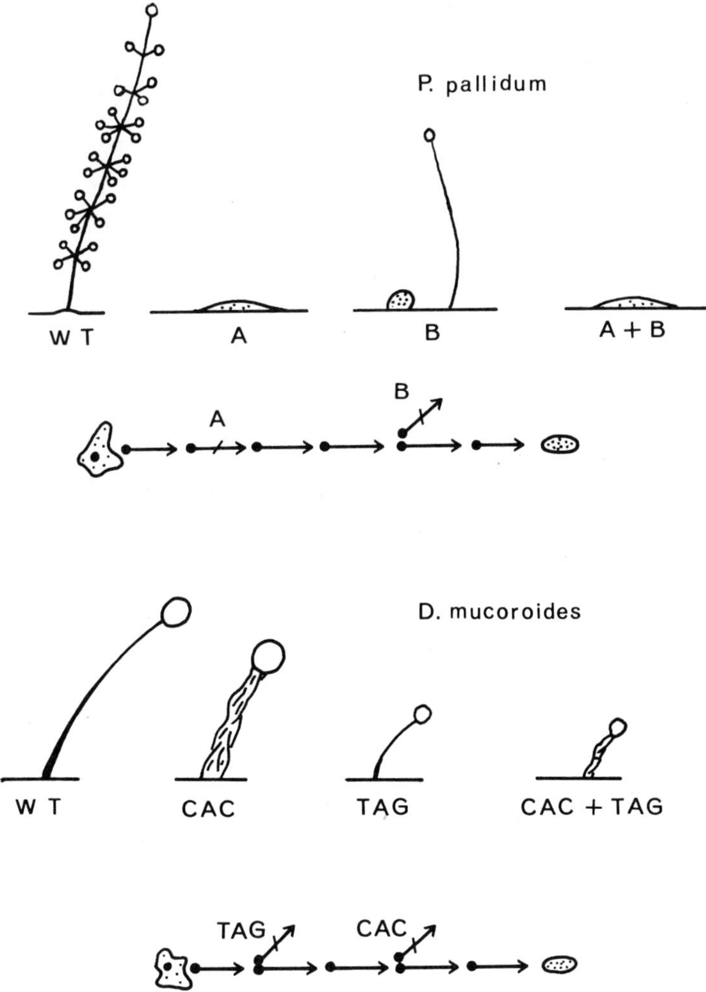

FIG. 4. Appearance of morphogenetic mutants and of constructed double mutants in *P. pallidum* and *D. mucoroides* (diagrammatic), and inferred locations of these mutations in the timing sequence. Further explanation is included in the text.

unable to express the B phenotype because the earlier active A does not allow development to proceed to the stage where B acts. Thus the A gene controls an essential event and is on the backbone of the timing sequence (Fig. 4). The location of the B gene is less certain. Since spores are produced in mutant B even though the fruiting bodies are abnormally shaped, one may imagine gene B as controlling a nonessential process dealing with fruiting body shape. This possibility is diagrammed in Fig. 4A by placing gene B on a side branch of the timing sequence backbone. An alternative possibility is that gene B is actually on the backbone, but that mutant B carries a leaky mutation and so is able to progress to spore formation, although abnormally.

The two morphogenetic mutants of *D. mucoroides* used by MacInnes and Francis [26] were also examined from this point of view. One of these, *tag*, operates at the aggregation stage and modifies development by making smaller-than-normal aggregates. In *cactus*, only the later stage of stalk formation is visibly affected. In the double mutant both mutations are expressed in a completely independent, additive fashion; the *cactus* gene produces the same effect on the very small stalks of *tag* as it did on the large stalks of wild-type. This result suggests that it is located on the side branch of the timing sequence. Alternatively, it may be a leaky mutation of a gene on the backbone, as with gene B of *P. pallidum*.

III. FINAL COMMENTS

In the cellular slime molds, two different sexual cycles are now operative. In the parasexual cycle, rare diploid cells occur during aggregation. They are more or less stable but eventually disintegrate via aneuploid intermediates to haploids. Selection procedures make it a routine matter to generate diploids and their haploid offspring. Because recombination is rare, linkage relationships are easily determined. Mapping within a linkage group is possible when appropriate selective markers are available.

In the macrocyst cycle frequent diploid zygotes are formed. These become enclosed in a resting macrocyst and undergo meiosis on germination of the cyst. Definition of linkage groups by this cycle will require a large set of markers because the high rate of meiotic recombination will obscure the linkage of all except closely-linked genes. If mapping of closely-linked genes is what is sought, however, no selective device is necessary to find recombinants. Studies using double-mutant haploids also are more favorably pursued using the macrocyst cycle, since only one cloning step is needed instead of two. In *Escherischia coli*, the analysis of regulatory units has required both haploid and diploid genetics. We expect that both the parasexual and macrocyst sexual cycles will be necessary for unravelling the regulation of development in cellular slime molds.

ACKNOWLEDGMENTS

Original work was supported by grants from the National Institutes of Health and the National Science Foundation, United States, and from the University of Delaware Research Foundation.

REFERENCES

1. Skupienski, R.-X., 1918. Comptes Rendus Acad. Sci., Paris 167: 960-962.
2. Wilson, C. M., 1952. Proc. Nat. Acad. Sci. USA 38: 659-662.
3. Wilson, C. M., 1953. Amer. J. Bot. 40: 714-718.
4. Huffman, D. M., 1967. J. Protozool. 14: 762-764.
5. Ross, I. K., 1960. Amer. J. Bot. 47: 54-59.
6. Sussman, R. R., and M. Sussman, 1963. J. Gen. Microbiol. 30: 349-355.
7. Sussman, M., and R. R. Sussman, 1962. J. Gen. Microbiol. 28: 417-429.
8. Loomis, W. F., Jr., 1969. J. Bacteriol. 99: 65-69.
8a. Gingold, E. B., 1974. Heredity 33: 419-423.
9. Katz, E. R., and M. Sussman, 1972. Proc. Nat. Acad. Sci. USA 69: 495-498.
10. Williams, K. L., R. H. Kessin, and P. C. Newell, 1974. J. Gen. Microbiol. 84: 59-69.

11. Warren, A. J., W. D. Warren, and E. C. Cox, 1975. Proc. Nat. Acad. Sci. USA 72: 1041-1042.

12. Brody, T., and K. L. Williams, 1974. J. Gen. Microbiol. 82: 371-383.

13. Sinha, U., and J. M. Ashworth, 1969. Proc. Roy. Soc. Ser. B 173: 541-555.

14. Gingold, E. B., and J. M. Ashworth, 1974. J. Gen. Microbiol. 84: 70-78.

15. Katz, E. R., and V. Kao, 1974. Proc. Nat. Acad. Sci. USA 71: 4025-4026.

16. Rothman, F. G., and E. T. Alexander, 1975. Genetics 80: 715-731.

17. Blackovics, J. C., and K. B. Raper, 1957. Biol. Bull. 113: 58-88.

18. Erdos, G. W., A. W. Nickerson, and K. B. Raper, 1972. Cytobiologie 6: 351-366.

19. Filosa, M. R., and R. E. Dengler, 1972. Develop. Biol. 29: 1-16.

20. Clark, M. A., D. Francis, and R. Eisenberg, 1973. Biochem. Biophys. Res. Comm. 52: 672-678.

21. Clark, M. A., 1974. J. Protozool. 21: 755-757.

22. Eisenberg, R., and D. Francis. J. Protozool., in press.

23. Erdos, G. W., K. B. Raper, and L. K. Vogen, 1975. Proc. Nat. Acad. Sci. USA 72: 970-973.

24. Nickerson, A. W., and K. B. Raper, 1973. Amer. J. Bot. 60: 190-197.

25. Nickerson, A. W., and K. B. Raper, 1973. Amer. J. Bot. 60: 247-253.

26. MacInnes, M. A., and D. Francis, 1974. Nature 251: 321-324.

27. Williams, K. L., and P. C. Newell, 1976. Genetics 82: 287-307.

28. Coukell, M. B., 1975. Molec. and Gen. Genetics 142: 119-135.

29. Clark, M. A., and S. E. Speight, 1973. J. Protozool. 20: 507.

30. Machac, M. A., 1975. Biology Senior Thesis, Princeton University.

31. O'Day, D. H., and K. E. Lewis, 1975. Nature 254: 431-432.

32. Loomis, W. F., 1975. *In* Isozymes, Vol. III, Developmental Biology, C. L. Markert (ed.), Academic Press, New York and London, p. 177.

33. Francis, D., 1969. Quart. Rev. Biol. 44: 277-290.

34. Free, S. J., and W. F. Loomis, Jr., 1974. Biochemie 56: 1525-1528.

35. Quance, J., and J. M. Ashworth, 1972. Biochem. J. 126: 609-615.

36. Coote, J. G., and G. Mandelstam, 1973. J. Bacteriol. 114: 1254-1263.

37. Francis, D., 1975. J. Gen. Microbiol. 89: 310-318.

38. Yamada, T., K. O. Yanagisawa, H. Ono, and K. Yanagisawa, 1973. Proc. Nat. Acad. Sci. USA 70: 2003-2005.

39. Francis, D., D. S. Salmon, and B. Moore, unpublished results.

INDEPENDENT AND DEPENDENT SEQUENCES IN DEVELOPMENT OF *DICTYOSTELIUM*

William F. Loomis, Randall L. Dimond, Stephen J. Free, and Sally S. White

Department of Biology
University of California, San Diego
La Jolla, California

I. INTRODUCTION

The sequential expression of specific genes is thought to underlie and determine the course of development in many organisms. Yet exactly how the sequence and timing of gene expression is regulated remains one of the unanswered questions of the biology of eucaryotes. In procaryotes we know of cascade inductions in which the product of one gene accumulates for a period before its activity triggers the expression of other genes [1-3], and in phage λ we know of a sequence in which there is a cascade of two tandem sequential inductions [4]. In the development of higher organisms we can often recognize many discrete stages which occur in an invariant temporal sequence, but techniques have not been available to analyze the detailed biochemical events which determine the stages and the transitions from one to the next.

In a general way, we can consider that the stages in development can be connected either in a dependent or an independent manner such that each stage depends on the attainment of prior stages or, alternatively, is independently triggered, either by the initial stimulus or by an underlying timing mechanism (Fig. 1). This concept is also pursued by David W. Francis in an earlier article in this volume. It is also possible that independent events are triggered by steps in a dependent sequence. To a certain extent our understanding of the connection of stages depends on the characteristics, either morphological or biochemical, by which we have chosen to characterize each stage. Only by analysis of the consequences of mutations affecting the components of these sequences can we hope to elucidate the interactions.

In the development of the cellular slime mold *Dictyostelium discoideum*, the temporal sequence of morphological changes and biochemical differentiations has been studied in some detail (for a review see Ref. 5). In this system, large numbers of growing amoebas can be induced to develop synchronously by removal of exogenous nutrients. Little cell division or growth occurs as the cells aggregate and develop into multicellular slugs, which culminate to form fruiting bodies consisting of a large number of spores supported on a cellular stalk (Fig. 2).

During the first 8 h after the initiation of development there is little or no morphological change, but the cells go through several modifications in preparation for aggregation [6,7]. The

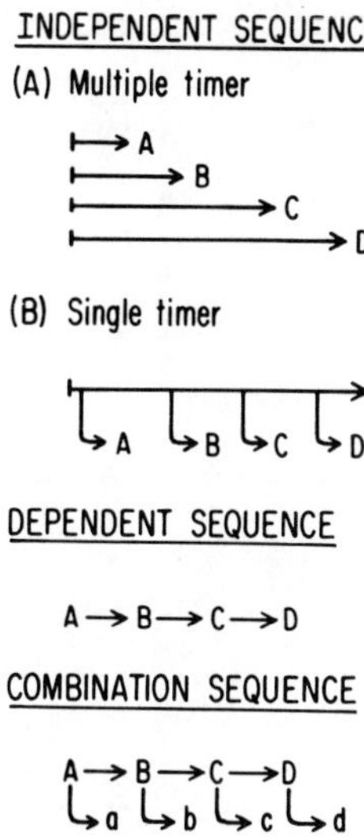

FIG. 1. Control of temporal sequences. Events controlled as an independent sequence may be timed independently or by a common mechanism. When the observed differentiations are arranged in a dependent sequence, each event requires all previous ones. Moreover, it is possible to have both dependent and independent events arranged in a combination sequence.

cells first acquire chemotactic sensitivity to cyclic adenosine-3',5'-monophosphate (cAMP), then become capable of relaying a cAMP signal by releasing cAMP themselves, and finally a few cells become capable of spontaneously releasing cAMP. Aspects of the aggregation phenomenon are detailed in a subsequent article by Anthony J. Durston. At this stage the cells start to aggregate into masses containing up to 10^5 cells. At about the same time, they insert a specific carbohydrate-binding protein into their plasma membranes and become mutually cohesive [8,9].

During the subsequent 8 h of development the cells become integrated into a slug by the deposition of an extracellular surface sheath surrounding the whole mass of cells and the formation of a tip where the sheath is distended at the anterior. Thereafter, the tip acts as an organizing region both during migration of the slug and during culmination. After about 18 h of development, a ring of cellulose is deposited below the tip and is extended through the

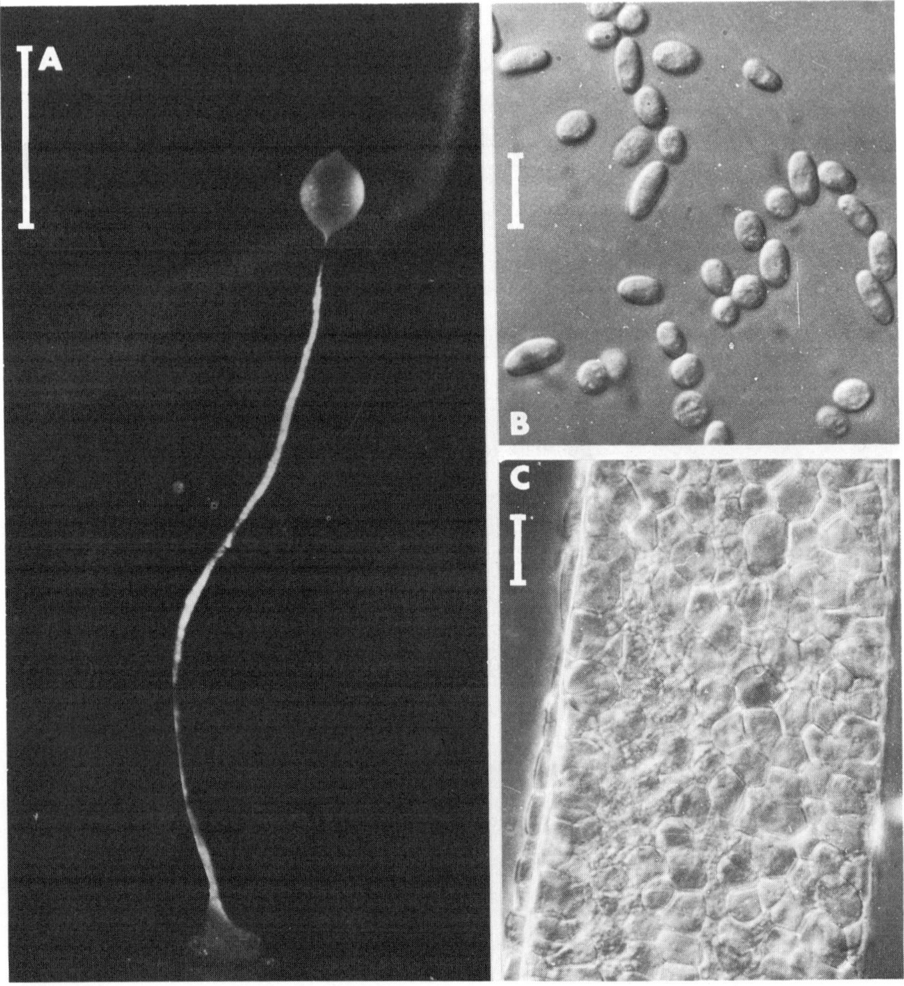

FIG. 2. Terminal differentiation in *D. discoideum*. About three-quarters of the cells become ellipsoid spores (B) while the remaining cells vacuolize and deposit a thick cellulose layer within the stalk tube (C). The spores are held in the sorus supported by the tapering stalk (A). The bar indicates (A) 0.5 mm, (B) 10 μm, and (C) 20 μm.

posterior cells forming the stalk tube. Cells included in the tube vacuolize and expand, lifting the mass of cells off the support. Further entry of anterior cells into the stalk tube and subsequent vacuolization of these cells continues to raise the posterior cells which proceed to encapsulate, forming spores. The final fruiting body is formed when all cells have either encapsulated or formed stalk cells. This sequence of morphological events is further detailed in Fig. 1 of Anthony J. Durston's contribution to this volume and is also summarized at the bottom of Fig. 3 below.

Although we can clearly distinguish four or five morphological stages in the development of *D. discoideum*, a large number of stages can be recognized when we analyze the pattern of accumulation of developmentally regulated enzymes. More than a dozen enzymes have

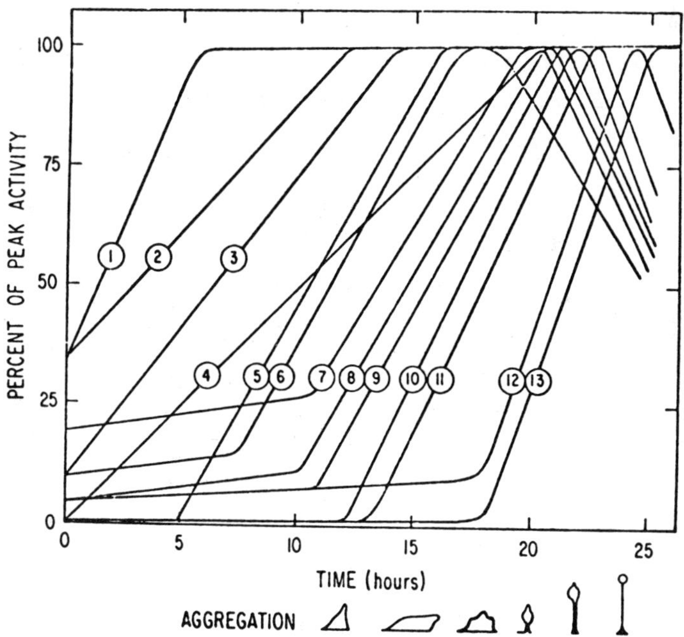

FIG. 3. Stage-specific enzymes. The kinetics of accumulation of 13 enzymes is presented along with a schematic representation of the morphological stages. The individual enzymes are (see Ref. 5 for review): (1) Leucine aminopeptidase, (2) alanine transaminase, (3) acetylglucosaminidase, (4) α-mannosidase, (5) trehalose-phosphate synthetase, (6) threonine-deaminase-2, (7) tyrosine transaminase, (8) UDPG pyrophosphorylase, (9) UDP galactose epimerase, (10) UDP galactose transferase, (11) glycogen phosphorylase, (12) alkaline phosphatase, and (13) β-glucosidase-2.

been found to accumulate during discrete stages and have been studied in some detail (Fig. 3). Each of these marker enzymes appears to be the product of unique genes and accumulates apparently as the result of de novo enzyme synthesis since the accumulation in each case has been shown to be inhibited by cycloheximide, a drug known to block protein synthesis in *D. discoideum*. Moreover, Franke and Sussman [10] have shown by radioimmunoprecipitation that accumulation of UDPG pyrophosphorylase is a direct consequence of an increase in the differential rate of synthesis of this enzyme. The increase in specific activity of each of the marker enzymes is inhibited within 1 to 2 h by actinomycin D alone or in conjunction with daunamycin [11,12], and therefore appears to depend on concomitant transcriptional events. Since these enzymes allow us to focus on the temporal expression of specific genes, we have studied the patterns of enzyme accumulation in various mutant strains selected either for loss of a specific enzyme or for aberrant morphogenesis.

II. CURRENT RESEARCH

A. Acid Hydrolase Mutants

Techniques for the screening of β-N-acetylglucosaminidase (acetylglucosaminidase), α-mannosidase, and β-glucosidase in tens of thousands of clones were developed so as to isolate mutations affecting these enzymes [13]. Cells suspended in broth medium from a mutagenized population of *D. discoideum* were deposited in multitest wells with a multipipette and allowed to grow into clones over a 2-week period. They were then assayed in the wells with the respective chromogenic substrates. Clones which failed to form the yellow nitrophenol product were analyzed, and a series of mutant strains lacking one or another of the acid hydrolases were recovered (Table 1) [14-16].

We isolated five independent strains lacking acetylglucosaminidase and four strains which formed thermolabile enzyme [16]. The absolute mutant strains aggregate normally but form slugs which are

TABLE 1

Peak-specific Activity[a]

Strain	NAG	MAN	TD-2	TT	UPP	Alp	β-G2
Wild type							
A3	250	40	5.5	30	100	25	30
NC-4	250	40	5.6	44	110	35	30
NAG Mutants							
DBL211	<u>2</u>	35	7.3	35	100	15	19
DBL230	<u>3</u>	49	5.3	32	nd	21	31
MAN Mutants							
M1	280	<u>0.05</u>	5.4	33	nd	26	33
M2	nd	<u>0.03</u>	nd	nd	100	24	21
β-Glu Mutants							
G1					nd	12	<u>0</u>
G2					nd	18	<u>3.1</u>
G4					100	24	<u>0</u>
Upp Mutants							
U1					<u>0</u>	63	8
UM1					<u>0</u>	15.7	4.9

[a]The activity of β-N-acetylglucosaminidase (NAG), α-mannosidase (MAN), threnine deaminase-2 (TD-2), tyrosine transaminase (TT), UDPG pyrophosphorylase (Upp), alkaline phosphatase (Alp), and β-glucosidase-2 (β-G2) was measured during development and the peak-specific activity (units/mg protein) determined; nd = not determined.

unable to migrate normally. The temperature-sensitive strains migrate at the permissive but not at the nonpermissive temperature. If the cells are allowed to develop under conditions which cause development to bypass migration they fruit normally. These strains accumulate each of the other marker enzymes measured (Table 1).

We have isolated four strains lacking α-mannosidase [14]. Cells of these strains develop normally under laboratory conditions and accumulate the other six marker enzymes (Table 1).

There are two isozymes of β-glucosidase, one of which is present in vegetative cells but is not synthesized during development and one which is synthesized late in development during culmination [17]. We have isolated seven strains which lack β-glucosidase in vegetative cells and, to our surprise, found that none of these accumulate the developmental isozyme [15]. Moreover, a strain, G2, was recovered which forms thermolabile vegetative isozyme and accumulates some developmental isozyme which is also thermolabile. Thus, it appears both isozymes share a common subunit. The loss of the late developmental isozyme did not affect accumulation of two other late enzymes, UDPG pyrophosphorylase or alkaline phosphatase (Table 1).

B. Mutants Lacking UDPG Pyrophosphorylase

A slightly different screening technique had to be used to isolate mutations lacking UDPG pyrophosphorylase since the cells had to by lysed before the enzyme could be assayed. We were able to replicate the clones into fresh multitest trays with a "bed-of-nails" replicator [13] and to assay the enzyme in detergent-lysed cell extracts by coupling the reaction to reduction of tetrazolium dye. Five strains were found which contain little or no measurable activity [18]. Four of these strains develop to the slug stage but fail to culminate and do not form spores or stalk cells (Fig. 4). These strains are incapable of forming cellulose, which makes up a large proportion of the spore wall and the stalk tube. One strain with little UDPG pyrophosphorylase activity, UN3, develops poorly but produces some encapsulated spores and cells with thick cellulose walls. It is likely that this strain forms UDPG pyrophosphorylase which is partially active in vivo but is rapidly inactivated when the cells are lysed and the reaction assayed in vitro. In

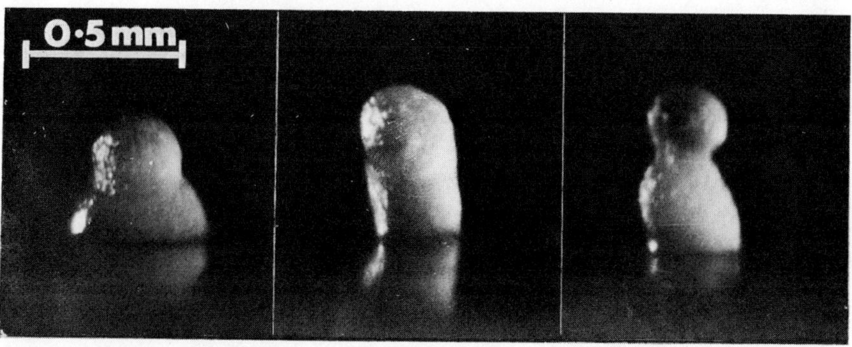

FIG. 4. Terminal differentiation in a strain lacking UDPG pyrophosphorylase. About 24 h after the initiation of development in strain U1 only mounds are formed.

fact, we have also isolated 10 strains in which considerable activity is measurable immediately upon lysis, but this activity decays within a few minutes following lysis [18].

The late stage-specific enzymes, alkaline phosphatase and β-glucosidase-2, accumulate in strains lacking UDPG pyrophosphorylase, but normal peak specific activity of β-glucosidase is not reached (Table 1). We confirmed by electrophoretic analysis that the activity present in developed cells of these strains was β-glucosidase-2 rather than remaining β-glucosidase-1. The reduced peak specific activity is probably a consequence of the autolysis and death of cells of these strains which occur during development, as judged by electron microscopy and determination of viable count (Fig. 5). It appears that even when terminal differentiation is aborted due to impaired synthesis of UDPG, the program which elicits synthesis of the stage-specific enzymes continues to function until late in development.

C. Independent Sequence of Events

The patterns of accumulation of stage-specific enzymes in mutants lacking one or another of them indicates that these gene

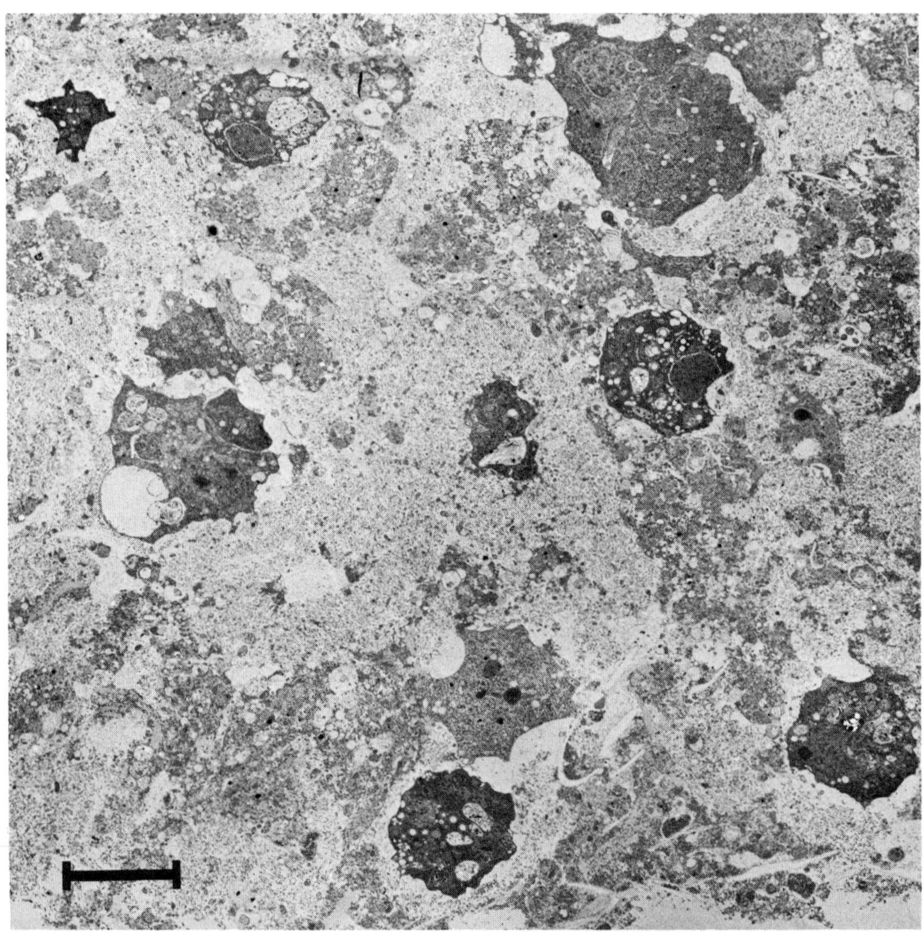

FIG. 5. Cells of a terminally differentiated structure of strain U1. No evidence of spores or stalk cells can be seen and considerable autolysis is apparent. Scale bar = 5 μm.

products accumulate independently of each other (Table 1 and Fig. 6). The individual genes may be controlled by multiple timing processes or could be triggered by a single timing device underlying the particular biochemical differentiations we have chosen to minitor. In either case, it is clear that development in *Dictyostelium* does not proceed through a series of stages in which progress from one to

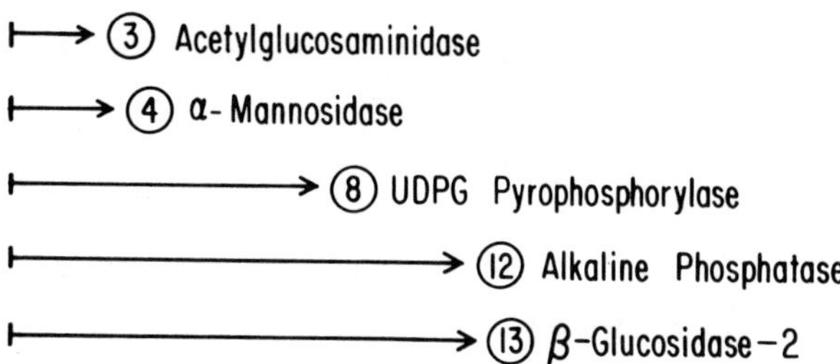

FIG. 6. Independent biochemical differentiations. The accumulation of these five enzymes occurs independently since mutations in any of them do not block accumulation of subsequent ones.

the next is dependent upon *all* of the previous biochemical differentiations.

D. Biochemical Differentiation in Morphological Mutants

Mutations which block morphogenesis at specific stages, such as aggregation, slug formation, culmination, or spore formation, can be isolated by visually screening the terminal developmental forms of a large number of clones derived from mutagenized populations [19,20]. When less than a hundred viable cells are spread on an agar plate carrying a bacterial lawn, the cells grow into discrete plaques within which they develop. By inspecting the plaques with a dissecting microscope, aberrant strains can be recognized and further studied. Since a large number of genes appear to be involved in the developmental phase, morphological mutants occur at a frequency of about 2×10^{-1} in a mutagenized culture. We have chosen seven morphological mutants to study in detail with respect to the marker enzymes [21].

Strains VA-5, DA-2, Agg206, and VA-4 are blocked early in development and fail to aggregate although they remain viable and continue protein synthesis. Strain TS2 is a temperature-sensitive strain which grows well at 22 or 27° but develops only at 22°. At 27° it remains as a smooth lawn of undifferentiated cells. Another temperature-sensitive strain, DTS6, aggregates at 27° but fails to form pseudoplasmodia. The temperature-sensitive strains were analyzed at both the permissive and nonpermissive temperatures. Strain KY3 develops to the slug stage but is blocked at culmination and so remains as migrating slugs for an extended period of time. The morphological phenotypes of these strains alone indicate a dependent sequence functions in development, since aggregation is a prerequisite of slug formation and the slug stage is a prerequisite of culmination.

When the pattern of accumulation of stage-specific enzymes is analyzed in these strains, it can be seen that the enzymes can be ordered such that the pleiotropic effects of the mutations give rise to a linear polarity (Fig. 7). When a morphological mutant fails

ENZYME		VA 5	DA 2	Agg 206	TS2 or VA 4	DTS6	KY3	NC4
3	NAG	−	+	+	+	+	+	+
4	MAN		+	+	+	+	+	+
6	TD		−	+	+	+	+	+
1	AT			−	+		+	+
2	LAP			−	+	+	+	+
5	TPS			−	+	+	+	+
7	TT		−	−	+	+	+	+
8	Upp				−		+	+
11	GP		−	−	−		+	+
12	Alp				−	−	−	+
13	β-Glu 2				−	−	−	+

FIG. 7. Biochemical differentiations in morphological mutants. The accumulation of 11 stage-specific enzymes was determined in various strains. The numbers refer to the enzymes described in Fig. 4. The accumulation was judged positive if a peak-specific activity of more than half that in wild-type cells was reached and was judged negative if the activity rose less than 20% above the basal level. Due to difficulties in retaining strains which fail to form spores, only strains Agg206, TS2, KY3, and NC4 still remain in our collection.

to accumulate an early enzyme, it was found to be also blocked in accumulation of all subsequent enzymes analyzed.

The dependent sequence differs somewhat from the actual temporal sequence of accumulation of the marker enzymes. This temporal discrepancy is most dramatic in strain Agg206. This strain accumulates acetylglucosaminidase, α-mannosidase, and threonine deaminase-2 in an essentially normal manner but does not accumulate the very early enzymes, alanine transaminase or aminopeptidase [21]. Alanine transaminase normally begins to accumulate immediately upon initiation of development and reaches peak-specific activity after 6 h, while threonine deaminase-2 does not normally accumulate until after 4 h of development and reaches peak-specific activity about 10 h later. The most likely explanation is that the processes necessary for accumulation of threonine deaminase-2 are triggered by a step which precedes that controlling the appearance of alanine transaminase and aminopeptidase, but that while these enzymes begin to accumulate almost immediately there is a delay before the threonine deaminase-2 gene is activated. Similar delays may also occur following the signals for acetylglucosaminidase and α-mannosidase.

E. Dependent Sequence of Events

The pattern of morphological aberrations and accumulation of the marker enzymes in these mutant strains can be used to define a linear dependent sequence of seven stages (Fig. 8; A-G). Mutant strains can be considered to have specific defects in the progression from one stage to the next if all of the differentiations characteristic of prior stages occur normally while differentiations characteristic of subsequent stages fail to occur. The stage-specific enzymes which we have used to define this sequence are listed in Fig. 7 in the order in which their accumulation begins. The developmental stages, on the other hand, are arbitrarily positioned in the table with respect to time since we have no idea at present what processes or functions comprise these stages and

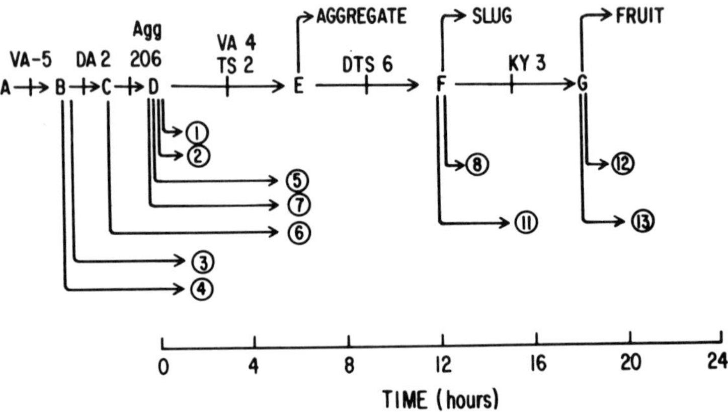

FIG. 8. Dependent sequence. The pattern of accumulation of stage-specific enzymes in various strains was used to define seven discrete steps (A-G) on which the accumulation of the enzymes depends. The individual enzymes are represented by the numbers used in Fig. 4. The mutant strains are given at their block points. Biochemical or morphological differentiations which depend on steps before the block point take place in the mutant strains while those that depend on subsequent steps do not. There is an uncertainty in assigning the step on which UDPG pyrophosphorylase (8) and glycogen phosphorylase (11) depend (it could be either E or F). The marker enzymes are given at the time at which accumulation begins.

therefore cannot directly determine when they occur. Since the stages are causally related (by definition) to the observed markers, all we can tell is that they occur before the biochemical differentiations themselves.

Although we have used only 11 marker enzymes and 3 clearly defined morphological criteria, 7 stages of the dependent sequence could be unequivocally defined by these mutant strains. It seems likely that there are more than 7 stages in the dependent sequence which might be recognized when more markers are analyzed in an expanded series of mutant strains. At present it appears that while the accumulation of the hydrolases and pyrophosphorylase are independent processes, they are controlled by a single timing mechanism consisting of the dependent sequence of stages.

F. Genetic Complexity

An estimation of the total number of genes which play an essential role in morphogenesis, either as part of the dependent sequence or as triggered independent genes, can be made by comparing the frequency of recovery of mutations in known genes to the frequency of recovery of morphologically aberrant strains. The mutations in strains with altered levels of specific enzymes are most likely in the structural genes for these enzymes, since about half produce enzymes with altered properties (stability, Km, pH optimum). Assuming the observed mutation frequency for specific genes (8.7×10^{-4}) represents the true single gene mutation fre-

TABLE 2

Genetic Complexity of Development in *Dictyostelium discoideum*

Specific enzyme	Number of mutants recovered	Mutation frequency[a]	Number of genes[b]
β-N-Acetylglucosaminidase	10	2.2×10^{-3}	1
α-Mannosidase	6	2.7×10^{-4}	1
β-Glucosidase	7	3.6×10^{-4}	2
UDPG Pyrophosphorylase	15	8.2×10^{-4}	1
	Average per gene:	8.7×10^{-4}	
Aggregation	165	1.1×10^{-1}	126
All developmental stages	310	2.0×10^{-1}	230

[a]The frequency was calculated as the total number of mutant strains recovered divided by the total number of clones screened. The number of clones screened in multitest wells was calculated assuming a random (Poisson) distribution of cells.

[b]The number of genes affecting the enzymes is based upon the known number of subunits in each, while the number of genes affecting morphogenesis was calculated from the observed frequency and the average mutational frequency per gene.

quency, we estimate from the frequency of recovery of aggregateless strains (1.1×10^{-1}) and the frequency of recovery of strains with aberrant multicellular morphogenesis (0.9×10^{-1}) that approximately 115 genes are essential for aggregation and about the same number of genes are essential for later morphogenesis (Table 2). This estimation concerns only those genes which, when mutated, give rise to aberrations which can be recognized in cultures which have developed under our standard laboratory conditions using a dissecting microscope and which are not essential for growth. Nevertheless, this analysis gives us some idea of the degree of genetic complexity of development in this system and indicates that the combination sequence may include several hundred genes.

III. FINAL COMMENTS

Biochemical analysis of mutant strains has been a highly fruitful approach to dissecting the control mechanisms which function in bacteria and viruses, and appears to be able to shed some light on the underlying temporal controls which function during development of *Dictyostelium*. It has also given some information on the physiological roles of certain of the stage-specific enzymes. Studies on strains lacking acetylglucosaminidase or UDPG pyrophosphorylase have shown that these enzymes play essential roles during the migration and culmination stages, respectively. These enzymes accumulate during the previous stages and so appear to be cases of predifferentiation at the molecular level as the cells prepare for subsequent metabolic demands. Preliminary evidence indicates that α-mannosidase and β-glucosidase play roles in the alternative developmental pathway leading to macrocyst formation [15].

The independence of the marker enzymes suggests that they are secondary consequences of a primary timing mechanism. When they are used to characterize morphological mutants, it can be seen that they are controlled by a single timing mechanism rather than by a series of independent timing mechanisms. However, the differential delay in the appearance of several of the marker enzymes suggests that independent secondary timing mechanisms may function along the specific pathway leading from the primary dependent sequence to the control of the various genes. Perhaps the analysis of strains in which one enzyme or another is expressed precociously or belatedly will be able to better define these temporal controls.

REFERENCES

1. Palleroni, N., and R. Stanier, 1964. J. Gen. Microbiol. 35: 319-334.
2. Stevenson, I., and J. Mandelstam, 1965. Biochem. J. 96: 354-362.
3. Hegerman, G., 1966. J. Bacteriol. 91: 1155-1160.
4. Kourilsky, P., M. Bourguignon, M. Bouquet, and F. Gros, 1970. Cold Spring Harbor Symp. Quant. Biol. 35: 305-314.
5. Loomis, W. F., 1975. Dictyostelium discoideum. Academic Press, New York and London.
6. Gerisch, G., D. Malchow, and B. Hess, 1974. *In* Biochemistry of Sensory Functions, L. Jaenicke (ed.), Springer-Verlag, New York, p. 279.
7. Durston, A., 1974. Develop. Biol. 37: 225-235.
8. Rosen, S., J. Kafka, D. Simpson, and S. Barondes, 1973. Proc. Nat. Acad. Sci. USA 70: 2554-2557.
9. Siu, C. H., R. Lerner, R. Firtel, and W. F. Loomis, 1975. *In* Cell and Molecular Biology, Vol. 2, D. MacMahon and C. F. Fox (eds.), pp. 129-134.
10. Franke, J., and M. Sussman, 1973. J. Mol. Biol. 81: 173-185.
11. Sussman, M., and R. Sussman, 1969. Symp. Soc. Gen. Microbiol. 19: 403-435.
12. Firtel, R., L. Baxter, and H. Lodish, 1973. J. Mol. Biol. 79: 315-327.
13. Brenner, M., D. Tisdale, and W. F. Loomis, 1975. Exp. Cell Res. 90: 249-252.
14. Free, S. J., and W. F. Loomis, 1975. Biochimie 56: 1525-1528.
15. Dimond, R., and W. F. Loomis, 1976. J. Biol. Chem. 251: 2680-2687.
16. Dimond, R., M. Brenner, and W. F. Loomis, 1973. Proc. Nat. Acad. Sci. USA 70: 3356-3360.
17. Coston, B., and W. F. Loomis, 1969. J. Bacteriol. 100: 1208-1217.

18. Dimond, R., and W. F. Loomis, 1975. *In* Cell and Molecular Biology, Vol. 2, D. MacMahon and C. F. Fox (eds.), pp. 533-538.

19. Sussman, M., 1955. J. Gen. Microbiol. 13: 295-309.

20. Yanagisawa, K., W. F. Loomis, and M. Sussman, 1967. Exp. Cell Res. 46: 328-334.

21. Loomis, W. F., 1975. *In* Isozymes III, Developmental Biology, C. Markert (ed.), Academic Press, New York, p. 177.

ENZYME TURNOVER IN *DICTYOSTELIUM*

Barbara E. Wright

Department of Microbiology and Molecular Genetics
Harvard Medical School
Boston Biomedical Research Institute
Boston, Massachusetts

David A. Thomas

Department of Developmental Biology
Boston Biomedical Research Institute
Boston, Massachusetts

I. INTRODUCTION

Over the past 10 years, a great deal of wishful thinking, speculation, and many negative experiments have been offered in support of the hypothesis that gene activation is the critical variable which triggers enzyme accumulation (and hence differentiation) in *Dictyostelium.* This hypothesis has now been examined in five cases: trehalose-6-phosphate (T6P) synthetase, uridine diphosphoglucose (UDPG) pyrophosphorylase, β-N-acetylglucosaminidase, α-glucosidase, and glycogen phosphorylase. In all cases, an increase in enzyme-specific activity occurs during differentiation and these increases are inhibited by high levels of actinomycin D. In only one of these cases, however, could gene activation be the critical variable limiting the rate of enzyme accumulation. This case, glycogen phosphorylase, will be the major subject of discussion. By way of introduction, investigations concerning the other four enzymes will be summarized briefly.

A. Trehalose-6-phosphate Synthetase

Trehalose-6-phosphate synthetase was originally reported to be absent in myxamoebas and to increase at least 150-fold between the 3rd and 16th hours of development [1]. However, careful attention to methodology with respect to enzyme preparation, stability, and assay resulted in the conclusion that enzyme is present from the beginning of differentiation and thereafter increases only a few fold in specific activity [2,3]. In this case, therefore, the striking increase in enzyme-specific activity reported originally did not correspond to the amount of enzyme protein present. In the following four cases, such a correspondence has been demonstrated.

B. UDPG Pyrophosphorylase

UDPG pyrophosphorylase is present and required during growth for the synthesis of UDP-glucose as the precursor of glycogen, which accumulates. To be maintained during cell division, the enzyme must be synthesized at a rate of about 20%/h [4,5]. Upon the initiation of starvation, two situations could exist, as indicated in Fig. 1: (1) the rate of enzyme synthesis could drop to

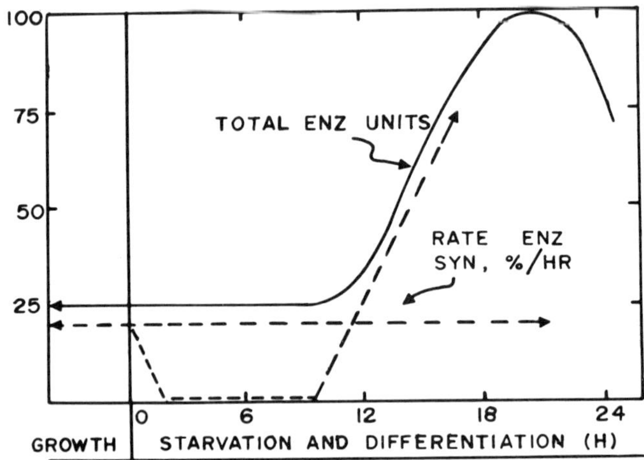

FIG. 1. A schematic representation of changes in the amount of UDPG pyrophosphorylase protein during the transition from the growth to the differentiation phase of the life cycle. The dashed lines represent two possible patterns with respect to the rate of enzyme synthesis.

zero, and be reinitiated coincident with the accumulation of enzyme during culmination; (2) alternatively, the rate of enzyme synthesis could continue at about 20%/h and enzyme degradation could decrease, allowing enzyme to accumulate. There is no biochemical basis or precedent for believing that enzyme synthesis would stop on the initiation of starvation. It was not surprising, therefore, that this possibility was clearly excluded by experiments such as that described by Fig. 2 [6]. As apparent from Fig. 1, there is no significant increase in the amount of UDPG pyrophosphorylase over the first 9 to 10 h of differentiation. However, it can be seen from Fig. 2 that enzyme is labeled rapidly during this period, providing a direct demonstration of active enzyme turnover in the absence of enzyme accumulation. (Although Franke and Sussman [7] were unable to detect significant radioactivity in amoeba enzyme, a careful examination of their methodology, data, and interpretations reveals a number of factors which apparently contributed to their negative results [8,9].) All our data, on the other hand, are consistent with the interpretation that, on the initiation of starvation,

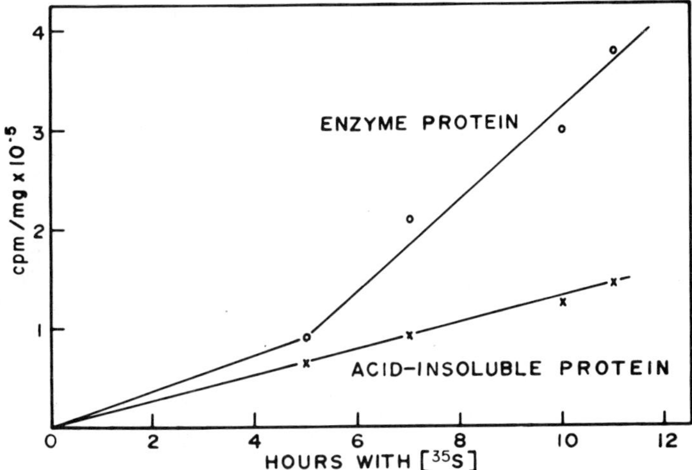

FIG. 2. The specific radioactivity of enzyme protein and of acid-insoluble protein during the first 11 h of differentiation. Cells were exposed to [^{35}S]methionine at 0 time and aliquots harvested at successive time periods beginning at 5 h [6].

enzyme turnover is very high; thereafter, synthesis remains fairly constant and degradation decreases, allowing enzyme to accumulate [5]. In the absence of evidence for an increase in the rate of enzyme synthesis, gene activation is not implicated.

The enzyme which accumulates during culmination is indistinguishable immunologically from the enzyme present in amoebas. However, a slight difference can be detected in mol wt, suggesting a posttranscriptional conversion during differentiation to a more stable form. That such a conversion occurs is also suggested by the fact that, in vitro, the enzyme present in culmination extracts is much more stable than the enzyme in amoeba extracts [10].

C. β-N-Acetylglucosaminidase and α-Glucosidase

Very recently, Every and Ashworth carried out similar immunological analyses for β-N-acetylglucosaminidase and α-glucosidase and arrived at conclusions bearing a striking resemblance to those summarized above for UDPG pyrophosphorylase [11]. These investigators have concluded that the rate of synthesis of these enzymes does not change in going from the growth to the differentiation phase of the life cycle, and that this rate of enzyme synthesis can fully account for the rate of enzyme accumulation. They further conclude that the increases in enzyme-specific activity at the beginning of development do not correspond to any change in the rate of enzyme synthesis or gene activation. Thus, the only difference between these enzymes and the pyrophosphorylase is that the latter, on the initiation of differentiation, is maintained by turnover at a low level for 9 to 10 h prior to enzyme accumulation. In all probability, the rate of enzyme synthesis will be found to be fairly constant during growth and differentiation for other enzymes present in both phases of the life cycle.

In each of the four cases discussed above, inhibitors blocking differentiation also interfere with increases in enzyme-specific activity--yet changes in the rate of enzyme synthesis are not involved. For many years, such inhibitor studies have been used as the basis for constructing genetic "maps" depicting specific periods of transcription and translation controlling the synthesis of enzymes such as T6P synthetase and UDPG pyrophosphorylase [1,12-15]. It is now clear that criticisms of these studies were justified [4,8,16,17].

D. Glycogen Phosphorylase

Glycogen phosphorylase was selected for turnover studies because this enzyme enjoys a unique position in the metabolism of *Dictyostelium*. In contrast to UDP-glucose pyrophosphorylase, phosphorylase would serve no function (and cannot be detected) during growth, when glycogen is being actively synthesized and stored. Also in contrast to the pyrophosphorylase, phosphorylase is in all probability rate-limiting during differentiation. Its activity (along with amylase) is essential for glycogen breakdown, which in turn provides a major source of precursors for the synthesis of cellulose and trehalose. The increase in specific enzyme activity during differentiation is adequate for, and correlated with, the in vivo increase in the rate of glycogen degradation. An analysis of kinetic parameters [18,19] and of substrate and effector concentrations in differentiating cells also indicates a rate-limiting role for glycogen phosphorylase [20-22]. Finally, this enzyme is unique in that, to date, it represents the most promising candidate for providing evidence of gene activation during development.

II. CURRENT RESEARCH

Glycogen phosphorylase activity cannot be detected during growth nor during the first 5 h of starvation. After 5 h its activity increases rapidly to a maximum at culmination (Fig. 3). The enzyme was purified to homogeneity from cells at the culmination stage of development and specific antisera was prepared [23]. During the first few hours of differentiation, the low level of antibody-precipitable protein present in the enzyme region of the gel is apparently nonspecific, judging from the following analysis of amoeba extracts. Figure 4 represents analytical gel analyses of phosphorylase protein at two stages of purification. The first gel shows purified enzyme from culminating cells; enzyme activity can be recovered by elution. The third gel represents partially purified culmination enzyme (diethyaminoethyl-cellulose fraction), and the middle gel indicates this same stage of purification for amoeba extracts. It is apparent that the enzyme band is absent in this case. The concentrated amoeba preparation was also subjected to Ouchterlony analysis, and no precipitin band was evident. Between aggregation and culmination the enzyme accumulates from 0.026% of the total cell protein to 0.11%, and then drops to about 0.07% of the total protein during sorocarp formation (24 h). After aggregation, the relationship between antibody-precipitable protein and enzyme activity indicates that phosphorylase does not increase in

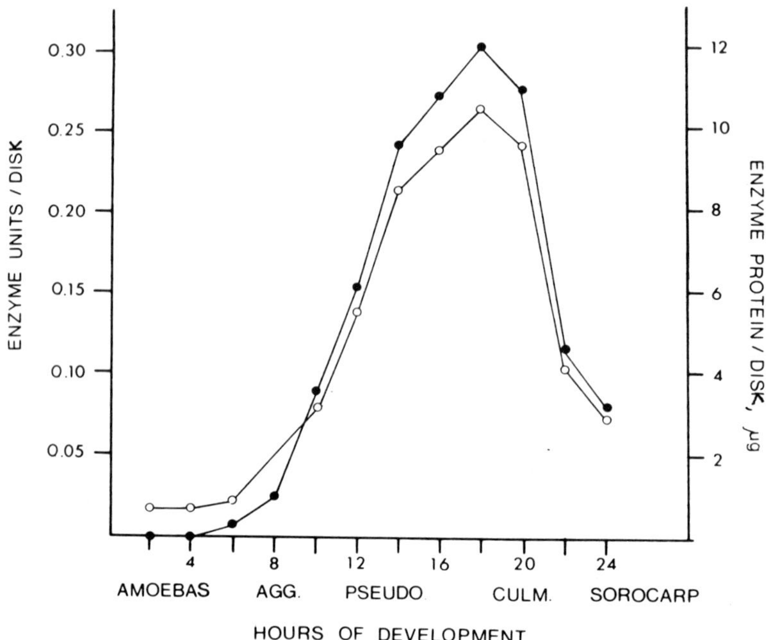

FIG. 3. The relationship between enzyme activity and phosphorylase protein during development. Cells of *D. discoideum* were spread on filter paper disks and allowed to develop at 22°. At the indicated time periods, cells from three filter disks were removed, combined in 4 ml of buffer, and frozen. Prepared cell extracts were assayed for enzyme activity (●) and quantitated for phosphorylase protein (○) by immunoprecipitation and SDS gel electrophoresis.

activity as a result of a modification of preexisting enzyme molecules, unless there exists a catalytically inactive precursor which is not immunologically reactive.

Isotope incorporation into enzyme and acid-insoluble (AI) protein: pulse and pulse-chase experiments were carried out in order to determine the rate of enzyme synthesis and degradation during development. Antisera precipitates prepared from cells exposed to [^{35}S]methionine were dissolved in sodium dodecyl sulfate (SDS) plus 2-mercaptoethanol and electrophoresed on 6% SDS acrylamide gels. After electrophoresis, the gels were stained, quantitated for protein, and sectioned for counting. Figure 5 represents the results

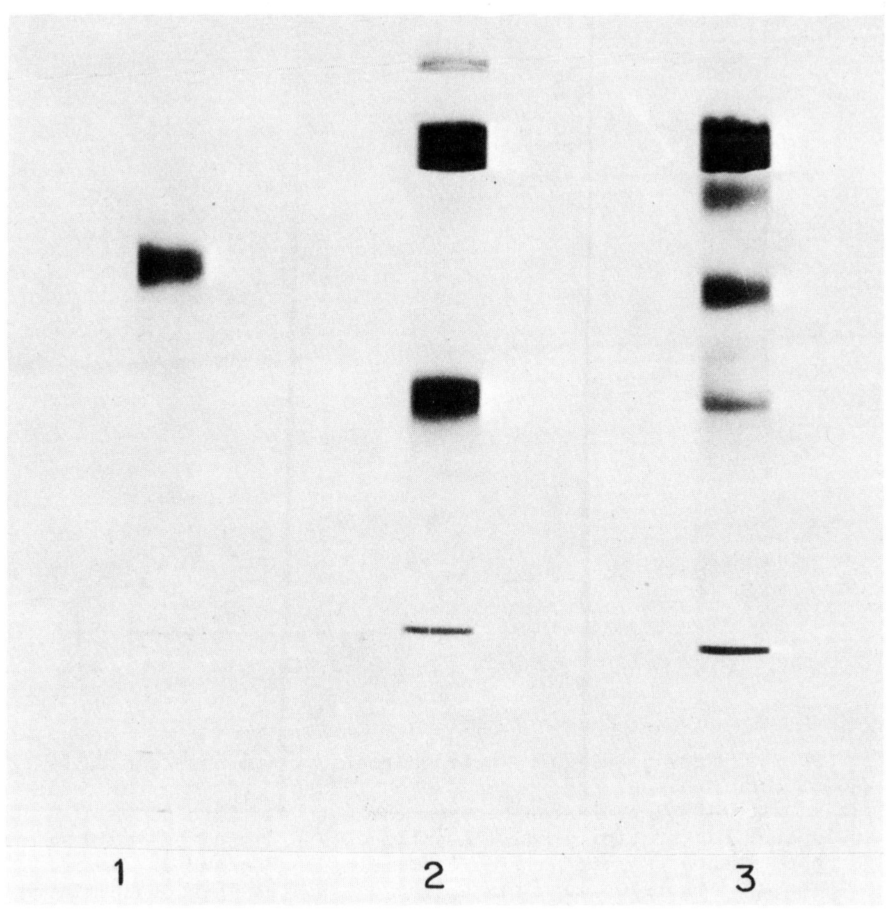

FIG. 4. Disk gel electrophoresis of glycogen phosphorylase. Gel 1: 45 μg of enzyme purified from culminating cells; gel 2: 175 μg of partially purified sample from amoeba cells; gel 3: 160 μg of partially purified enzyme from culminating cells.

from one such experiment, in which enzyme protein was precipitated from cell extracts labeled for 3 h. A comparison of the electrophoretic pattern of precipitated slime mold enzyme (lower gel) with purified rabbit muscle phosphorylase (upper gel) shows that the main radioactive band coincides with the subunit band of the rabbit muscle enzyme, which has the same mol wt (95,000). The major pro-

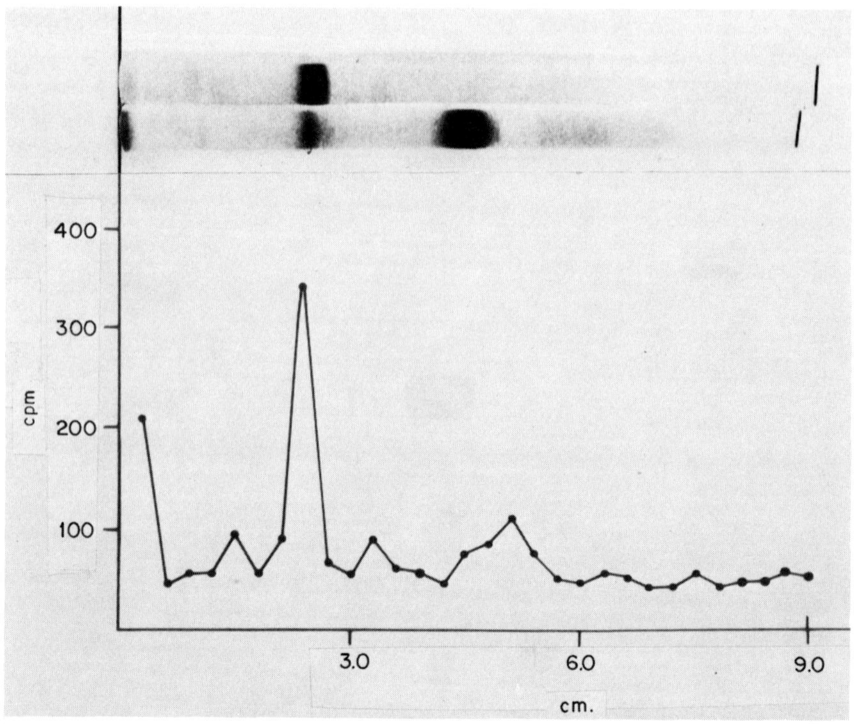

FIG. 5. SDS-acrylamide gel electrophoresis of immunoprecipitated [^{35}S]methionine-labeled glycogen phosphorylase. Enzyme from cells labeled with [^{35}S]methionine between the 12th and 15th hours of development was precipitated with antisera and dissolved in 10 mM sodium phosphate buffer, pH 7.0, containing 1% SDS and 1% 2-mercaptoethanol. After electrophoresis, the gel was stained, quantitated for protein, and sectioned for counting. A photograph of the original gel was aligned with the corresponding gel sections (lower gel). A control gel containing 30 μg of purified rabbit muscle phosphorylase was also electrophoresed following the above procedure (upper gel).

tein band in the slime mold gel is heavy chain γ-globulin.

Figure 6 shows the results of an isotope incorporation experiment performed when enzyme units per cell were increasing at a maximal rate. During this period of development (14 to 16 h), linear rates of label incorporation into both enzyme and AI protein were observed. Based on specific radioactivity, incorporation of label

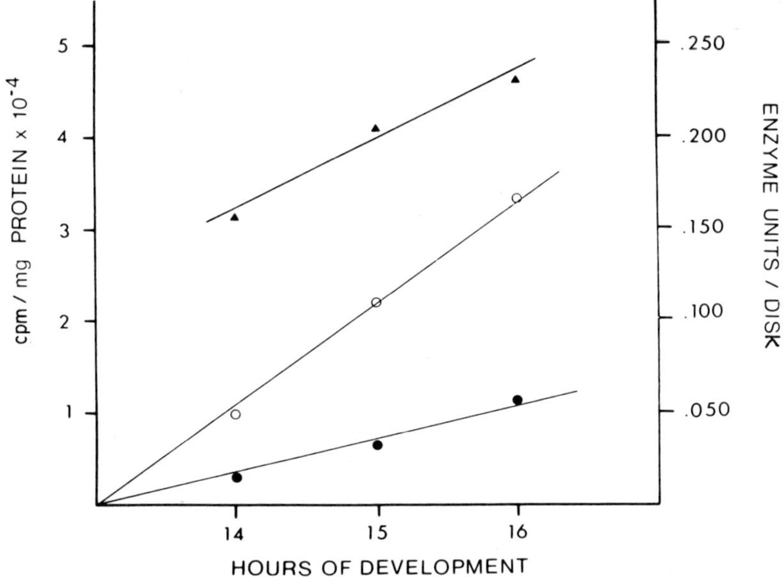

FIG. 6. Incorporation of [^{35}S]methionine into phosphorylase and acid-insoluble protein during preculmination (14 to 16 h). Cells on filter paper disks were exposed to [^{35}S]methionine beginning at the 13th hour of development. At the designated time intervals cell extracts were prepared for the isolation of enzyme and AI protein. Results are expressed as specific radioactivity (cpm/mg protein) incorporated into phosphorylase protein (○) and acid-insoluble protein (●). The change in phosphorylase units per filter paper disk (▲) over the period of label exposure is also represented.

into enzyme protein is about three times faster than that incorporated into AI protein. While enzyme is accumulating rapidly, the concentration of AI protein does not change significantly over this period of time. In previous studies, the rate of AI protein synthesis and degradation (i.e., turnover) was determined, based on the specific radioactivity of endogenous [^{35}S]methionine (Table 1). This was essential because of 10-fold changes in permeability which occur during development [24]. Due to these permeability barriers, it is not possible to do meaningful chase experiments at the earlier stages of differentiation [24,25]. However, chase experiments can be carried out during culmination when permeability is greatest.

TABLE 1

Protein Turnover[a]

Stage	Hours of differentiation	No. exps.	% Turnover/h
Amoebas	2-6	4	18 ± 4
Agg-pseudo	8-14	5[b]	10 ± 3
Preculm-culm	14-18	3	7 ± 2
Young sorocarp	20-24	2	8 ± 1

[a]Taken from Wright and Anderson [24].
[b]One value 3 × average of other 6 omitted from this calculation.

Knowing the relationship between enzyme units and micrograms of enzyme protein (Fig. 3; 1 μg enzyme = 0.028 units), it has been possible to estimate the rates of enzyme synthesis and degradation at various stages of development.

Estimates of rates of enzyme synthesis and degradation show that, in the case of UDPG pyrophosphorylase and the glycosidases in *D. discoideum* [5,7,11], enzyme is not degraded during the period of its maximal rate of accumulation. This also appears to be the case for glycogen phosphorylase between the 10th and 16th hours of differentiation. In the experiment described by Fig. 6, at the preculmination stage of differentiation, the rate of average protein turnover is about 8%/h (Table 1). Thus, assuming no enzyme degradation, phosphorylase is being synthesized at 24%/h. In fact, the increase in enzyme protein from 5.6 to 8.2 μg/h over the course of the experiment can be almost exactly accounted for by a rate of synthesis of 24%/h. A rate of enzyme synthesis calculated from a comparable experiment carried out at the aggregation-pseudoplasmodium stages of development (hours 10 to 14) could again fully account for the increase in enzyme protein (assuming no degradation).

Chase experiments were carried out during culmination, as indicated in Fig. 7. Cells were incubated with label from the 12th to 15th hour of development and then transferred to cold methionine

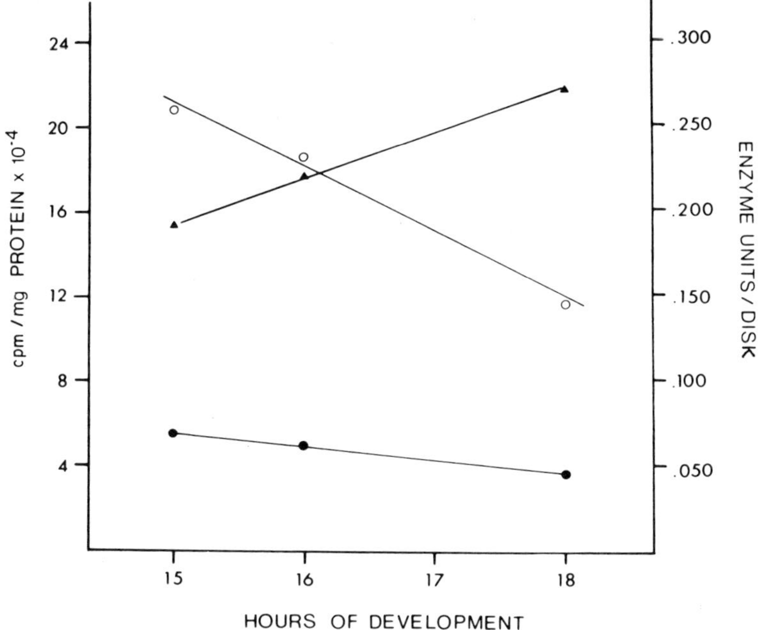

FIG. 7. Chase of [^{35}S]methionine from prelabeled phosphorylase and acid-insoluble protein during culmination (15 to 18 h). Cells on filter paper disks were incubated with [^{35}S]methionine from the 12th to the 15th hour of development. After the 15th hour, the cells were transferred to cold methionine and chased through hour 18. At the designated time intervals cell extracts were prepared for the isolation of enzyme and AI protein. Results are expressed as specific radioactivity (cpm/mg protein) in phosphorylase protein (○) and acid-insoluble protein (●). The change in phosphorylase units per filter paper disk (▲) over the chase period is also represented.

and chased through hour 18. Calculations from the half-life of AI protein gives a value of about 11%/h turnover, in rough confirmation of the earlier studies in which turnover was calculated based on the specific radioactivity of endogenous methionine (Table 1). As this correspondence indicates that the chase was effective in this experiment, the loss in total counts should indicate the extent of enzyme degradation. A degradation rate based on total counts in enzyme gave a value of about 12%/h. As net enzyme accumulation (synthesis minus degradation) is occurring at a rate of 1 μg/h, the addition of 12%/h of the amount of enzyme present (to compensate for enzyme degradation) gives a rate of enzyme synthesis of about 1.8 μg/h.

Following culmination, when the enzyme level decreases, a chase (from hours 21 to 23) and a pulse (from hours 22 to 26) experiment were done. During the chase experiment enzyme decreased from 4.3 to 1.5 μg in 2 h, or 18%/h. A degradation rate based on total counts in enzyme gave a value of 40%/h, necessitating a rate of enzyme synthesis of 22%/h or roughly (using an average of 3 μg enzyme present) 0.6 μg/h. The fact that some synthesis must occur was also indicated by a decrease in enzyme-specific radioactivity. During the pulse experiment, from hours 22 to 26 (Fig. 8) enzyme-

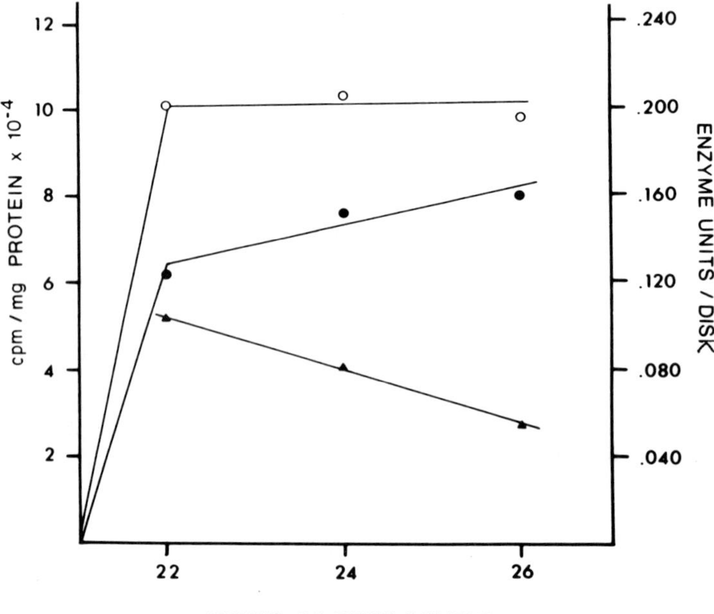

FIG. 8. Incorporation of [^{35}S]methionine into phosphorylase and acid-insoluble protein during sorocarp (22-26 h) formation. Cells on filter paper disks were exposed to [^{35}S]methionine beginning at the 20th hour of development. Results are expressed as specific radioactivity (cpm/mg protein) incorporated into phosphorylase protein (○) and acid-insoluble protein (●). The change in phosphorylase units per filter paper disk (▲) over the period of label exposure is also represented.

TABLE 2

Estimates of Enzyme Synthesis and Degradation

Stage	Hours	Rate synthesis (μg/h)	Rate degradation (%/h)	Net accumulation or decrease (μg/h)
Amoeba-aggregation	4-8	∿0.3	?	∿+0.3
Aggregation-pseudoplasmodium	10-14	0.6	0	+0.6
Preculmination	14-16	1.3	0	+1.3
Culmination	15-18	1.8	12	+1.0
Postculmination	21-23	0.6	40	-1.4
Young sorocarp	22-26	0	13	-1.8

specific radioactivity did not change, and the amount of enzyme fell by 13%/h. Thus, it appears that insignificant enzyme synthesis occurred. Estimates of enzyme synthesis and degradation are summarized in Table 2. The value for hours 4 to 8 (amoeba aggregation) is based on small amounts of enzyme protein and hence represents very rough estimate. Glycogen phosphorylase cannot be detected by enzymatic nor by immunological analyses prior to 5 h of development. Between aggregation and culmination, the rate of enzyme synthesis in μg/h increases about six-fold, then drops to an insignificant level in young sorocarps. However, expressed as %/h, the rate of enzyme synthesis is quite constant (about 25%/h) between aggregation and late culmination. During the maximum rate of enzyme accumulation, enzyme degradation is negligible, then increases to a peak value of 40%/h after culmination when enzyme levels are falling rapidly.

III. FINAL COMMENTS

If we really wish to analyze the mechanism of differentiation, the mechanism of enzyme accumulation is not nearly as important as the role that enzyme plays in the complex metabolic networks underlying differentiation unless, of course, one is so naive as to equate the mechanism of differentiation with the mechanism of enzyme accumulation. For years it has been apparent that there is a poor correlation indeed between in vitro changes in enzyme-specific activity and enzyme activity in vivo. Furthermore, assuming that the five enzymes discussed in this presentation are representative, the great majority of changes in enzyme-specific activity are not attributable to gene activation. Let us assume that new mRNA synthesis is responsible for the initiation of glycogen phosphorylase synthesis during differentiation. Nevertheless, this fact will have relevance to the mechanism of differentiation only if the enzyme is essential, unique, and rate-limiting.

In this regard, the study of so-called "developmental mutants" will be of minimal value in understanding mechanism [17]. Such mutants will, for the most part, probably fall in the class of being vital but not unique; that is, the majority of enzymes essential to differentiation are probably not unique to the transformation in question, as in the case of the morphological mutants in *Neurospora* with abnormalities in enzymes such as glucose-6-phosphate dehydrogenase. A mutant impaired in the activity of any enzyme essential to differentiation will of course interfere with the process, but tell us nothing as to whether the enzyme is unique to differentiation, rate-limiting, or present but unused prior to the time of its expression in vivo. Nor can a mutant help reveal the mechanism(s) by which enzyme activity is normally expressed during the course of differentiation. Such fundamental questions can only be answered through an in-depth study of the normal differentiating system. To reveal rate-limiting events, such studies must encompass and integrate the complexities and interdependencies of a variety of cellular components, as they exist in the intact cell. An analysis of the mechanism of a change in enzyme-specific activity is a far cry from an analysis of the mechanism of differentiation.

As an illustration of the kind of data very relevant to the mechanism of differentiation, let us examine some new aspects of the role of glycogen phosphorylase in the intact cell. Through the use of kinetic models, it is possible to calculate or predict the activity of an enzyme over the course of differentiation [20-22]. The basis for this so-called "enzyme activation function" is its compatibility with reaction rates determined in vivo, the accumulation patterns of relevant metabolites and end products, and the enzyme mechanisms and kinetic constants determined in vitro. The predicted enzyme activity (V_v) is expressed as mM of product per minute per ml of cell volume. This value can be compared to enzyme activity (V_{max}) which is determined in vitro and also based on cell volume. A correspondence between V_v and V_{max} suggests that we have taken into account all variables affecting the enzyme in vivo when we measure its activity in vitro. We have made such comparisons for glycogen phosphorylase and found that the values correspond at aggregation. However, at culmination (where two cell types exist), V_v has only one-half the value of V_{max}, suggesting that the enzyme is inhibited or is perhaps more substrate-limited in vivo than was assumed in calculating V_v. It is therefore very interesting that Dr. Charles Rutherford has found that inorganic phosphate is in fact concentrated in the stalk cells at culmination at a level about four-fold higher than in the spore cells [27]. In contrast, glycogen and glycogen phosphorylase are mainly localized in the spores and in the presumptive spore and stalk cells at this stage of differentiation. Such an association between phosphorylase and glycogen has been observed in other systems [26]. Thus, the divergence between V_v and V_{max} at culmination can be rationalized: the predicted value, V_v, was based on the assumption of no compartmentalization between enzyme and substrates. When the two cell types are modeled separately, a higher enzyme level will be required in order to overcome the limiting levels of inorganic phosphate and still achieve the rate of glycogen degradation depicted in the model (and actually measured in vivo). Localization of phosphate in the stalk cells would help to insure the complete utilization of glycogen in this cell type, which dies. However, such a localization would in turn require excessive enzyme levels at the site of glycogen breakdown (at the apical tip, in presumptive spore and stalk cells). In view of all of these observations, it is probable that glycogen phosphorylase levels are limiting the rate of glycogen degradation in vivo [20-22]. The fact that insignificant glycogen levels and low enzyme levels are found in stalk cells [27] is incompatible with our observation that the presence of glycogen stabilizes glycogen phosphorylase.

Thus, insight into the complex mechanisms operating during development can be obtained through the integration of many different kinds of information: (1) flux values obtained with isotopes in intact cells; (2) enzyme activities and mechanisms determined in vitro; (3) enzyme activities predicted in vivo by using kinetic models; and (4) information on the distribution of substrates and enzymes in the two cell types during differentiation.

ACKNOWLEDGMENTS

This investigation was supported by Grant P1B1765 from the National Science Foundation, by Public Health Service Research Grant HD05357 from the National Institute of Child Health and Human Development, and by General Research Support Grant 5S01RR05711. David A. Thomas is a Postdoctoral Fellow of the National Institute of General and Medical Sciences (5F02GM53288).

REFERENCES

1. Roth, R., and M. Sussman, 1968. J. Biol. Chem. 243: 5081-5087.
2. Killick, K. A., and B. E. Wright, 1972. J. Biol. Chem. 247: 2967-2969.

3. Killick, K. A., and B. E. Wright, 1975. Arch. Biochem. and Biophys. 170: 634-643.

4. Gustafson, G. L., and B. E. Wright, 1972. CRC Crit. Rev. Microbiol. 1: 453-478.

5. Gustafson, G. L., W. Y. Kong, and B. E. Wright, 1973. J. Biol. Chem. 248: 5188-5196.

6. Gustafson, G. L., and B. E. Wright, 1973. Biochem. Biophys. Res. Comm. 50: 438-442.

7. Franke, J., and M. Sussman, 1973. J. Mol. Biol. 81: 173-185.

8. Killick, K. A., and B. E. Wright, 1974. Ann. Rev. Micro. 28: 139-166.

9. Wright, B. E., and K. A. Killick, 1975. *In* Spores VI, R. N. Costilow, H. L. Sadoff, and P. Gerhardt (eds.), American Society for Microbiology, Washington, D.C., p. 13.

10. Wright, B. E., and D. Dahlberg, 1968. J. Bacteriol. 95: 983-985.

11. Every, D., and J. M. Ashworth, 1975. Biochem. J. 148: 169-177.

12. Ashworth, J. M., and M. Sussman, 1967. J. Biol. Chem. 242: 1696-1700.

13. Roth, R., J. M. Ashworth, and M. Sussman, 1968. Proc. Nat. Acad. Sci. USA 59: 1235-1242.

14. Telser, A., and M. Sussman, 1971. J. Biol. Chem. 246: 2252-2257.

15. Loomis, W. F., Jr., 1970. J. Bacteriol. 97: 375-381.

16. Wright, B. E., 1972. Develop. Biol. 28: f13-f20.

17. Wright, B. E., 1975. ASM Conference on Cell Differentiation and Communication, M. Dworkin and L. Shappiro (eds.), American Society for Microbiology, Washington, D.C.

18. Jones, T. D. H., and B. E. Wright, 1970. J. Bacteriol. 104: 754-761.

19. Firtel, R. A., and J. Bonner, 1972. Develop. Biol. 29: 85-103.

20. Wright, B. E., 1973. Critical Variables in Differentiation, Prentice Hall, Englewood Cliffs, N.J.

21. Wright, B. E., and G. L. Gustafson, 1972. J. Biol. Chem. 247: 7875-7884.

22. Wright, B. E., and D. J. M. Park, 1975. J. Biol. Chem. 250: 2219-2226.

23. Thomas, D. A., and B. E. Wright, 1975. J. Biol. Chem. 251: 1258-1263.

24. Wright, B. E., and M. L. Anderson, 1960. Biochem. Biophys. Acta 43: 67-78.

25. Gustafson, G. L., and B. E. Wright, unpublished data.

26. Meyer, F., L. M. G. Heilmeyer, R. H. Haschke, and E. H. Fischer, 1970. J. Biol. Chem. 245: 6642-6648.

27. Rutherford, C., 1976. J. Embryol. Exp. Morph. 35: 335-343.

A MODEL FOR CONTROLLED AUTOLYSIS DURING DIFFERENTIAL MORPHOGENESIS OF FISSION YEAST*

Byron F. Johnson

Division of Biological Sciences
National Research Council of Canada
Ottawa, Ontario, Canada

G. B. Calleja†

University of the Philippines
Diliman, Quezon City, The Philippines

Bong Y. Yoo

Department of Biology
University of New Brunswick
Fredericton, New Brunswick, Canada

---

*N.R.C.C. Publication Number 14837.

†Current affiliation: Division of Biological Sciences, National Research Council of Canada, Ottawa, Ontario K1A 0R6

I. INTRODUCTION

It would be delightful if one could describe differential morphogenesis of any cellular organelle, if one could list all the genes directly involved with its biogenesis and with its subsequent differentiation, if one could detail the interrelationships among those genes and their products down to the assembled differentiating organelle, and finally, if one could have insight into the overall controlling mechanisms involved. Such an analysis seems possible if the organelle one studies is large enough to allow easy recognition of its different stages by light microscopy of living cells, if some of the enzymes involved in elaborating the organelle are discontinuously operative, if temperature-sensitive mutants fail in organellogenesis at their restrictive temperatures, if it can be shown that the relevant enzymes act in recognizable sequences, and if activation of a differentiating sequence can be simply induced by the investigator. These characteristics apply to some or all of the differential morphogenesis of the fission yeast cell wall.

We hope that the following description of those processes will encourage others to consider using the fission yeast cell wall system for their own experiments in differential morphogenesis.

The fission yeast *Schizosaccharomyces pombe* is a sausage-shaped cell which grows by extension [1-4] at its end(s) and divides by fission midway between the ends [1,5-8]. Under certain circumstances the cells may pair off, form conjugation tubes, fuse their cell walls together, and dissolve away those portions of the walls which separate their cytoplasms. Karyogamy, meiosis, and ascospore formation then take place in the fused cytoplasm; all of these are intensely interesting but none pertains directly to our theme here. Eventually the fused walls may open to release the ascospores. To the best of our knowledge, all these differential changes of the cell walls involve controlled autolysis, which is manifested in a variety of ways.

A. Autolysis and Controls

Yeast cells have very strong walls whose characteristic shape and strength are usually ascribed to their glucan component [9,10]. In addition, yeast cells also contain glucanolytic enzymes, both endoglucanase [11,12] and exoglucanase [13,14], whose natural substrate is the glucan in the wall. Hence, yeast cells should be unstable, which leads us to consider our titular use of the word, control.

As mentioned above, we are ignorant of the overall controlling mechanisms. Thus we must discuss *control* in a more restricted sense in which tight controls are implied by the existence of high stability in a potentially unstable circumstance. In any autolytic system, poor controls predicate lysis--but for vegetatively growing yeast cells, spontaneous lysis is so uncommon as to merit a small descriptive literature of its own [15,16]. The "controls" must be highly effective!

II. CURRENT RESEARCH

Most of the work reviewed here is based upon experiments with *Schizosaccharomyces pombe* Lindner, N.C.Y.C. 132 (A.T.C.C. 26192, originally obtained from Prof. J. M. Mitchison) or a subclone (360-2) characterized by enhanced flocculation and conjugation capabilities. Wherever necessary, attention will be specifically drawn to usage of another strain, 968 h^{90} (obtained from Dr. A. Nasim). Culturing conditions have been reviewed [17], and our experimental procedures are detailed in earlier reports from our laboratories.

A. Cell Extension

Fission yeasts extend at one or both ends, depending upon the strain and growth conditions. Direct morphometric analysis [1,3,18] and autoradiographic studies [2,4] have together yielded a reasonable description of the phenomenon. Based upon this and upon variable lytic responses to 2-deoxyglucose, a model has been proposed for extension of yeast cell walls by molecular intussusception [19].

As shown digrammatically (Fig. 1), the wall is thought to extend by having glucan synthetase activity so closely coordinated with endoglucanase activity (hypothetical, when proposed) that the cellular turgor pressure can cause only limited stretching in the weakened region of the broken glucan molecule(s). The broken glucan molecules are rejoined by covalent addition of glucose or oligoglucan [20] across their breaks before the wall can be completely ruptured.

Since the model was proposed, no obvious contrary evidence has appeared. On the other hand, an endo-β-glucanase having the proper sorts of affinities and activities has been described [11,12] and that description was subsequently confirmed [14]. In addition, endoglucanase activity has been found [21] in vesicles and from electron microscope observations. The activity was thought to be near the ends of yeast cells. In contrast, exoglucanase isolated from yeast cells [13,14] has no measurable hydrolytic activity against intact yeast cell walls. This suggests that native glucan molecules are too long to provide a substrate for exoglucanase enzymes. Compatible with the latter suggestion is the complete failure to find tetramethyl endgroups after methylation of the glucan (α-linked, in this case) isolated from *Penicillium patulum* [22].

In summary, the extension process seems to involve endoglucanase activity very closely coordinated with, but obviously precedent to, glucan synthetase activity. Both enzymes are operative on the same glucan molecules (Table 1A). As we proceed, we shall show how

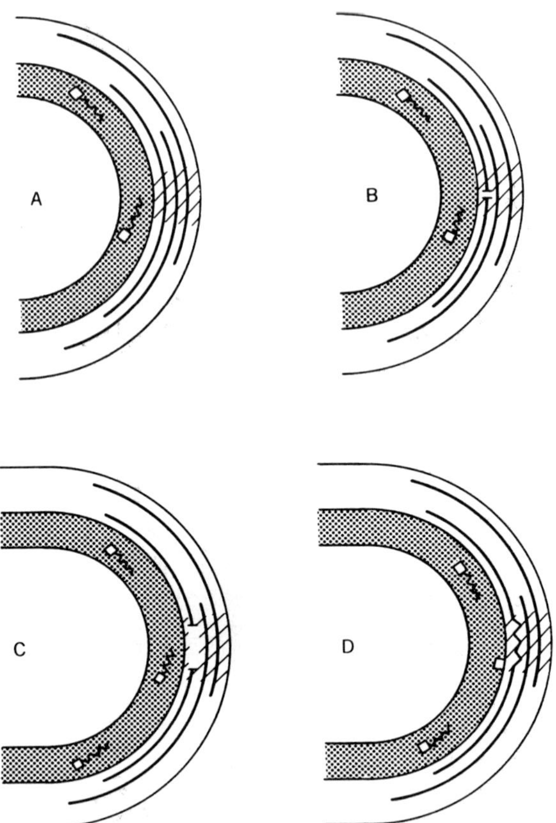

FIG. 1. Model for extensile growth of yeast cell walls. The dotted periplasmic layer contains glucan precursors (⁄\/\/\) attached to glucan synthetase molecules (□). The cytoplasm, internal to the periplasmic layer, is depicted as "empty," and is marked only by the sequence letters (A-D). The cell wall, external to the periplasmic layer, is mostly shown as empty; preexisting β-glucan molecules (probably branched in reality) are shown as heavy, smoothly curved, concentrically oriented lines; crosshatching within the wall is the site of the purported endo-β-glucanase activity. In (A), the system is depicted before the endoglucanase attack upon preexisting glucan molecules. In (B), endoglucanase activity has hydrolysed the innermost glucan molecule near its midpoint, weakening the wall. (C) Depicts the weakened wall as having stretched in response to turgor pressure. A glucan synthetase-glucan precursor (glucose or oligoglucan) complex is seen in the periplasm near the broken glucan molecule in the wall. In (D), the once-broken glucan molecule is again intact, albeit longer. (Adapted from Ref. 19.)

TABLE 1

Differential Deployment of Putative Glucan Hydrolases and Glucan Synthetases During Morphogenesis of Fission Yeasts

Morphogenetic process	Activation coordination: Sequence	Activation coordination: Timing	Comment
A. Cell extension	Endoglucanase Glucan synthetase	Tight	Requires pre-existing glucan molecules
B. Cell division			
1. Primary septum (template)	Template synthetase		
2. Secondary septum			
a. Elaboration	Glucan synthetase		Remote from pre-existing glucan molecules
b. Bonding to old cell wall	Glucan synthetase		
3. Hydrolytic attack			
a. Old cell wall	Endoglucanase Exoglucanase	Loose	
b. Primary septum	Template lyase	Loose	
C. Sexual phenomena			
1. Copulation	Endoglucanase Glucan synthetase	Tight	
2. Conjugation tube extension	Endoglucanase Glucan synthetase	Tight	Compare with A
3. Cross wall resorption	Endoglucanase 28 + exoglucanase (?)		Compare with B 3 a + b
4. Conjugant lysis			Probable failure of controls in C 3
D. Spore liberation			
1. 968 h^{90}	Endoglucanase Exoglucanase		Compare with B 3 a
2. NCYC 132 (Isolate 360-2)			Spore liberation a rare event

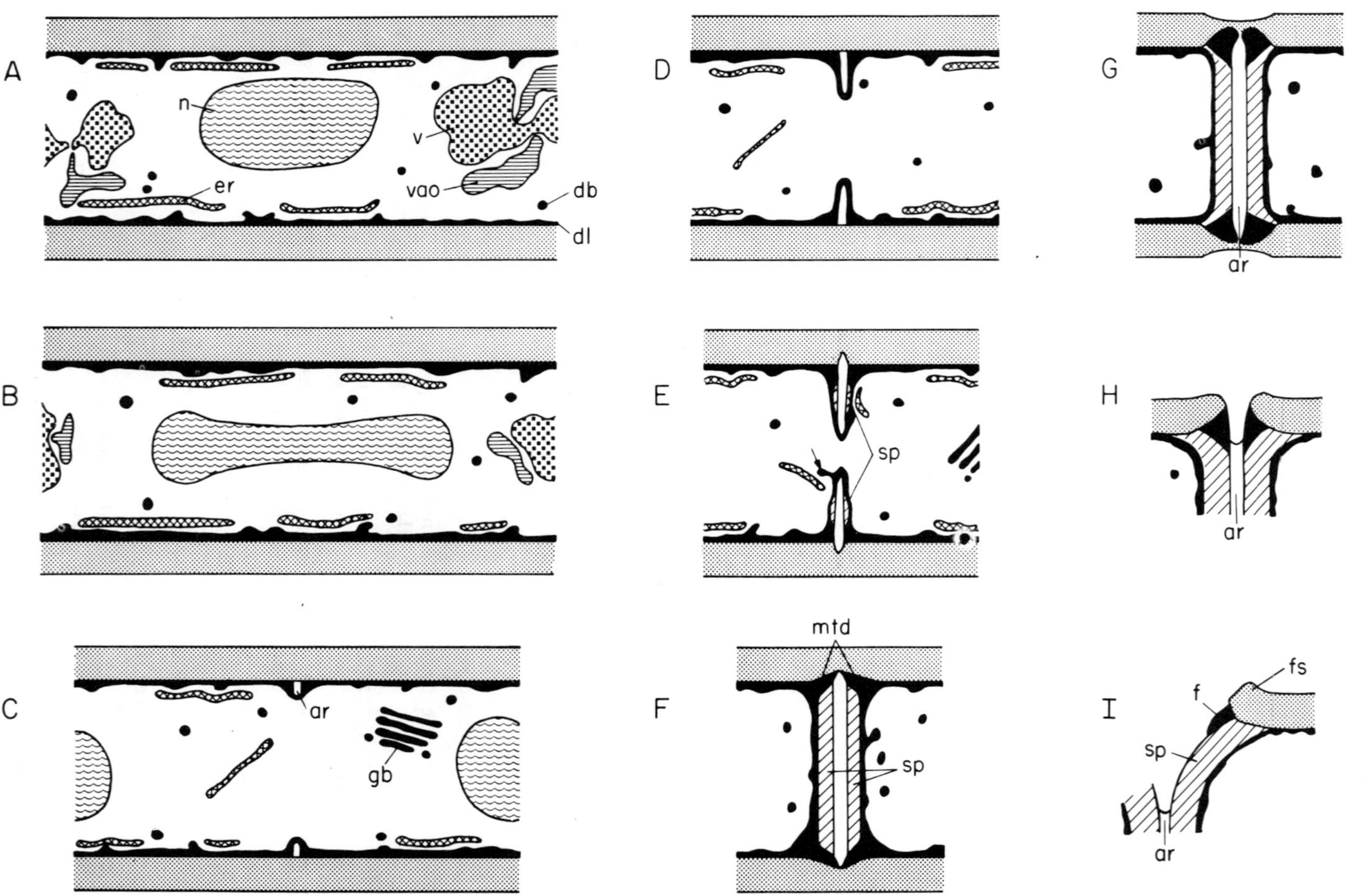
A
n
v
er
vao
db
dl
D
G
ar
B
E
sp
H
ar
C
ar
gb
F
mtd
sp
I
fs
f
sp
ar

FIG. 2. Cell division of *S. pombe*. (A-C), Mitosis and initiation of primary septum (AR). (D), Establishment of dense fillets at base of primary septum. (E), Initiation of secondary septum (SP); fusion of dense body (DB) with dark periplasmic layer (DL, arrow). (F), Migration of dense fillet into wall to become *material triangulaire dense* (MTD [6]); closure of pore. (G), Growth of secondary septum to make contact and bind to old cell wall; initiation of fission by hydrolytic erosion of old cell wall. (H), Outward movement of MTD to become fuscannel (F); fission scar (FS) ridge becomes apparent; material of primary septum disappears at edge of layer. (I), Fission almost complete; F at final disposition; secondary septum constitutes new end of daughter cell; material of primary septum nearly all lost. Nucleus (N); endoplasmic reticulum (ER); vacuole (V); vacuole-associated organelle (VAO); golgi body (GB). This figure and legend have been adapted from [7] and are reproduced with permission of the American Society for Microbiology.

other morphogenetic activities of the fission yeast cell may involve these and/or similar enzymatic activities in comparable and contrasting manners.

B. Cell Division

Electron microscope observations [6,7] of cell division of the fission yeast have provided a fair basis for understanding the sequence of morphogenetic events. In addition, a combined fluorescence microscope-electron microscope study [8] allows the integration of observations on living cells with the ultrastructural description. Mitosis is usually completed before cell division begins [7,23].

Cell division in *S. pombe* consists of (1) generation of a primary septum (newly adapted terminology from Shannon and Rothman [24]) which first appears as an electron-transparent annular rudiment in contact with the old cell wall at about the midpoint of the cell; this primary septum (its chemical nature is unknown [8]) serves as a template for generation of the secondary septa (Fig. 2); (2) de novo generation of secondary septa on both sides of the primary septum, originally remote from the old cell wall and growing mostly centripetally toward the center, but also outward toward the old cell wall (Fig. 2E, F); (3) more or less simultaneous fusion of the new secondary septa to the innermost surface of the old cell wall and erosion of the old cell wall between the sites of attachment of the secondary septa; and finally (4) complete dissolution of the old cell wall between the secondary septa and complete dissolution of the primary septal template, allowing separation of the daughter cells.

It is notable that the glucan synthetic and hydrolytic activities are found in a different sequence and that the times and sites of activation are in contrast with extension (Table 1B; Fig. 3); hence the controls if not the enzymes themselves must be quite different. Note in particular that the glucan synthetase elaborates secondary septum remote from the glucan molecules in the old cell

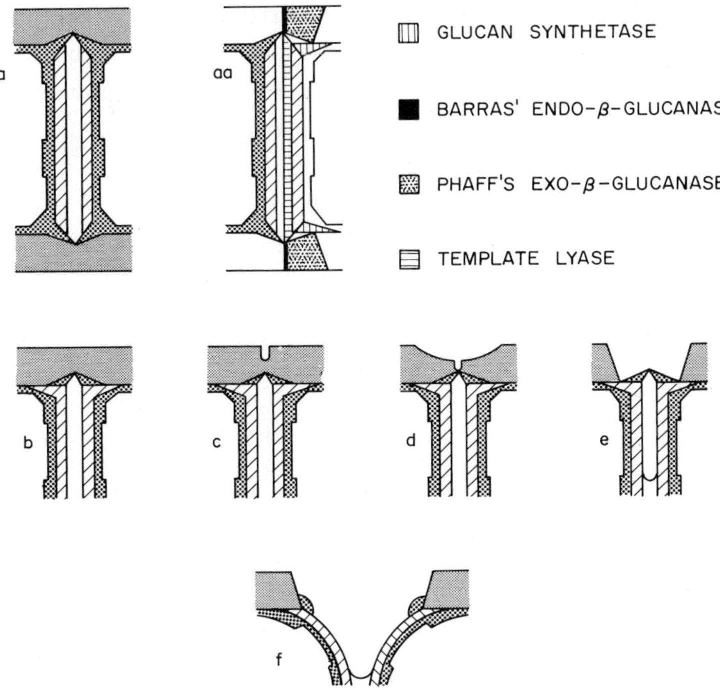

FIG. 3. Suggested enzyme cytochemistry of fission. In (a-f), changes in structure of assembled cell plate (primary septum, secondary septum, and associated old cell wall) are shown; aa (left side) repeats (a) and Fig. 2F; aa (right side) demonstrates where obvious enzymatic activities must occur. In (a), the primary and secondary septa are shown as completely elaborated except for the final fusion of the secondary septum to the old cell wall, shown in (b) as the function of a glucan synthetase. (c) Illustrates the initiation of erosion of the old cell wall between the sites of attachment of secondary septa. Endoglucanase activity would be required to generate a "cut" through the old glucan molecules. In (d), both the deactivation of the endoglucanase at the primary septum and the activation of the exoglucanase can be seen. The exoglucanase would not have had substrate for its hydrolytic activity until after endoglucanase activity had generated glucan "ends." (e) Demonstrates both the termination of exoglucanase activity and the effect of template lyase activity. (f) Merely demonstrates the effect of turgor pressure in rounding up the new cell ends.

wall. Also, it is clear that glucanase activities during cell division must be under very strict control for large amounts of old wall are dissolved, but wall regions proximal to those dissolved (perhaps including even the same glucan molecules which have already been attacked) are not further attacked. The hydrolytic activity abruptly ceases. It is difficult to see how any portion of the extension model can be applied to the cell division process. Only some of the labels of the enzyme activities seem the same, and that coincidence might well be spurious. The contrast between the processes of cellular extension and cell division could hardly be greater (Table 1A,B; Fig. 3).

C. The Sexual Process

Cultures of *S. pombe* exhibit flocculation (cohesion of cells mediated by noncovalent bonds), copulation (cohesion by covalent bonds) and conjugation tube formation, conjugation, meiosis and ascus formation, and spore liberation [25,26], all in response to a simple induction stimulus. That simple stimulus is stationary phase aeration following anaerobic log phase growth. Many genes which are repressed in normal cell cycles seem operative during this sequence of events. Genetic analysis of the sequence has been initiated [27] and continues [28,29]. Obviously it will be easier to identify the genes than to characterize their products and their controls. Portions of the sequence have been described ultrastructurally [26], and more will be published elsewhere. Again, light microscopy of living cells is useful, with phase contrast, Nomarski, and fluorescence techniques all finding application. On the other hand, some of these events, e.g., flocculation, seem to have little influence upon the appearance of the wall.

About 75% of a culture will flocculate in response to induction. This is a necessary preliminary to copulation, the establishment of covalent bonds between paired cells. Regardless of the nature of the initial covalent linkage, there can be no doubt that eventually glucan-glucan bonds are established. If these bonds are mediated by an endoglucanase-glucan synthetase system, then bonding could be merely a consequence of the first stages of conjugation tube generation. Conjugation tube extension (Fig. 4B) seems closely comparable with cellular extension of the normal cell growth cycle. Eventually the conjugation tubes are bonded together over a large area (Fig. 4C) before an opening appears at their midpoint (Fig. 4D). Resorption of the crosswalls is apt to involve both endo- and exoglucanases but might also include some participation from a glucan synthetase to keep things bonded together. Some indirect evidence [29] suggests that only endoglucanase activity may be present. It seems inappropriate to compare resorption of crosswalls with any aspect of cell division. At any rate, the final junction between

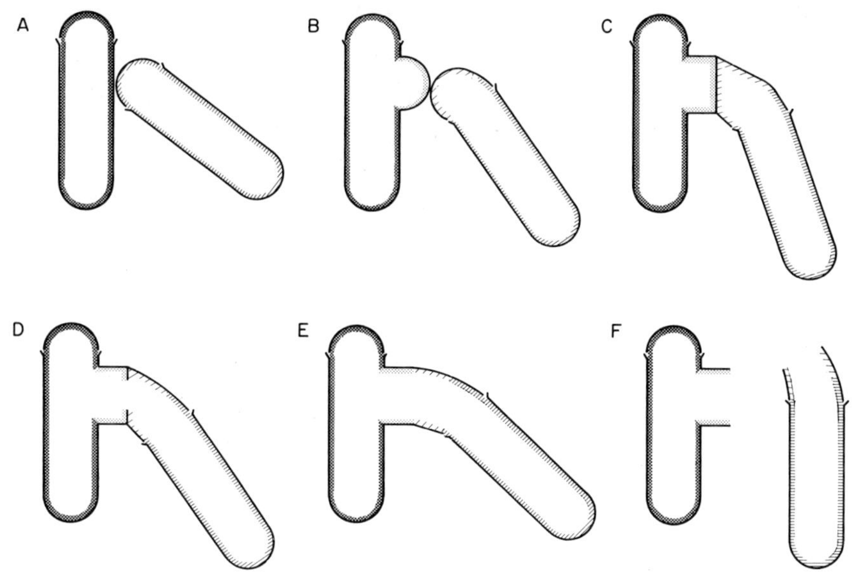

FIG. 4. Morphogenetic events in copulation, conjugation, sporulation, and exporulation of *S. pombe*. Old cell walls shown as dark, new cell walls (conjugation tubes) shown as light. (A) Paired, flocculated cells. (B) Copulating cells forming conjugation tubes. (C) Late copulation with cytoplasms still completely separated by conjugation tube cross wall. (D) Early conjugation with fused cytoplasm, but still some remnants of conjugation tube cross wall visible. (E) Late conjugation and ascus formation with smooth, perfectly fused conjugation tube; cross walls completely resorbed. (F) Exporulation (only in 968 h^{90}; a blocked step in N.C.Y.C. 132).

the conjugants is smooth and featureless by all the methods of observation available to us (Fig. 4E).

The late copulation-conjugation period is fraught with danger for the paired cells. Often the controls over the hydrolytic system seem to fail: lysis of *S. pombe* conjugants is frequent (Fig. 5) [29]. This is in marked contrast to the stability of vegetatively growing cells. Other yeasts also lyse frequently at conjugation [30,31]. An indirect assay of autolytic activity [29] established a correlation "with the formation of conjugation tubes (late copulation) rather than with the dissolution of the separating walls." However, visual evidence (Fig. 5) suggests that lysis per se, as opposed to carbohydrate release [29], is associated with crosswall resorption. Hence we conclude that lysis during late copulation and early conjugation constitutes a form of "inadequately" controlled autolytic activity in differential morphogenesis (Table 1C).

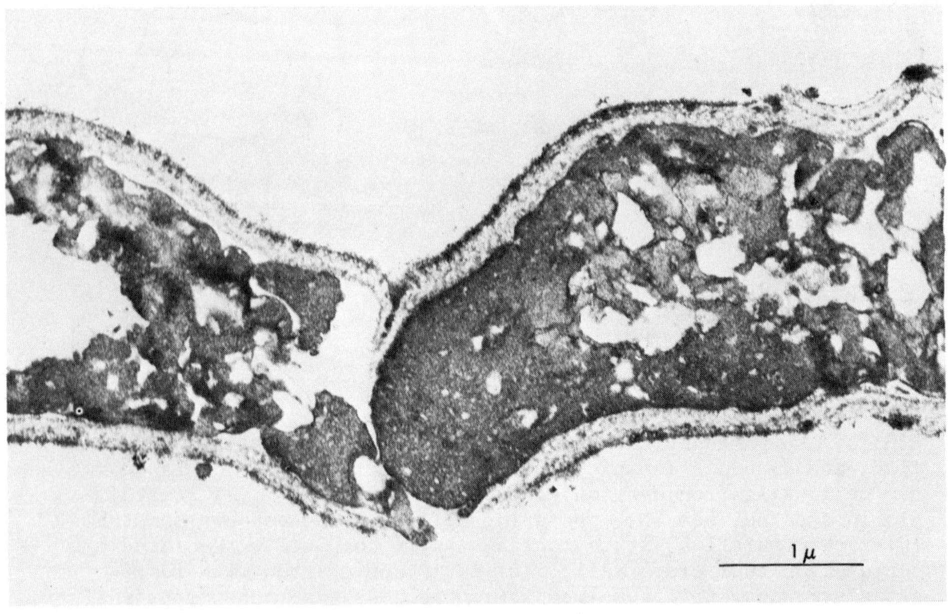

FIG. 5. Electron micrograph of fission yeast conjugants lysing (isolate 360-2). Some of the crosswall remains visible across from the site of lysis, but resorption of the crosswall is advanced. The cytoplasm of one partner is badly disintegrated, but the quality of the other suggests that it has just begun to lyse.

D. Spore Liberation

The perfectly fused walls of a pair of cells constitute an ascus. Karyogamy, meiosis, and sporulation all take place within that ascus. Karyogamy and meiosis lie outside the terms of reference of this report, and little is known about spore wall formation. However, because the spore walls are generated at sites remote from the old ascus walls [26], the entire process might be more akin to generation of the secondary septum at cell division than to extensile processes. At any rate, the final product is a mature ascus typically containing four ascospores. The last morphogenetic event in the sexual sequence is spore liberation. Biologically, it is the thing to do. Thus it is curious that isolate 360-2 does not exhibit spore release. Its ascus simply does not open up to liberate the spores. We shall return to this point.

Spores differ in shape and refractility from vegetative cells, hence are easily recognized by light microscopy, whether they are within the ascus or are liberated. Both light and electron microscopy (Fig. 6) show that the asci (of strain 968 h^{90}) open transversely at approximately the site of crosswall resorption. The spores eventually slip out of the half-asci, leaving the empty ascus walls behind (Figs. 4F and 6). However, the walls continue to disintegrate long after the spores have gone, so that one eventually sees only the highly refractile free spores and miscellaneous debris.

The transverse split in the ascus wall is highly reminiscent of the hydrolytic attack upon the old wall at cell division between the fused secondary septa (line 3a of Table 1B) and probably involves similar enzyme activities (endoglucanase followed by exoglucanase; Table 1D). Because the ascus walls eventually disintegrate, it would seem that the negative controls which keep a dividing cell from continuing to autodigest its wall do not operate here. This event would be similar to cell wall digestion during germination of *Polysphondylium pallidum* as described by Danton O'Day in this volume. By contrast, because the asci of isolate 360-2 do not open to

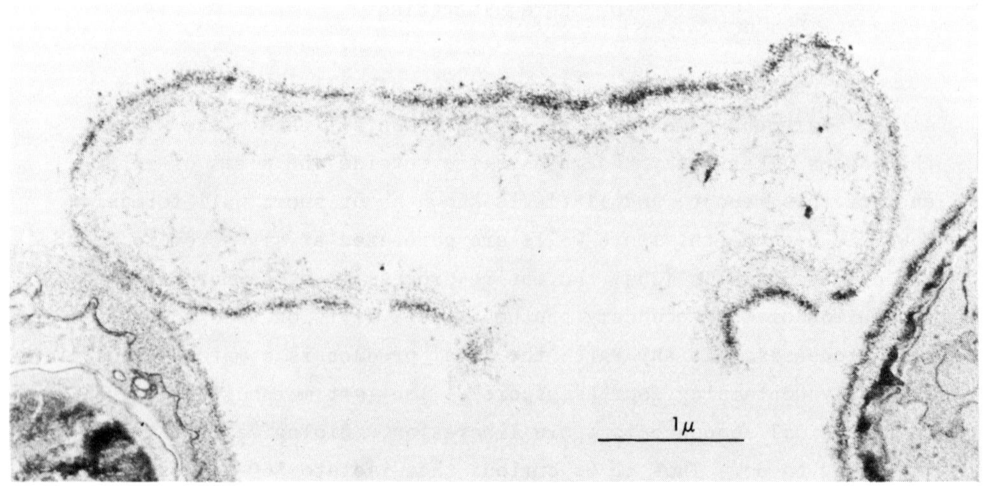

FIG. 6. Electron micrograph of half-ascus (968 h^{90}) after spore liberation. Many signs of disintegration are apparent.

liberate their spores, it would seem that the enzymes for spore liberation do not exist, or fail to become activated in the ascus. This seems to be an intriguing chance to do comparative study of morphogenetic controls.

III. FINAL COMMENTS

Morphogenetic processes during the vegetative cell cycle of the fission yeast include extension and cell division. Both involve polysaccharide biosynthesis, but ultrastructural [5-7] and fluorescent [8] studies of cell division have demonstrated marked differences between those morphogenetic processes. Direct comparison of the two would seem naive now. Nevertheless, it does not seem too naive to draw comparisons between vegetative morphogenetic processes and morphogenetic processes which occur during and consequent to sexual pairing of fission yeast cells. Perhaps one should be cautious, and not suggest that the endo-β-glucanase and glucan synthetase activities imputed to vegetative extension are identical to those apparently involved with extensile activity during conjugation tube formation. The same degree of caution should apply to other "easy" comparisons, such as the hydrolytic attack upon the old cell wall late in cell division and during spore liberation. It is noteworthy that N.C.Y.C. 132 divides readily but its derivative, isolate 360-2, fails to liberate its spores. That failure emphasizes the need for interpretive caution.

The current position for the proposed different roles of autolytic and synthetic enzymes in differential morphogenesis of the fission yeast is unproven, but hardly any other explanations seem reasonable (if only because of the refractory nature of yeast cell walls). The model proposed for extension has received some acceptance [32-34]. It obviously could apply to hyphal extension in fungi by simply substituting for glucan the pertinent "backbone" polysaccharide, and such substitution has been made [35]. Obviously, we are far from understanding the nature of the controls of differential morphogenesis, but genetic analysis should indicate something of the number of genes and their integration. Having reasonable information on the various structural changes and their timing, and models of activation and inactivation of relevant enzymes, we hope soon to have a reasonable description of the relevant genetics, so that characterization of their derivative enzymes may begin.

ACKNOWLEDGMENTS

We thank Donna McLeish for assistance through various phases of the work discussed here, and Dr. R. Egel for transmitting some of his results before they were published. G. B. Calleja was a Canadian International Development Agency-National Research Council

Research Associate in 1973 and 1974. Bong Y. Yoo was supported by N.R.C.C. Grant No. A 3651.

REFERENCES

1. Mitchison, J. M., 1957. Exp. Cell Res. 13: 244-262.
2. Johnson, B. F., 1965. Exp. Cell Res. 39: 613-624.
3. Streiblová, E., and A. Wolf, 1972. Z. Allg. Mikrobiol. 12: 673-684.
4. Biely, P., J. Kovařík, and Š. Bauer, 1973. Arch. Mikrobiol. 94: 365-371.
5. Streiblová, E., I. Málek, and K. Beran, 1966. J. Bacteriol. 91: 428-435.
6. Oulevey, N., J. Deshusses, and G. Turian, 1970. Protoplasma 70: 217-244.
7. Johnson, B. F., B. Y. Yoo, and G. B. Calleja, 1973. J. Bacteriol. 115: 358-366.
8. Johnson, B. F., B. Y. Yoo, and G. B. Calleja, 1974. *In* Cell Cycle Controls, G. M. Padilla, I. L. Cameron, and A. M. Zimmerman (eds.), Academic Press, New York, p. 153.
9. Phaff, H. J., 1963. Ann. Rev. Microbiol. 17: 15-30.
10. Bacon, J. S. D., 1973. *In* Yeast, Mould and Plant Protoplasts, J. R. Villanueva, I. Garcia-Acha, S. Gascon, and F. Uruburu (eds.), Academic Press, London, p. 61.
11. Barras, D. R., 1969. Antonie van Leeuwenhoek 35: Supplement I17-I18.
12. Barras, D. R., 1972. Antonie van Leeuwenhoek 38: 65-80.
13. Fleet, G. H., and H. J. Phaff, 1973. *In* Yeast, Mould and Plant Protoplasts, J. R. Villanueva, I. Garcia-Acha, S. Gascon, and F. Uruburu (eds.), Academic Press, London, p. 33.
14. Fleet, G. H., and H. J. Phaff, 1974. J. Biol. Chem. 249: 1717-1728.
15. Necas, O., 1956. Nature 177: 898-899.
16. Johnson, B. F., 1967. J. Bacteriol. 94: 192-195.
17. Mitchison, J. M., 1970. *In* Methods in Cell Physiology, Vol. IV, D. M. Prescott (ed.), Academic Press, New York, p. 131.
18. Johnson, B. F., 1968. Exp. Cell Res. 49: 59-68.
19. Johnson, B. F., 1968. J. Bacteriol. 95: 1169-1172.

20. Kjosbakken, J., and J. R. Colvin, 1975. Can. J. Microbiol. 21: 111-120.

21. Cortat, M., P. Matile, and A. Wiemken, 1972. Arch. Mikrobiol. 82: 189-205.

22. Perry, M. B. Personal communication.

23. McCully, E. K., and C. F. Robinow, 1971. J. Cell Sci. 9: 475-507.

24. Shannon, J. L., and A. H. Rothman, 1971. J. Bacteriol. 106: 1026-1028.

25. Calleja, G. B., and B. F. Johnson, 1971. Can. J. Microbiol. 17: 1175-1177.

26. Yoo, B. Y., G. B. Calleja, and B. F. Johnson, 1973. Arch. Mikrobiol. 91: 1-10.

27. Bresch, C., G. Müller, and R. Egel, 1968. Molec. Gen. Genet. 102: 301-306.

28. Egel, R., and M. Egel-Mitani, 1974. Exp. Cell Res. 88: 127-134.

29. Kröning, A., and R. Egel, 1974. Arch. Microbiol. 99: 241-249.

30. Shimoda, C., and N. Yanagishma, 1972. Arch. Mikrobiol. 85: 310-318.

31. Lee, E.-H., C. V. Lusena, and B. F. Johnson, 1975. Can. J. Microbiol. 21: 802-806.

32. Bartnicki-Garcia, S., and I. McMurrough, 1971. *In* The Yeasts, Vol. II, A. H. Rose and J. S. Harrison (eds.), Academic Press, London, p. 441.

33. Ingram, M., 1969. Antonie van Leeuwenhoek 35: Supplement 7-29.

34. Matile, P., 1973. Ber. Deutsch. Bot. Ges. 86: 241-255.

35. Bartnicki-Garcia, S., 1973. *In* Microbial Differentiation, Symp. Soc. Gen. Microbiol. #23, J. M. Ashworth and J. E. Smith (eds.), Cambridge University Press, London, p. 245.

PART II

Cell Communication and Morphogenesis

EDITORS' INTRODUCTION

The induction of morphogenetic changes in response to specific cellular effector molecules is an area that has been of special interest to developmental biologists during the last half century. Questions such as "how do cells communicate with one another?" and "how can extremely small amounts of specific 'chemical messengers' produce dramatic metabolic changes leading to specific alterations in patterns of cell differentiation?" probe at the very basis of our understanding of development. The first of these questions, while proving to be quite challenging in many systems, has been answerable. The second question is extremely more complex and, while pieces of the puzzle are beginning to fit together for certain effectors, an understanding of the total picture is still a long way from being realized in any eucaryotic system. Because of their simplicity and their many attributes, mentioned time and again by authors of this volume, eucaryotic microbes are proving invaluable as models to formulate an understanding of how cells can communicate with one another.

Chemical effectors are known to exist in almost all the major groups of eucaryotic microbes. While it is beyond the scope of this section to discuss the multitude of ways in which eucaryotic microbes can communicate with one another, four interesting systems are discussed which utilize four different types of chemical effectors: glycoproteins, terpenoid hormones, steroid hormones, and cyclic nucleotides. All of these classes of compounds are recognized to have profound effects on morphogenesis throughout the eucaryotic kingdom.

GLYCOPROTEINS

Among the glycoproteins (proteins known to have side chains of sugars) are an amazing number of regulatory molecules in eucaryotes. These include enzymes, hormones, antiviral compounds (interferon), and molecules associated with nerve transmission and memory [1]. Probably the most interesting group of glycoproteins are those found on the surfaces of cells. Although in number they only are a minor constituent of the membrane components, they are important because they appear to be partly responsible for communicating the individuality of the cell. Membrane glycoproteins serve as antigenic determinants, as markers of cellular identity, and as virus receptors. When cells become malignant, often the carbohydrate components of certain membrane glycoproteins become altered [1]. Glycoproteins have also been shown to act as inducers of sexual development in the colonial green alga *Volvox* [2,3]. In this section Gary Kochert discusses the glycoprotein sexual inducer of *Volvox carteri f. wiesmannia*. Specific attention is given to the isolation and characterization of the inducer and its mode of action.

TERPENOID HORMONES

Terpenes are minor lipid components of cells which are constructed of multiples of isoprene, a five-carbon hydrocarbon. Multiples of the basic isoprene unit are important in the biosynthesis of steroid hormones and of vitamin A. Furthermore, at least two classes of plant hormones--the gibberellins and abscisic acid--are terpenoids. For several years, trisporic acid has been considered the sex hormone for the Mucorales, an order of fungi in the class of Zygomycetes [4]. Richard F. Sutter in this part reviews the data implicating trisporic acid and suggests some interesting alternative hypotheses. He presents some convincing observations which suggest that the chemical precursors to trisporic acid are the real sex hormones in the Mucorales.

STEROID HORMONES

Probably the only completely characterized steroid sex hormones outside the animal kingdom are antheridiol and oogoniol, the male and female sex hormones of the aquatic fungus *Achlya*. Antheridiol (hormone A) was first extracted and purified by Raper and Hagen-Smit in 1942 [5]. Recently, the chemical structure and characterization of oogoniol (hormone B) were reported by McMorris and several collaborators [6]. In this part, P. A. Horgen discusses some of the biochemical effects of antheridiol on the development of the male sex organ initial in *Achlya*. He specifically discusses the effects of the hormone on nucleic acid metabolism, protein metabolism, and chromosomal protein modification.

CYCLIC NUCLEOTIDES

Cyclic-3'5'-adenosine monophosphate (cAMP) is a recognized regulator molecule throughout the eucaryotic kingdom (and also in procaryotes). Among the many functions served by cAMP in man and other animals is its ability to act as a chemical messenger that regulates the enzymatic activities within cells that store fats and sugars. cAMP is involved in controlling the activity of genes in the nucleus. Furthermore, a condition for the development of one of the various kinds of cancerous growth in animals appears to be an inadequate supply of cAMP [7]. Aggregation in the cellular slime mold *Dictyostelium discoideum* occurs because certain cells begin secreting a signal molecule, initially called "acrasin," which has been shown to be cAMP [8]. These specific cells produce pulses of cAMP, which act as a signal causing the individual amoeba cells to form aggregation centers [9]. Anthony J. Durston, in this part, discusses the signal-relaying mechanisms during aggregation in *D. discoideum*. Communication between individual cells is important whenever component cells of an organism are assembled during embryogenesis and pulsating signals appear to be involved in morphogenesis throughout the eucaryotic kingdom. Durston discusses the physiology and biophysics associated with wave pattern formation in *Dictyostelium*.

REFERENCES

1. Sharon, N., 1974. Sci. Amer. 230: 78-86.
2. Kochert, G., and I. Yates, 1974. Proc. Nat. Acad. Sci. USA 71: 1211-1214.
3. Starr, R., and L. Jaemicke, 1974. Proc. Nat. Acad. Sci. USA 71: 1050-1054.
4. Gooday, G., 1974. Ann. Rev. Biochem. 43: 35-49.
5. Raper, J., and H. Hagen-Smit, 1942. J. Biol. Chem. 143: 311-320.
6. McMorris, T., R. Seshadri, G. Weithe, G. Arsenault, and A. Barksdale, 1975. J. Amer. Chem. Soc. 97: 2544-2545.
7. Pastan, I., 1972. *In* Current Topics in Biochemistry, C. B. Anfinsen, R. F. Goldberger, and A. N. Schechter (eds.), Academic Press, New York, p. 65.
8. Konijin, T., C. Van de Meene, J. Bonner, and D. Barkley, 1967. Proc. Nat. Acad. Sci. USA 58: 1152-1154.
9. Robertson, A., and J. Grutsch, 1975. Life Sci. 15: 1031-1043.

SEXUAL HORMONES AND CELL DIFFERENTIATION IN *VOLVOX CARTERI*

Gary Kochert

Department of Botany
University of Georgia
Athens, Georgia

I. INTRODUCTION

The potential of the green alga *Volvox* for developmental studies has long been realized by developmental biologists. The feature of the organism which was attractive to early workers in the field was chiefly that it was a very simple organism yet it showed a clear differentiation of two very distinct cell types. Many splendidly detailed studies of the morphology, life history, and phylogenetic relationships of *Volvox* species were carried out with field collections (for a review see Ref. 1), but until the late 1950s it was not possible to grow *Volvox* in axenic laboratory cultures. The development of a defined medium for growth of such cultures [2] has made possible the present resurgence of interest in *Volvox* as an experimental tool.

Individuals of all *Volvox* species are composed of two cell types: somatic and reproductive. The cells are arranged in a surface layer over the spheroid and are embedded in a sheath of transparent matrix material (Fig. 1). The composition of the matrix has not been established in detail, but it contains large amounts of a protein in which 22% of the amino acid residues are hydroxyproline [3]. Typically, asexual individuals are composed of a few thousand somatic cells and about 10 asexual reproductive cells. Somatic cells are biflagellate and provide the propulsive force which renders the organisms actively motile. In some species, somatic cells of mature spheroids are linked by cytoplasmic bridges. In addition to flagella, each somatic cell contains a single nucleus and chloroplast. Asexual reproduction is accomplished by a series of divisions of the reproductive cells to form miniature "daughter" spheroids. After release of mature daughter spheroids, the somatic cells of the parent organism undergo senescence and eventually die. In this way more than 99% of the cells present die each generation.

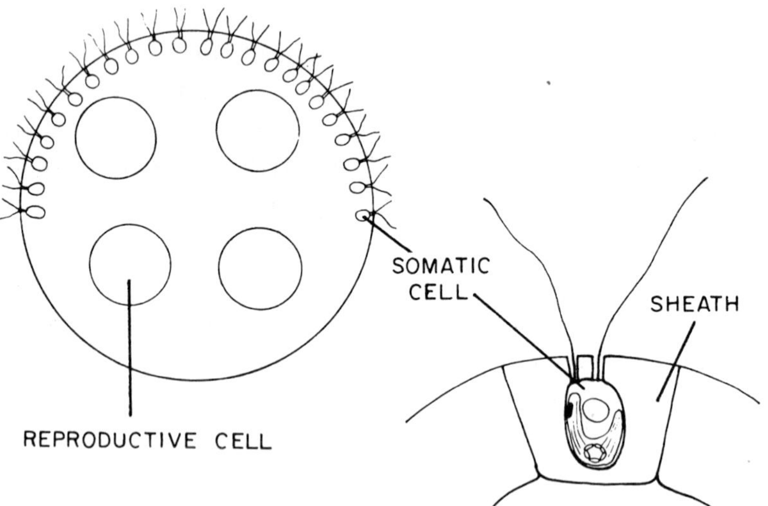

FIG. 1. Diagram of the structure of an asexual spheroid of *Volvox carteri*. Not all the somatic cells are depicted.

This is an unusual feature in a microorganism, and suggests *Volvox* as a model system for the study of cellular senescence.

All *Volvox* spheroids display a marked degree of polarity. A definite anterior-posterior axis is present with the anterior end being always directed forward when swimming. Reproductive cells are usually present only in the posterior portion of the organism. In addition, the somatic cells are not quite the same in all parts of the organism. Those in the anterior portion have a large eyespot which is thought to be important in phototaxis. The eyespot becomes progressively smaller in posterior cells and, in the posterior half of the spheroid, somatic cells usually have no eyespot at all.

Sexual reproduction and its control have been the aspects of *Volvox* development most emphasized in recent studies. The initial contribution in this field was made by Darden [4] when he discovered that formation of sexual colonies in *V. aureus* could be induced by a substance released into the culture medium by male spheroids. The inducer was reported to be a glycoprotein [5]. Since Darden's work, several other *Volvox* species have been demonstrated to produce sexual inducers (for reviews see Refs. 6 and 7). In most cases the inducer is produced by male spheroids, released into the culture medium, and induces the formation of female spheroids or of more male spheroids. Each inducer seems to be species-specific or even strain-specific. None will elicit the formation of sexual spheroids in another species.

II. CURRENT RESEARCH

A. Control of Gonidial Differentiation

Thus far *V. carteri* has been the species which has received the most attention and its development will be emphasized in the present article. One aspect of its development which has interested us has been the factors controlling differentiation of the gonidia in developing asexual spheroids. Mature asexual spheroids show a very regular arrangement of gonidia. These are placed in two or three tiers of four gonidia in planes normal to the longitudinal axis of the spheroid. Since this pattern expresses itself early in development, we reasoned it might be programmed into the gonidium before cleavage, perhaps during the period of expansion before divisions begin. To test this idea we utilized an experimental design frequently used in developmental studies of cytoplasmic localizations in animal eggs. In these cases it has been conclusively shown that regions of cytoplasm in the uncleaved egg affect the pattern of differentiation of nuclei which are placed in contact with them during cleavage (for a review see Ref. 8). One way of showing this is to irradiate an uncleaved egg unilaterally with ultraviolet light (UV) in the hope of destroying a cytoplasmically localized morphogenetic substance. Specific abnormalities in later development are then taken as evidence of such destruction.

By these and other methods many animals have been shown to localize a germ plasm in specific areas of the developing egg. This material is essential for the differentiation of germ cells, apparently exerting its effect by inducing nuclei which come in contact with it to form the series of gene products necessary for normal germ cell differentiation. If the germ plasm is inactivated by UV, no viable germ cells are formed. No germ line of cells has been demonstrated in higher plants, where germ cells are formed directly from somatic cells.

Many animals, then, effectively differentiate their germ cells early in development, and only specific cells have the capacity to form germ cells. The strains of *V. carteri* that we use also differentiate their germ cells early in development and were originally called *V. weismannia* in honor of the early proponent of a soma-germ theory. Is this early differentiation of reproductive cells in *V. carteri* also controlled by the localization in the cytoplasm of a specific morphogenetic substance? We have shown that if one UV irradiates mature but undivided gonidia with low UV doses that do not affect viability, abnormal spheroids are produced when these gonidia cleave [9]. Very specific sorts of abnormalities are produced in which one or more gonidia are missing from one side of the organism (Fig. 2). Somatic cells in the area of the missing gonidia are almost always normal in morphology. Moreover, the remaining gonidia in the spheroid produce organisms with normal numbers and positioning of reproductive cells on cleaving. This would seem to eliminate the possibility of nuclear mutation and is consistent with the idea of destruction of a germ plasm.

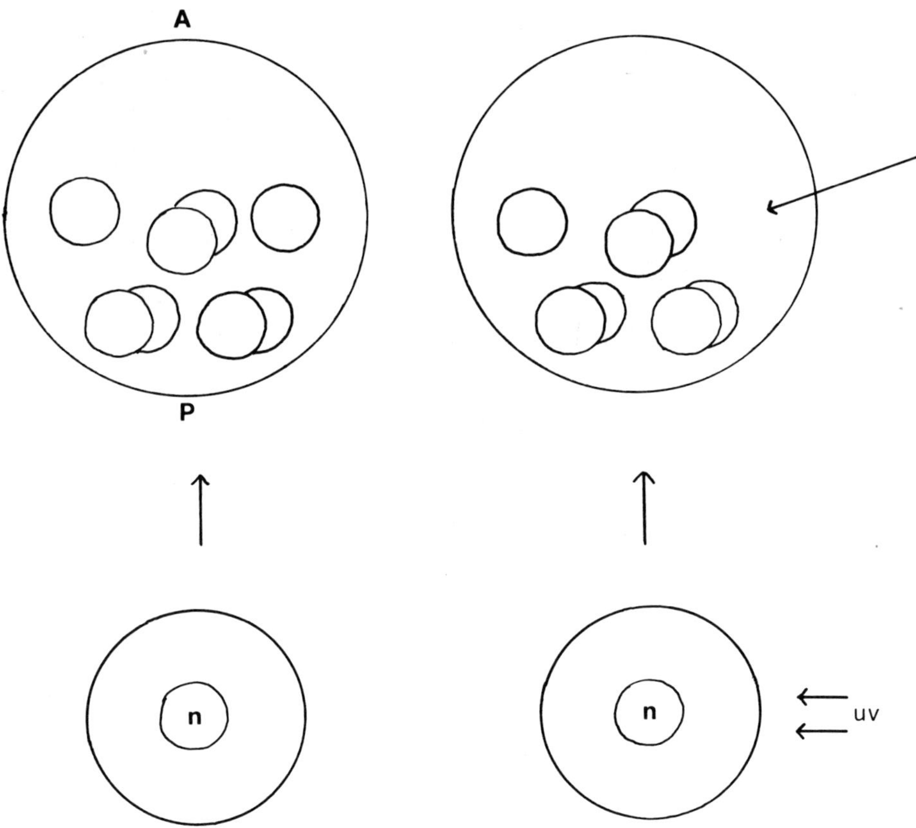

FIG. 2. Diagram showing the production of an abnormal spheroid from a UV-irradiated gonidium. The control (left) produces two tiers of four gonidia. The UV-irradiated gonidium produces a spheroid which lacks one gonidium (upper arrow); n = nucleus of gonidium.

B. Control of Spheroid Development

Factors controlling the appearance of sexual spheroids are also very interesting. The organisms are haploid and two types of clones can be isolated from nature (Fig. 3). Both reproduce asexually, but each clone produces only male or female spheroids and never both. A mature asexual spheroid contains approximately 3,000 somatic cells

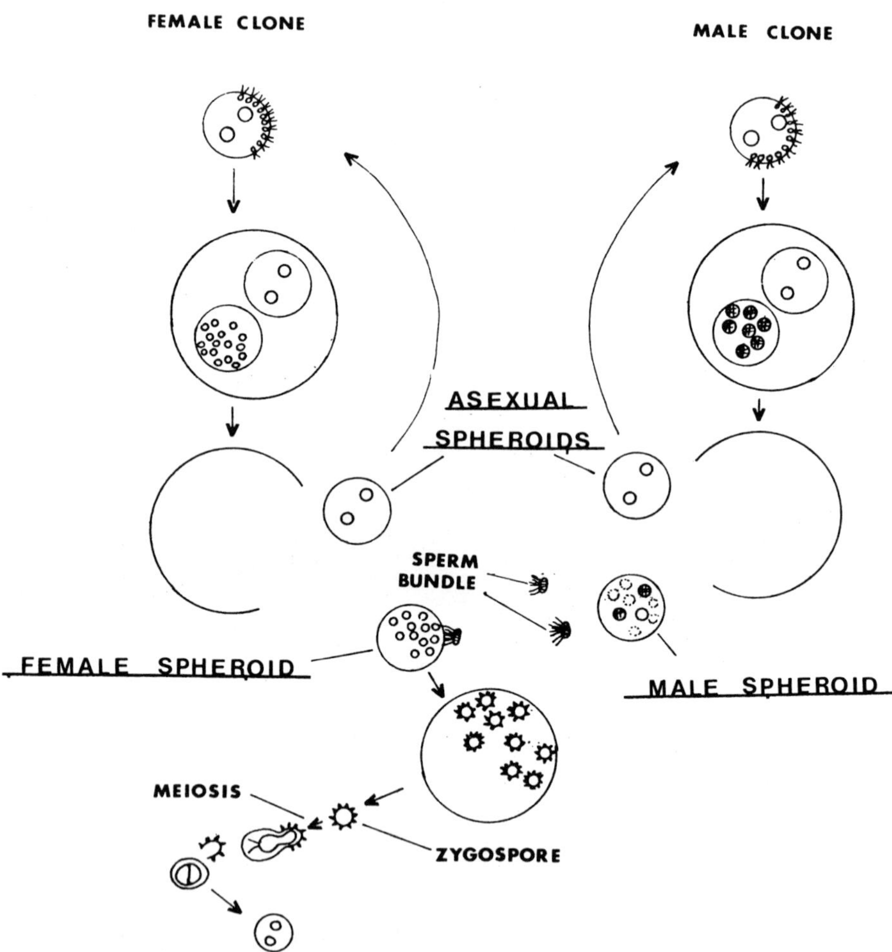

FIG. 3. Life cycle of *Volvox carteri*.

and 8 to 16 reproductive cells (gonidia). Each reproductive cell cleaves to form a miniature spheroid inside the parent. A gonidium from the male clone can cleave to form either another asexual spheroid or a male spheroid. Male spheroids differ from asexual spheroids in that they are smaller and they contain fewer somatic cells and more reproductive cells. All of the approximately 50 reproductive cells of the male spheroid cleave to form packets of sperm. Asexual and male spheroids both appear to be formed by a true cleavage process; the cells become progressively smaller as the divisions proceed. It is not as yet clear what controls the initial appearance of male spheroids in male clones. In practice we induce the formation of males by transfer to fresh medium. Male spheroids are then formed for two or three generations after which the culture produces only asexual spheroids.

Female spheroids, however, never appear spontaneously in female clones and only asexual spheroids are produced. I initially discovered from growing mixed populations of male and female clones that female spheroids were formed only in the presence of an inducing substance released in the culture medium by male clones [10]. Apparently the inducing substance is produced in the male clone only by male spheroids and not by asexual spheroids. It is not known whether the sexual inducer is actively secreted or is a breakdown product. Also it is not known whether the substance is produced by somatic cells in the male spheroid or by the sperm bundles themselves. However, the highest concentrations of inducer are found in the culture medium only after the sperm bundles have been released from the male colonies and the individual sperm have lysed.

C. Characterization of the Sexual Inducer

We have isolated and partially characterized the sexual inducer from male conditioned media of *V. carteri f. weismannia* [11]. For production of inducer, light-dark synchronized cultures [12] were used. Large numbers of male spheroids are produced synchronously when these cultures are transferred to fresh medium. After the male spheroids have formed and disintegrated, the remaining asexual spheroids are removed by filtration. The conditioned medium is then adjusted to pH 5.0 with acetic acid and the inducer is adsorbed batchwise onto carboxymethyl (CM)-Sephadex. After recovery of the CM-Sephadex particles, the inducer is eluted with NaCl and then dialyzed against H_2O. The final concentration step is accomplished by lyophilization, after which the inducer is stored frozen in H_2O.

The purified inducer elutes from CM-Sephadex columns as a single peak with protein, carbohydrate, and biological activity coincident. When the purified inducer is electrophoresed in urea or sodium dodecyl sulfate-urea polyacrylamide gels, a single band is formed. The band stains with both Coomassie Blue and Periodic Acid Schiff reagent, indicating the active material is a glycoprotein. When sliced gels are bioassayed, inducer activity migrates with the glycoprotein band. When electrophoresed in sodium dodecyl sulfate gels with mol wt markers, the material migrates with an apparent mol wt of 32,000. Since some glycoproteins behave anomalously in sodium dodecyl sulfate gels, one must be cautious in interpreting results of this sort, but we believe this is near the true mol wt. The molecule contains approximately 20% carbohydrate, but we do not as yet know how many carbohydrate chains are present per molecule or which sugars are present. The inducer is very stable, as befits a molecule designed to be released into the capricious environment of a pond or lake. It can be boiled for short periods of time without loss of biological activity, and it is resistant to such enzymes as trypsin and chymotrypsin, although pronase will inactivate it. It is stable almost indefinitely at room temperature.

Inducers from other *Volvox* species also appear to be proteins or glycoproteins [5,13,14], but the only other inducer which has been characterized in detail is that from Japanese strains of *V. carteri f. nagariensis* [15]. The overall properties of the molecule are similar to the *V. carteri f. weismannia* inducer, but the inducers do not cross react. The constituent sugars present in the inducer from the Japanese strains have been analyzed; the principal components are xylose and glucose.

D. Mode of Action of the Sexual Inducer

To learn more about the mode of inducer action we have performed experiments in which inducer was added at various stages of the *V. carteri* life cycle. When a gonidium of *V. carteri* begins cleavage to form a daughter spheroid, no visible cell differentiation occurs through the first few cleavages. If the developing embryo is to become an asexual spheroid, the reproductive cell initials will be differentiated from the somatic cells by unequal divisions at the cleavage from 16 to 32 cells. If the gonidium has been induced, the unequal divisions are delayed until division of the 32- or 64-cell embryo occurs. Eggs instead of gonidia are formed and the eventual result is, of course, a female rather than an asexual spheroid. In *V. carteri*, however, a gonidium cannot be induced to form a female spheroid by adding inducer during cleavage or to the mature, uncleaved gonidium. To obtain full induction the gonidium must be placed in an inducing medium near the beginning of the lengthy precleavage growth period. Thus spheroids containing mature gonidia or developing daughter spheroids produce only asexual spheroids the first generation after being placed in inducing medium; females are not formed until the second generation.

Long exposure to the inducer is also necessary to achieve full induction. If spheroids are placed in inducing medium and then removed after various periods, "pulses" of inducer can be given. Pulses of 12 h induce only a small percentage of gonidia to form females, and the inducer must be continuously present to achieve full induction. Even if gonidia are left in inducing medium until just before cleavage begins, the percentage of females will be reduced considerably by washing in fresh medium.

Irradiation with UV also has a profound effect on induction of females. If induced, mature gonidia are irradiated just before cleavage or even during the initial stages of cleavage, the number of females formed is greatly reduced. This inhibition of induction is accomplished without any loss of viability or any evident morphological anomalies [16]. The UV-mediated repression of sexual induction is also stage-dependent. If induced gonidia are irradiated early in their precleavage enlargement period, they apparently are able to recover and nearly normal numbers of females are formed (Table 1).

The above observations seem to indicate that, although gonidia must be placed in inducing medium much prior to the actual visible differentiation stages, the inducer may in fact exert its ultimate effect during cleavage. It is difficult to assess the role of permeability in experiments of this sort since we know nothing about where the inducer exerts its effects and hence little about the sorts of permeability barriers that must be traversed for induction to occur. The results of the pulse experiments would indicate, however, that the inducer is not irreversibly taken up by the target organisms.

The UV experiments may be interpreted in several ways. It may be that the inducer must be present during the critical cleavage stages and that the UV (by destroying the inducer) effectively "washes" the gonidia and prevents female induction. It could also be that the primary action of the inducer is to repress the formation of the asexual reproductive cells and then sexual reproductive cells appear as a consequence of this repression. It is clear that the first recognizable effect of the inducer is to prevent the formation of the gonidia at the stage when they normally appear. The UV could then be destroying a repressor of the expression of asexual loci; by destroying the repressor, asexual characters could express

TABLE 1

UV Irradiation of Induced Gonidia

Developmental stage of the irradiated gonidium	% Females
In inverting spheroid	90
In newly released spheroid	68
Mature gonidium	26
Beginning cleavage (2-4 cells)	16

themselves. In this case UV would be acting as it does in the bacteriophage lambda system. With lambda a protein is produced which prevents expression of the loci necessary for cell lysis and production of virus particles. Treatment with UV causes the destruction of this repressor and the expression of the previously repressed viral genes [17].

We have also begun to investigate the uptake of purified inducer by target spheroids. One would suppose that the organism must possess a well-developed mechanism for binding inducer from the surrounding medium, since the normal environment of the organism is ponds or lakes where one would suspect that inducer concentrations would remain low.

To study inducer binding we have labeled the inducer in vitro with ^{125}iodine (^{125}I) by the chloramine-T method originally devised by Greenwood [18]. This procedure does not cause any decrease in biological activity detectable by the standard bioassay. Specific activities of 2 to 6 μCi/μg are obtained. To measure binding, labeled inducer is added to a culture of asexual spheroids from the female clone. Aliquots of the target spheroids are then removed at intervals, collected on filters cut from nylon screen cloth (Tobler), washed with culture medium, and assayed for bound radioactivity. With these methods, binding of labeled inducer can be demonstrated. The time course apparently reflects a rapid saturation of binding sites (Fig. 4). Even very long-term incubation does not increase the amount of labeled inducer bound. Very little of the bound inducer can be removed by repeated washes with culture medium. From the specific activity of the labeled inducer it can be calculated that each spheroid binds 2 to 4 × 10^6 inducer molecules.

We do not as yet know whether the inducer binding we observe is of biological significance or represents a nonspecific adsorption or uptake of inducer. The number of inducer molecules bound is in the general range of that required for biological activity. Some induction of females is detectable in our *Volvox* strains at levels of 10^{-13} M. This corresponds to about 2 × 10^5 molecules per spheroid present in the assay.

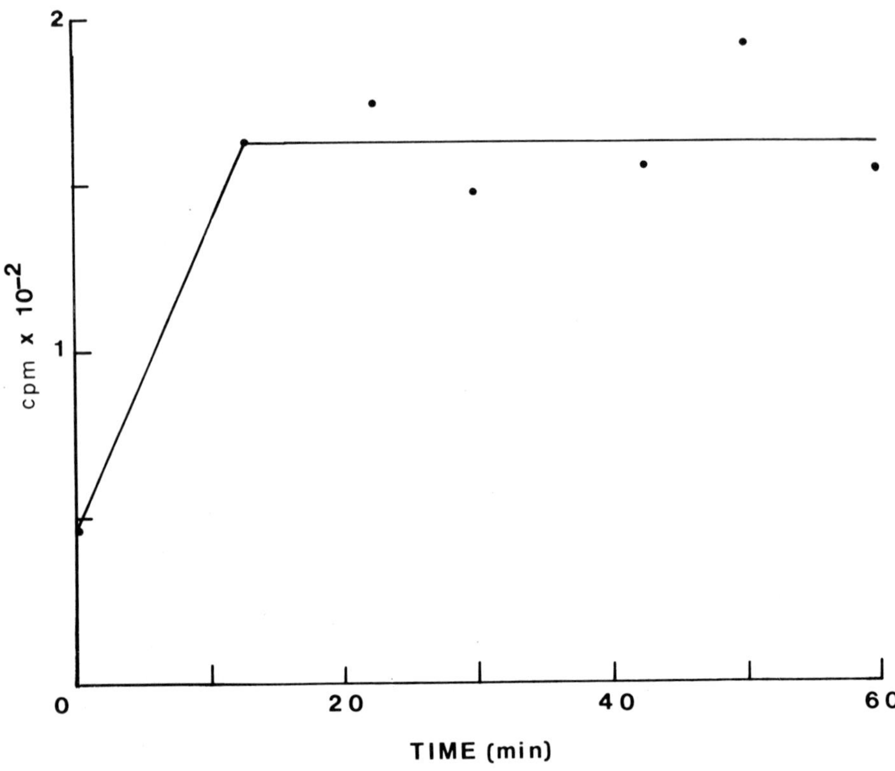

FIG. 4. Binding of radioactively-labeled *Volvox carteri* sexual inducer. Approximately 0.5 μg of ^{125}I-labeled inducer was added to a 100-ml culture (170 spheroids/ml). 5-ml samples were collected on filters at timed intervals.

It will be recalled that the *Volvox* inducers so far investigated have proven to be species-specific in their biological activity. We have investigated the binding of ^{125}I-labeled sexual inducer from *V. carteri* to other species of *Volvox*. When spheroids of *V. aureus* or *V. rousseletii* (both of which produce inducers of their own) are used in the assay, very little binding occurs (Table 2). This may mean that each *Volvox* species possesses receptors specific for its own inducer. Hopefully it also means that in our binding studies with *V. carteri* we are dealing with a specific sort of binding rather than a nonspecific adsorption of inducer molecules.

TABLE 2

Binding of Radioactively-labeled Inducer to *Volvox* Spheroids

Target organism	Inducer bound ($cpm/10^3$ spheroids)
V. carteri	210
V. rousseletii	20
V. aureus	16

III. FINAL COMMENTS

One of our primary areas for future research is further characterization of the sexual inducer and of its mode of action. We know the inducer to be a glycoprotein but we know nothing of the number or the arrangement of carbohydrate residues on the molecule and very little about the biological functions of the carbohydrate and protein portions of the inducer. All we know in this respect is that the protein portion is necessary for biological activity, since Pronase treatment of inducer destroys its ability to induce females. The molecule is very resistant to degradation by trypsin and chymotrypsin. Is this stability of the molecule imparted by the carbohydrate portion? Is the carbohydrate portion involved in specific binding of inducer to target spheroids?

We also know nothing about the series of events involved in the biological action of the inducer. The presumed target cells are the gonidia, and these are located inside the organism beneath the layer of spheroid sheath and have no obvious contact with the surrounding medium. This may be significant if the colony matrix is a permeability barrier to inducer entry. From our volume calculations we believe that it must be, since if one removes spheroids from long-term incubations with ^{125}I-labeled inducer and isolates them on filters with no washing, they contain less than the amount of inducer expected if free equilibrium of inducer exchange existed inside and outside the spheroid.

Thus we do not know whether the inducer acts at the spheroid surface, penetrates the spheroid and acts at the gonidium surface, or enters the gonidium to exert its effect. We will be studying the localization of inducer effect in several ways, with the aim of solving the questions outlined above. Several approaches are possible, but these are complicated by the fact that so few inducer molecules are apparently required for biological activity. In the most sensitive system, Starr has reported that as few as 1,800 molecules per gonidium may suffice for a 25% response in the bioassay [15]. In our binding studies we achieve saturation at only a few million molecules per spheroid. With the specific activity levels of labeled inducer that we are presently able to achieve, autoradiographic methods of localization are limited in utility. We may be able to perform localization experiments by using ferritin or peroxidase antibodies against the inducer. In collaboration with Dr. F. P. Inman, we have now produced antibodies against purified inducer and will attempt to use these for localization experiments.

We have also used the antibodies to devise an agar plate diffusion assay for the inducer (Fig. 5). The sensitivity of this assay is limited, however, and we are attempting to modify one of the existing radioimmunoassay techniques for use with the *V. carteri* inducer. Elimination of the cumbersome bioassay would make many sorts of experiments more feasible.

We also hope to clarify the question of when during the *Volvox* life cycle the inducer exerts its effect. To induce a gonidium to form a female rather than an asexual spheroid, the spheroid containing the gonidium must be placed in inducing medium while the gonidium is still very small. This would be consistent with the idea that the inducer exerts its effect by altering the pattern of cytoplasmic localization in the growing gonidium. Mature gonidia form only asexual spheroids if placed in inducing medium. This is consistent with the idea that the pattern of subsequent differentiation has already been "programmed" into the gonidium, as our UV experiments would seem to indicate. However, we find that induced gonidia will form asexual rather than female spheroids if the

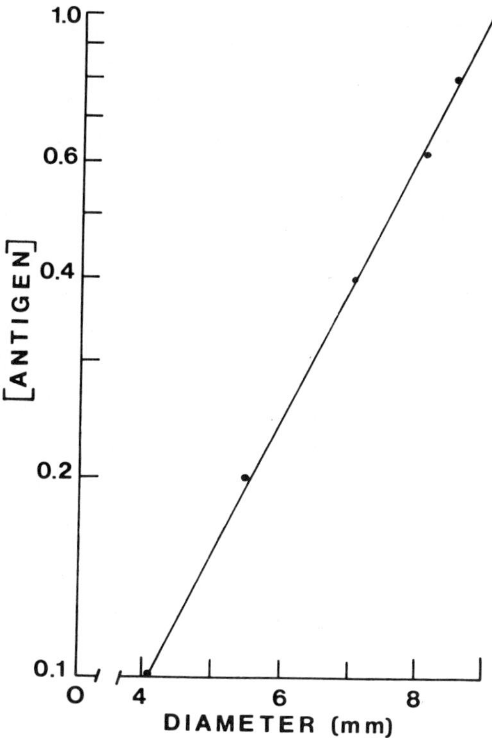

FIG. 5. Radial diffusion immunoassay of the *Volvox carteri* sexual inducer. 3-mm holes were cut into a 0.8% agarose plate containing 2% rabbit antiinducer serum. Inducer concentration is expressed in mg/ml.

inducer is washed away or if a short UV pulse is given while maintaining the organisms in inducing medium. This indicates that the actual time of inducer action may be during the cleavage process. Early application of inducer may be necessitated by permeability barriers, or it may be that the inducer must be continuously present for long periods to effect female induction.

It will be recalled that the main morphological manifestation of inducer action is a change in the cleavage pattern exhibited by gonidia during colony formation. Induced gonidia delay differentiation of reproductive cells for one or more cleavages and then form eggs rather than gonidia. We would like to know how the inducer affects the gonidial cleavage pattern. Is the cleavage pattern programmed into the gonidium before cleavage? It is possible that the inducer acts by repressing formation of gonidia, and that if no gonidia form, eggs are differentiated as a secondary response.

In summary then, *Volvox* is an organism that possesses distinct advantages for certain sorts of studies of cell differentiation. First of all, the organism shows true cell differentiation as a routine part of its life cycle. Two very distinct cell types are formed, different in both morphology and function. Cells in the mature spheroids are embedded in the sheath and not connected to each other; thus they are separable and comparative biochemical studies can be done [12]. Huskey and Starr have shown that many interesting developmental mutants can be generated [6,19]. Finally, a sexual induction system exists where one can specifically control a developmental pathway. The inducer involved has been isolated and characterized. A very simple procedure suffices to isolate this glycoprotein and it is a very stable molecule. We believe the currently increasing interest in *Volvox* as a research organism justifies our confidence that basic mechanisms of eucaryote cell differentiation will be particularly amenable to study in this fascinating organism.

REFERENCES

1. Smith, G. M., 1944. Trans. Amer. Microsc. Soc. 63: 265-310.
2. Provasoli, L., and I. J. Pintner, 1959. Spec. Publ. No. 2, Pymantuning Lab. of Field Biology, Univ. of Pittsburg: 84-96.
3. Lamport, D., personal communication.
4. Darden, W. H., Jr., 1966. J. Protozool. 13: 239-255.
5. Ely, T. H., and W. H. Darden, Jr., 1972. Microbios. 5: 51-56.
6. Starr, R. C., 1970. Develop. Biol. Supp. 4: 59-100.
7. Kochert, G., 1975. Develop. Biol. Supp. 8: 55-90.
8. Davidson, E., 1968. Gene Activity in Early Development, Academic Press, New York.

9. Kochert, G., and I. Yates, 1970. Develop. Biol. 23: 128-135.

10. Kochert, G., 1968. J. Protozool. 15: 438-452.

11. Kochert, G., and I. Yates, 1974. Proc. Nat. Acad. Sci. USA 71: 1211-1214.

12. Yates, I., and G. Kochert, 1975. Cytobios 12: 211-223.

13. McCracken, M., and R. C. Starr, 1970. Arch. Protistenk. 112: 262-282.

14. VandeBerg, W. J., and R. C. Starr, 1971. Arch. Protistenk. 113: 195-219.

15. Starr, R. C., and L. Jaenicke, 1974. Proc. Nat. Acad. Sci. USA 71: 1050-1054.

16. Crump, W., and G. Kochert, unpublished results.

17. Roberts, J. W., and C. W. Roberts, 1975. Proc. Nat. Acad. Sci. USA 72: 147-151.

18. Greenwood, F. C., W. M. Hunter, and J. S. Glover, 19 . Biochem. J. 89: 114-118.

19. Sessoms, A. H., and R. J. Huskey, 1973. Proc. Nat. Acad. Sci. USA 70: 1335-1338.

REGULATION OF THE FIRST STAGE OF SEXUAL DEVELOPMENT IN *PHYCOMYCES BLAKESLEEANUS* AND IN OTHER MUCORACEOUS FUNGI

Richard P. Sutter

Department of Biology
West Virginia University
Morgantown, West Virginia

I. INTRODUCTION

The first stage of sexual development in mucoraceous fungi includes zygophore formation, sex-stimulated carotenogenesis, and trisporic acid biosynthesis. Trisporic acids, metabolites of vitamin A, are synthesized by way of mating type-specific precursors. Although it is not known whether trisporic acids or their mating type-specific precursors regulate the first stage of sexual development, I speculate that the precursors are the regulators and discuss recent observations in relation to this speculation.

A. Organisms and Developmental Events

There are approximately 300 species of mucoraceous fungi belonging to the class zygomycetes [1,2]. Most mucorales are terrestrial saprophytes. Each species has two life cycles: an asexual or vegetative cycle involving large numbers of sporangiospores for efficient dispersal of the organism, and a sexual cycle involving highly resistant zygospores, presumably designed to withstand unfavorable seasons. Some species (including *Phycomyces blakesleeanus*, *Blakeslea trispora*, and *Mucor mucedo*) are heterothallic, meaning they require a mate to form zygospores; other species are homothallic, meaning they form zygospores by themselves [3].

P. blakesleeanus contains about seven times more DNA per nucleus than *E. coli* [4]. Its asexual cycle involves mycelial and vegetative spore stages (Fig. 1). The mycelium is a mass of multibranched, nonseptate (coenocytic) hyphae which grow terminally in, at, and above the surface of solid medium. The hyphae are yellow due to the presence of β-carotene [5,6]. The hyphal cell wall contains chitin, poly-N-acetylglucosamine [7]; its synthesis apparently involves the interaction of chitin synthetase, several proteases, and inhibitors of proteases [8], as has been described for yeast [9]. The initiation of sporangiophore development occurs throughout the mycelium when cultures are grown in continuous light or continuous

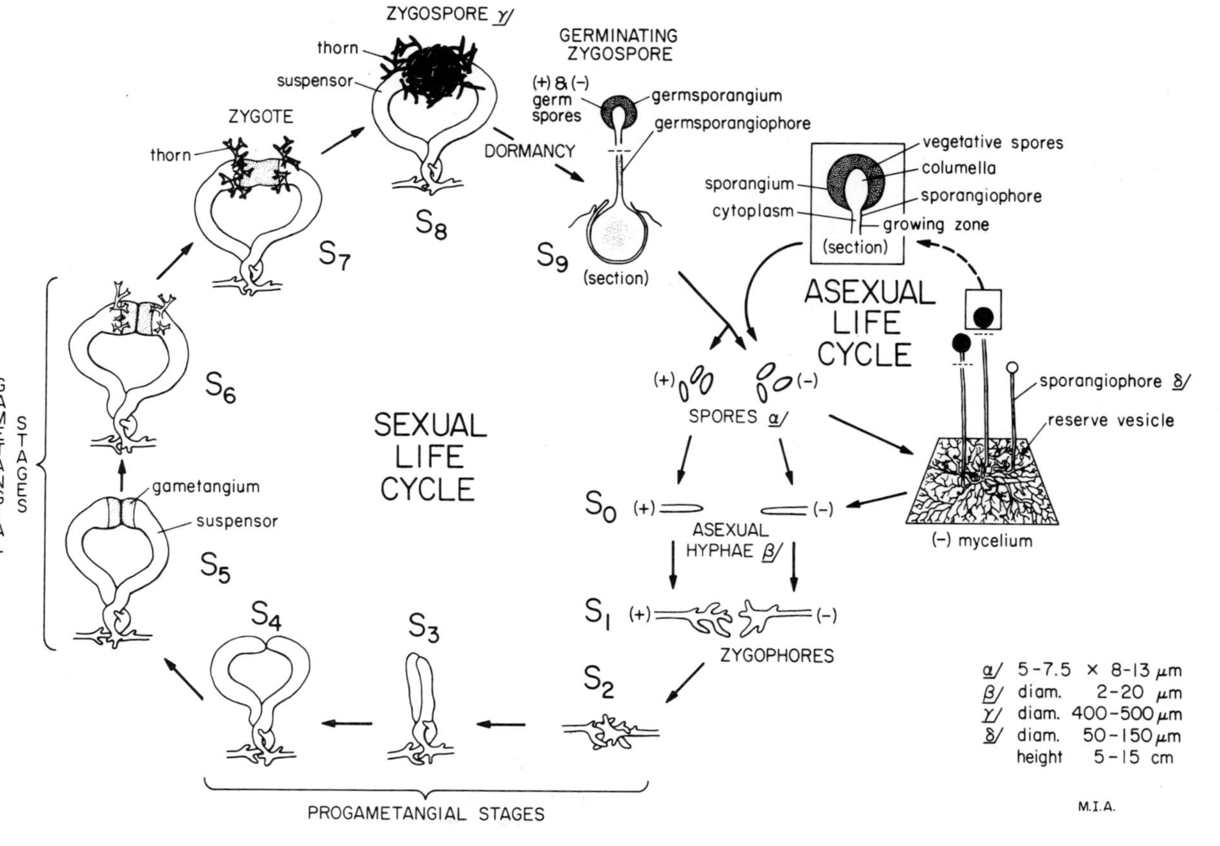

FIG. 1. Life cycle of *Phycomyces blakesleeanus*.

dark but, in a light-dark cycle, only where the mycelium grew in the light [10]. When a sporangiophore is 1 to 3 cm tall, elongation ceases temporarily and its tip swells to form a sporangium. Nuclei, which apparently are transported from the ground mycelium by protoplasmic streaming, migrate to the outer edge of the sporangium [11]. Spore walls enclose groups of one to six (haploid) nuclei within the sporangium [12] after a cell wall delimits the columella. Elongation of the sporangiophore then resumes with the growing zone about 0.1 to 2.5 mm below the sporangium [14]. Sporangiophores grow toward light [15] and avoid objects [16]. Spores, numbering between 50,000 and 100,000, are not released from a sporangium until contacted by another object [2]. Vegetative spores germinate readily on rich medium; on minimal medium, spore germination usually requires prior activation by incubation at 48° for 15 min [17] or treatment with chemicals [18]. The complete vegetative cycle takes 3 to 4 days.

The sexual cycle begins when vegetative hyphae of opposite mating types, designated (+) and (-), grow near each other (Fig. 1). Zygophores--short, stubby, multibranched hyphae--develop underneath the surface of solid medium prior to physical contact between the two sexes (S_1) [19]. Zygophores of (+) and (-) sexes are morphologically indistinguishable. Zygophores of opposite sexes grow toward each other, sometimes changing the direction of growth from that followed by the former asexual hyphae. As soon as zygophores come into contact, the short, multibranched hyphae interlock to form the first progametangial stage (S_2). The paired zygophores then become enlarged, breaking through the surface of solid medium, appearing as knobby knots. Enlarged paired zygophores develop stout, yellow, paired pillars which project vertically above the surface of the medium (S_3). The paired pillars separate, except at their base and apex, and enlarge to form a loop (S_4). Septa form in both halves of the loop to delimit gametangia and suspensor cells (S_5). Thorns, translucent initially, develop on the suspensor cells (S_6). Minute aerial hyphae from the ground mycelium surround the paired gametangia [20]. The cell walls separating gametangia dissolve from the center outwards to form a translucent zygote, surrounded by dark thorns (S_7). Each gametangium contributes thousands of haploid nuclei to the zygote [21]. The zygote enlarges and develops a thick, hardened black wall to become a zygospore, surrounded by black thorns (S_8). Zygospore dormancy, a polygenic trait [22], varies with strains [23]. Zygospores of crosses between UBC21(+) and NRRL1555(-) germinate about 60 days from the beginning of the sexual cycle to form germsporangia (S_9) [22]. A single germsporangium contains 7,000 to 15,000 germspores, each with one to six haploid nuclei. The bulk of the germspores are homocaryotic with respect to sex; however, about 4% are heterocaryotic. Heterosexual heterocaryons do not produce zygospores by themselves, but do produce pseudophores, abnormal zygophores, which may sometimes develop into sporangiophores [24]. Analysis of 84 fertile germsporangia from a four-factor cross involving two auxotrophic, one color, and sex marker indicates that germspores of a single germsporangium are derived from 1, 2, and 3 meioses in 87, 21, and 1% of the cases [25]. The four markers are on separate chromosomes; each marker is located about 15 map units from its centromere. The total number of chromosomes in *Phycomyces* is unknown [11,26].

B. Historical Background

Blakeslee, who discovered sexuality in mucoraceous fungi in 1903, studied sexual development in over 35 heterothallic species, testing between 10,000 and 20,000 isolates or races [3,27]. Sexual vigor varied among cultures, causing some races to appear neutral. However, evidence for sex intergrades was not found. When Blakeslee and Cartledge [28] tested every possible cross between (+) and (-) mating types of 25 species, they found that (1) zygospores developed in all of the (+/-) intraspecies crosses; (2) incomplete development (zygophores and progametangia) occurred in 306 of the (+/-) interspecies crosses; (3) no sexual reaction occurred in 294 (+/-) interspecies crosses; and (4) no sexual reaction occurred in any of the interspecies crosses between the same mating types. The extremes of behavior among the strains tested were (1) the (-) mating type of *Mucor H*, which reacted with the (+) mating type of every species; and (2) the (+) mating type of *P. blakesleeanus*, which failed to react with the (-) mating type of any species except *P. blakesleeanus*, *Mucor H*, and *Mucor mucedo*. These observations suggest that the initial stages of sexual development may be common to all species of mucoraceous fungi.

Burgeff [29] reported that (+) and (-) cultures of *M. mucedo*, separated by a cellodin membrane, developed zygophores which grew toward their counterpart of the opposite sex. His experiment demonstrated that diffusible and/or volatile substances pass between the sexes to stimulate the formation of zygophores and to direct their growth. His experiment provided the first evidence for fungal sex hormones.

Both Blakeslee [3] and Burgeff [29] observed that the sexual structures of a number of species, including *P. blakesleeanus*, were more yellow than the asexual hyphae. Barnett et al. [30] reported that (+/-) cultures of *Choanephora cucurbitarum* accumulate 15 to 20 times more β-carotene than separate (+) and (-) cultures; and sex-stimulated carotenogenesis occurs in both mating types even though separated by a cellophane membrane. Soon afterwards Hesseltine and Anderson [31] found that all members of the family choanephoraceae, including *B. trispora*, exhibit sex-stimulated carotenogenesis. Prieto and co-workers [32] reported in 1964 that trisporic acids stimulate carotenogenesis in separate mating type cultures of the choanephoraceae. They found trisporic acids in the culture medium of mated, but not unmated, cultures of *B. trispora*. Trisporic acids are 18-carbon atom derivatives of vitamin A containing three double bonds conjugated with a keto group. (Trisporic acid C is 1,3-dimethyl-4-oxo-2-(3'-methyl-7'hydroxy-octa-1',3'-dienyl)cyclohex-2-en-1-oic acid [33]. Trisporic acid B has a keto group instead of a hydroxyl group at 7' [34]. The unsaturation in each acid may be either 1'-trans-3'-trans or 1'-trans-3'-cis [35,36]. The absolute stereochemistry of trisporic acid C is 1 S, 7'R [35,37]. Trisporic acids are unstable: they lose biological activity and exhibit altered UV absorbance spectra when exposed to UV radiation and/or oxygen [33,38-40]. Trisporic acids, which are biosynthesized from β-carotene and retinol [41], have been synthesized chemically [42, 43].)

In the meantime, Plempel [44,45], a student of Burgeff's, found that the (+) mating type of *M. mucedo* makes a "gamone" (hormone) which stimulates zygophore formation in the (-) mating type, and the (-) mating type makes a "gamone" (hormone) which stimulates zygophore formation in the (+) mating type. Plempel attempted to purify the two hormones from (+/-) culture medium of *M. mucedo*. Instead, he isolated a single compound in crystalline form which stimulates zygophore formation in both mating types. Van den Ende [46,47] then demonstrated that trisporic acid B and trisporic acid C, when tested separately in the *Mucor* bioassay, stimulated zygophore formation equally well in both (+) and (-) mating types. (In contrast, the methyl esters of trisporic acids stimulated zygophore formation selectively in the (-) mating type [35,36].) Trisporic acid B was three times more active than trisporic acid C [47]; the more polar isomer of cis-trans trisporic acid B was twice as active as the less polar isomer [36]; and as little as 20 ng of an unresolved mixture of trisporic acids (about 90% trisporic acid C) stimulated zygophore formation in both mating types of *M. mucedo* [48, 49].

II. CURRENT RESEARCH

A. Trisporic Acid Biosynthesis in *B. trispora*

How do the two mating types of *B. trispora* cooperate to form trisporic acids? Preliminary experiments, in which combined (+/-) cultures were grown 5 days on a gyratory shaker in flasks with liquid medium containing potato extract and glucose, revealed (1) mated cultures accumulated 130 mg of trisporic acids (and 14 mg of β-carotene) per liter of medium [39,50,51]; (2) trisporic acid accumulation is directly proportional to the amount of mycelium of either mating type in cultures containing excess mycelium of the opposite mating type [47,50,52]; (3) physical contact between mycelia of opposite mating types greatly increases the rate of trisporic acid accumulation [50]; and (4) cycloheximide, an inhibitor of protein synthesis, prevents the initiation of trisporic acid biosynthesis in young cultures and completely stops ongoing trisporic acid biosynthesis within 10 to 15 h in older (solid mat) cultures [51].

Do both mating types synthesize trisporic acids? Yes. Trisporic acids are made when mycelium of one mating type is incubated with filter-sterilized culture medium of the opposite mating type [49,53,54]. The trisporic acids can be detected by (1) *Mucor* bioassay analysis of the culture medium (or acid fraction); (2) UV absorbance analysis of the acid fractions, but not culture medium (Table 1). The validity of these screening procedures was verified by extracting the putative trisporic acids from large quantities of culture medium and purifying them by LH-20 Sephadex, DEAE-Sephadex, and silica gel chromatography. The putative and authentic trisporic acid, cochromatographed as a single spot in four separate solvent systems, exhibited identical UV absorbance spectra and similar biological activities in the *Mucor* bioassay [49].

Are the trisporic acid-stimulating (TAS) and zygophore-stimulating components, detected in the medium of separate mating type cultures, soluble in organic solvents? Yes, the biological activity is extracted with neutral, not acidic, compounds. Of the total TAS activity in filter-sterilized (+) culture medium, 90% is recovered in the neutral fraction if the pH of the culture medium is not adjusted prior to extraction [55], but only 40% of it is adjusted to pH 2 [49].

The amount of TAS activity in the neutral fractions isolated from 5-day cultures is shown in Table 2. TAS activity is measured as trisporic acid B and trisporic acid C after incubating the neutral fractions with cultures of the opposite mating type, and is expressed as a percentage of the trisporic acids isolated from the medium of 5-day (+/-) cultures. Although separate (+) and (-)

TABLE 1

Bioassay and UV-absorbance Analyses
of the Medium Isolated from Cultures of *B. trispora*

Incubation[a]		*Mucor* bioassay[b] zygophores produced		UV analyses[c] (A_{325} units/ml)	
Culture medium	Mycelium added	(+)	(-)	CM	AF
(+)	None	No	Yes	5.6	1.2
(+)	(-)	Yes	Yes	6.0	5.8
(-)*	(-)	No	No	---	---
(-)	None	Yes	No	5.4	0.8
(-)	(+)	Yes	Yes	5.4	1.3
(+)*	(+)	No	No	---	---
(+/-)	None	Yes	Yes	19.6	230

[a]Culture media (125 ml) were incubated for 0.5 days with and without mycelia. Culture media marked with * were from 1.5-day cultures; rest from 5-day cultures. Mycelia were from 1.5-day cultures.

[b]Tenth milliliter of filter-sterilized culture medium tested [49,56].

[c]CM, culture medium; AF, acid fraction (in 5 ml ethanol) [53].

TABLE 2

Trisporic Acids Isolated from the Culture Medium of *B. trispora*

Cultures[a]	Additions[b]	Trisporic acids[c]
(+/-)*	None	100%
(-)	(+)NF	1%
(-)	None	Nil
(+)	(-)NF	0.4%
(+)	None	Nil

[a]Cultures: 1.4-day except *, 5-day. [b]Neutral fractions from 5-day culture media adjusted to pH 2 prior to extraction. [c]50 mg trisporic acid C and trisporic acid B per liter of (+/-) culture medium [49].

cultures synthesize 1% or less of the trisporic acids synthesized by (+/-) cultures, the (-) cultures, for example, make over 25 times more trisporic acid per milliliter than is needed for a minimal response in the *Mucor* bioassay.

Are the TAS components in the neutral fractions precursors of trisporic acids? Yes, there are four types of evidence for this conclusion. First, cycloheximide does not inhibit trisporic acid biosynthesis when neutral fractions are added to opposite mating type cultures [49,53,54]. This observation also implies that the enzymes which convert the precursors to trisporic acids are constitutive. Second, the amount of trisporic acids formed in (-) cultures is directly proportional to the amount of (+) neutral fraction added [49]. Third, TAS components disappear from the medium when trisporic acids are formed as judged both by thin layer chromatography [53] and bioassay [49] data. Fourth, labeled trisporic acid

TABLE 3

Trisporic Acid Biosynthesis in *B. trispora* By Way of Mating Type-Specific Precursors

Culture	Additions[a]	Trisporic acid analyses[b]	
		mg	cpm
(+)	C^{14}-(+)NF	Nil	Nil
(-)	C^{14}-(+)NF	1.25	21,875
None	C^{14}-(+)NF	Nil	Nil
(+)	C^{14}-(-)NF	0.59	7,434
(-)	C^{14}-(-)NF	Nil	Nil
None	C^{14}-(-)NF	Nil	Nil

[a]Neutral fractions, partially purified, were isolated from the medium of cultures grown in [^{14}C]glucose.

[b]Separated trisporic acid B and trisporic acid C were analyzed: 20% TAB, 80% TAC; specific activities were identical [55].

B and trisporic acid C are made by separate (+) and (-) cultures incubated with labeled extracts isolated from the medium of cultures of the opposite mating type (Table 3); unlabeled trisporic acids are made when the cultures are incubated with [^{14}C]glucose and unlabeled extracts. Over 100-fold less trisporic acids are made when cultures are incubated with labeled extracts isolated from the medium of cultures of the same mating type. Thus, the precursors are mating type-specific.

Trisporic acid biosynthesis by way of mating type-specific precursors is comprehensible by assuming that several of the steps between β-carotene and trisporic acids are sex-linked while the remaining steps are common to both sexes. This pathway implies that the enzymes in trisporic acid biosynthesis have group rather than molecule specificity. These considerations, the common mechanism of sexuality in mucoraceous fungi [3,28], the inability of caroteneless mutants of *P. blakesleeanus* to stimulate zygophore formation in a wild-type mate (see below), and the incorporation of retinyl acetate into trisporic acids in mated cultures of *B. trispora* [57], led to the schema for trisporic acid biosynthesis in mucoraceous fungi shown in Fig. 2 [58].

The structures of the mating type-specific precursors of trisporic acids in *M. mucedo* were determined by Nieuwenhuis and van den Ende [59] and are shown in Fig. 3. Trisporins (M) and 4-hydroxymethyltrisporates (P) stimulate zygophore formation and trisporic acid biosynthesis in (+) and (-) cultures, respectively. (Bu'Lock and co-workers [60], using the *Mucor* bioassay to identify mating type-specific compounds, had previously isolated and chemically characterized trisporins and trisporols from (-) cultures of *B. trispora* and 4-hydroxymethyltrisporates from (+) cultures. Trisporols, however, were reported to stimulate trisporic acid biosynthesis in the (-) mating type of *B. trispora* over 50% as well as in the (+) mating type [36].)

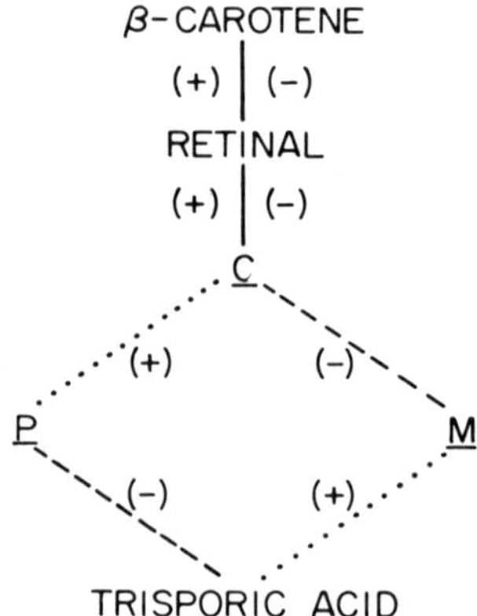

FIG. 2. Hypothetical schema for trisporic acid biosynthesis in mucoraceous fungi. The solid lines represent enzymatic reactions common to both mating types; the dotted and dashed lines represent enzymatic reactions unique to (+) and (-) mating types, respectively.

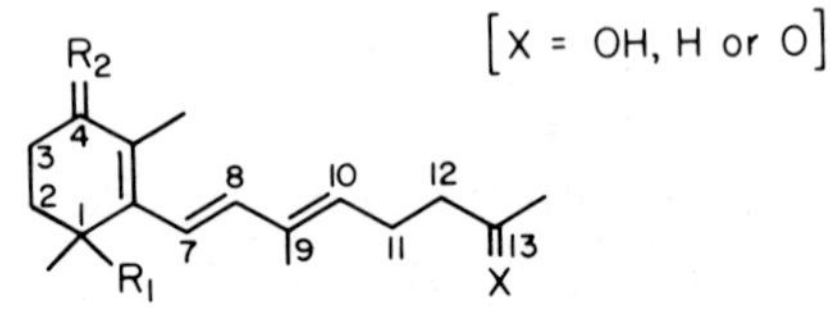

[X = OH, H or O]

COMPOUND	R_1	R_2
C	CH_3	OH, H
P	$COOCH_3$	OH, H
M	CH_3	O
TRISPORIC ACID	COOH	O

FIG. 3. Structures of P (4-hydroxymethyltrisporates) and M (trisporins) isolated from *Mucor mucedo* [59], C (a hypothetical intermediate made by both mating types), and trisporic acids. The numbering system of carbon atoms follows the β-carotene system.

B. Regulation of Sexual Development: A Working Hypothesis

It is not known whether trisporic acids or their mating type-specific precursors regulate the first stage of sexual development in mucoraceous fungi.

I speculate that the mating type-specific precursors (or their pretrisporate metabolites) regulate zygophore formation, sex-stimulated carotenogenesis, and trisporate biosynthesis. According to this viewpoint, trisporic acids are inactive metabolites of the regulatory compounds. This viewpoint is appealing because (1) regulatory molecules, such as cyclic AMP [61] and acetylcholine [62], are converted to inactive compounds during the regulatory process; and (2) the accumulation of large amounts of trisporic acids in the medium of cultures, such as *P. blakesleeanus* and *B. trispora* (see below), during normal sexual development becomes understandable. The organisms which are about to form zygospores, a dormant state, are discarding inactive, unneeded molecules. This viewpoint implies that cultures such as *M. mucedo*, which form zygophores in response to low concentrations of trisporic acid, have enzymes which convert trisporic acids to the regulatory compounds. See Sec. II.C for some recent observations in relation to this working hypothesis.

C. Critiques

Zygophore formation

Each mating type makes a chemical messenger which stimulates zygophore formation in the opposite mating type. Recent observations supporting this conclusion come from: crosses of wild-type and *car* mutants of *P. blakesleeanus* in which the mutant is unable to stimulate zygophore formation in its mate (Table 4); crosses of *M. mucedo* in which (1) (-) mutants, unable to form zygophores in response to trisporic acids, form zygophores when crossed with (+) wild type [63]; and (2) (+) and (-) wild types form zygophores when separated by an air space, suggesting that the volatile mating type-

TABLE 4

First Stage of Sexual Development in *P. blakesleeanus*

	Zygophores[b]		Carotenes[c]		Trisporic acids[d] (mg/2 liter)	
Cultures[a]	(+)	(-)	β-Carotene	Lycopene	Medium	Mycelium
wt × wt	Yes	Yes	2.0	Nil	12.0	0.24
wt × m	No	Yes	0.3	3.3	0.8	<0.01
m × wt	Yes	No	3.1	Nil	0.7	<0.01
m × m	No	No	0.6	0.4	Nil	Nil

[a]Wild types, wt; mutants, m.

[b]See [19] for mutants tested.

[c]Levels of carotenes in mating zone relative to the asexual mycelium, which is assigned a value of 1. There were 30, nil, 1470 μg β-carotene and nil, 1060, nil micrograms lycopene per gram mycelium in wild types, *carR21*(-), and *car-43*(+), respectively [70].

[d]Cultures [wild types, *carA12 carR27 mad-48*(-), and *car-43*(+)] were grown for 3.5 days at 20°.

specific precursors, rather than the nonvolatile trisporic acids, initiate their formation [64]. (Note: trisporins and 4-hydroxymethyltrisporates are *not* the zygotropic agents in *M. mucedo* [65].)

It has been assumed that trisporic acids (once formed from their precursors) stimulate zygophore formation in all mucoraceous fungi because a common regulatory mechanism appears to control the early stages of sexual development in all mucorales, and nanogram quantities of trisporic acids stimulate zygophore formation in both (+) and (-) mating types of *M. mucedo* [66-69]. However, 1,000 to 10,000 times more trisporic acids are needed to stimulate zygophore formation in both (+) and (-) mating types of *P. blakesleeanus* [19, 51]. *P. blakesleeanus* either lacks the enzymes required to convert trisporic acids to the zygophore-initiating compounds or has a hyphal membrane which is impermeable to trisporic acids. Similar explanations are applicable to (-) mutants of *M. mucedo* which do not make zygophores in response to trisporic acids but do form zygophores when crossed with the (+) wild type [63].

Sex-stimulated carotenogenesis

Studies with wild types and *car* mutants of *P. blakesleeanus* demonstrate that each mating type makes a chemical messenger which stimulates carotenogenesis in the opposite mating type (Table 4). Although Prieto and co-workers [32] reported that trisporic acids stimulate carotenogenesis in separate mating type cultures of the choanephoraceae, subsequent investigators [46,50,71,72] found that trisporic acids stimulate carotenogenesis selectively in the (-) mating type of *B. trispora*. (The significance of this selectivity is not known because isogenic mating types were not used in any of the studies.) What is more interesting, however, is the report of Sebek and Jaeger [71] that trisporic acids do not stimulate carotene synthesis in other mucoraceous fungi (such as *P. blakesleeanus*, *M. hiemalis*, *Cunninghamella blakesleeana*, and *Absidia glauca*). Whitaker [73] has confirmed that trisporic acids do not stimulate carotenogenesis in *P. blakesleeanus*.

Trisporic acid biosynthesis

Large quantities of trisporic acids (>100 mg/liter) accumulate in the medium of (+/-) cultures of *B. trispora* grown 6 days in (defined) liquid medium on a gyratory shaker at 25 ± 3°. Negligible quantities of trisporic acids (<1 mg/liter) accumulate in the medium of (+/-) cultures of other mucoraceous fungi, such as *P. blakesleeanus* and *M. mucedo*, grown under the same conditions. However, 20 mg/liter of trisporic acids accumulate in the medium of (+/-) cultures of *P. blakesleeanus* [74] and *B. trispora* [75] when the cultures are grown in liquid medium in petri dishes.

Both mating types in (+/-) cultures must be able to synthesize precursors or the yield of trisporic acids drops to 10% or less. This has been demonstrated with crosses of wild-type and *car* mutants of *P. blakesleeanus* (Table 4) and with fluorouracil pretreated wild-type mycelia of *B. trispora* [76].

Trisporic acids stimulate (derepress) M and P syntheses 10- to 15-fold in separate mating type cultures of *B. trispora* [55,77] and *M. mucedo* [78]. However, large excesses of trisporic acids were added to cultures for the amounts of precursors formed. For example, 16 μg of trisporic acids were added to (-) *B. trispora* for every microgram of M synthesized [55]. These facts suggest that trisporic acids are not the active molecules in derepressing P and M syntheses, or/and P and M inhibit their own syntheses.

The accumulation of P and M in the medium of separate (+) and (-) cultures of *B. trispora* is dependent on the composition of the medium in which the mycelium is grown. For example, we obtained 1 mg of P per liter of (+) culture medium with one lot of potato extract [49] but only 0.1 mg with the next lot [51]. To insure reproducible results, I developed a defined medium (see [19] for composition, except 0.2% MSG with 0.02% $aspNH_2$). One mg of P accumulates when (+) *B. trispora* is grown in this medium [51]. Less than 0.05 mg of P accumulates if either 0.2% yeast extract or 0.2% asparagine is added to the medium. These observations suggest that Plempel's progamones [45] may be artifacts of experimental design because the cultures were grown on a nitrogen-rich medium, and a critical set of controls was neglected. Furthermore, data leading to the speculation that π and μ stimulate M and P synthesis are also explicable by assuming that P stimulates M synthesis and M stimulates P synthesis [49].

Although trisporic acids are formed when (+) and (-) mating types of *B. trispora* are separated by a membrane filter, the extent of synthesis was not reported [46]. Recently, Jelinek [75] found that filter-separated (+) and (-) mycelia of *B. trispora* synthesize only 1.5% of the trisporic acids synthesized by combined (+/-) cultures. Similar percentages were obtained when the cultures were incubated for 72 and 168 h even though the absolute amount of trisporic acids in (+/-) cultures increased 2.4-fold with time. Furthermore, the medium of controls of separate (+) and (-) cultures contained precursors for 0.5% of the trisporic acids synthesized by combined (+/-) cultures. (It was not possible to use cycloheximide to determine if the 1% differential in trisporic acid biosynthesis between filter-separated mycelia and control cultures was due to derepression of P and M biosyntheses because cycloheximide also inhibits "basal level" P synthesis [79].) The absence of extensive derepression of trisporic acid biosynthesis in cultures with mating types separated by a membrane filter suggests that other components, which cannot pass through a 0.45-μm pore-size Gelman Acropor filter, may be required for derepression of trisporic acid biosynthesis, or/and P and M--present in low levels--are inactivated by conversion to trisporic acids before they are able to derepress M and P biosyntheses, respectively. (Others had speculated that each mating type of (+/-) *B. trispora* synthesizes trisporic acids de novo after derepression [80]. The speculation is based, in part, on the finding that 5-day (+) cultures of *B. trispora*, grown without additives from (-) cultures, accumulate 0.1% of the trisporic acids made by (+/-) cultures [49,54]. Although the speculation is now considered unlikely [69], it has not been disproved experimentally.)

III. FINAL COMMENTS

We plan to isolate mutants of *P. blakesleeanus*, which are blocked in trisporate biosynthesis between the mating type-specific precursors and trisporic acids. With the mutants we plan to demonstrate which compounds--the mating type-specific precursors or trisporic acids--regulate zygophore formation, sex-stimulated carotenogenesis, and trisporate biosynthesis. This demonstration will complete the long-developing story on which compounds regulate the first stage of sexual reproduction in mucoraceous fungi.

These studies may also shed light on the role of secondary metabolites in the life of microorganisms and plants. The raison d'être of secondary metabolites is unknown [81]. Trisporic acids are secondary metabolites. If trisporic acids are inactive products in the regulation of the first stage of sexual development in mucoraceous fungi, it is not unreasonable to suggest that secondary metabolites in other organisms may be inactive end products of as yet unrecognized regulatory mechanisms.

Studies on the mechanism of action of compounds regulating the first stage of sexual development in mucoraceous fungi may shed light on the role of vitamin A in higher organisms. Besides its role in vision, vitamin A is necessary for normal growth, development, and sexual reproduction [82]. How vitamin A affects growth, development, and reproduction is not known. Nor is the biologically active form of vitamin A mediating these effects known. The recent discovery of "trisporate-like" compounds in the urine of rats fed high doses of vitamin A [83] makes studies on the mechanism of action of compounds regulating the first stage of sexual development in mucoraceous fungi all the more exciting and important.

ACKNOWLEDGMENTS

I am grateful to my co-workers at WVU over the last 7 years for their enthusiasm and assistance. They are D. A. Capage, R. N. DeHaven, G. Galasko, T. L. Harrison, B. G. Jelinek, P. A. Kaufman, W. A. Keen, D. A. Mastrorocco, D. L. Petrone, J. M. Swadley, and J. P. Whitaker. Our research has been supported in part by NSF.

This paper is dedicated to Max Delbrück and Virgil Greene Lilly, two outstanding teachers.

REFERENCES

1. Hesseltine, C. W., and J. J. Ellis, 1973. *In* The Fungi, Vol. 4B, G. C. Ainsworth, F. K. Sparrow, and A. S. Sussman (eds.), Academic Press, New York, p. 187.
2. Ingold, C. T., 1971. Fungal Spores: Their Liberation and Dispersal. Clarendon Press, Oxford.
3. Blakeslee, A. F., 1904. Proc. Amer. Acad. Arts Sci. 40: 205-319.
4. Dusenbery, R. L., 1975. Biochem. Biophys. Acta 378: 363-377.
5. Garton, G. A., T. W. Goodwin, and W. Lijinsky, 1951. Biochem. J. 48: 154-163.
6. Lilly, V. G., H. L. Barnett, and R. F. Krause, 1960. West Virginia University Agricultural Experiment Station Bulletin 441T.

7. Bartnicki-Garcia, S., 1968. Ann. Rev. Microbiol. 22: 87-108.

8. Thomson, K. S., and E. P. Fisher, personal communication.

9. Holzer, H., H. Betz, and E. Ebner, 1975. *In* Current Topics in Cellular Regulation, Vol. 9, B. L. Horecker and E. R. Stadtman (eds.), Academic Press, New York, p. 103.

10. Bergman, K., 1972. Planta 107: 53-67.

11. Bergman, K., P. V. Burke, E. Cerdá-Olmedo, C. N. David, M. Delbrück, K. W. Foster, E. W. Goodell, M. Heisenberg, G. Meissner, M. Zalokar, D. S. Dennison, and W. Shropshire, 1969. Bacteriol. Rev. 33: 99-157.

12. Heisenberg, M., and E. Cerdá-Olmedo, 1968. Mol. Gen. Genet. 102: 187-195.

14. Ortega, J. K. E., R. I. Gamow, and C. N. Ahlquist, 1975. Plant Physiol. 55: 333-337.

15. Delbrück, M., and W. Reichardt, 1956. *In* Cellular Mechanisms in Differentiation and Growth, D. Rudnick (ed.), Princeton University Press, New Jersey, p. 3.

16. Cohen, R. J., Y. N. Jan, J. Matricon, and M. Delbrück, 1975. J. Gen. Physiol. 66: 67-95.

17. Rudolph, H., 1960. Planta 55: 424-437.

18. Delvaux, E., 1973. Arch. Mikrobiol. 88: 273-284.

19. Sutter, R. P., 1975. Proc. Nat. Acad. Sci. USA 72: 127-130.

20. Schröder, W. H., personal communication.

21. Cutter, V. M., 1942. Bull. Torrey Botan. Club 69: 592-616.

22. Eslava, A. P., M. I. Alvarez, P. V. Burke, and M. Delbrück, 1975. Genetics 80: 445-462.

23. Goodell, E. W., personal communication.

24. Blakeslee, A. F., 1906. Ann. Mycol. 4: 1-28.

25. Eslava, A. P., M. I. Alvarez, and M. Delbrück, 1975. Proc. Nat. Acad. Sci. USA 72: 4076-4080.

26. Cerdá-Olmedo, E., 1974. *In* Handbook of Genetics, Vol. 1, R. C. King (ed.), Plenum Press, New York, p. 343.

27. Blakeslee, A. F., J. L. Cartledge, D. S. Welch, and A. D. Bergner, 1927. Bot. Gaz. 84: 27-50.

28. Blakeslee, A. F., and J. L. Cartledge, 1927. Bot. Gaz. 84: 51-57.

29. Burgeff, H., 1924. Botan. Abh. 4: 1-135.

30. Barnett, H. L., V. G. Lilly, and R. F. Krause, 1956. Science 123: 141.

31. Hesseltine, C. W., and R. F. Anderson, 1957. Mycologia 49: 449-452.

32. Prieto, A., C. Spalla, M. Bianchi, and G. Biffi, 1964. Commun. Internatl. Fermentation Symp., London, Abstract, p. 38.

33. Caglioti, L., G. Cainelli, B. Camerino, R. Mondelli, A. Prieto, T. Salvatori, and A. Selva, 1966. Tetrahedron Suppl. 7: 175-187.

34. Cainelli, G., P. Grasseli, and A. Selva, 1967. Chim. Ind. 49: 628-629.

35. Reschke, T., 1969. Tetrahedron Letters 39: 3435-3439.

36. Bu'Lock, J. D., D. Drake, and D. J. Winstanley, 1972. Phytochemistry 11: 2011-2018.

37. Bu'Lock, J. D., D. J. Austin, G. Snatzke, and L. Hruban, 1970. Chem. Commun.: 255-256.

38. van den Ende, H., 1967. Nature 215: 211-212.

39. Sutter, R. P., 1970. J. Gen. Microbiol. 64: 215-221.

40. Kaufman, P. A., 1970. M. S. Thesis, West Virginia University.

41. Bu'Lock, J. D., B. E. Jones, D. Taylor, N. Winskill, and S. A. Quarrie, 1974. J. Gen. Microbiol. 80: 301-306.

42. Edwards, J. A., V. Schwarz, J. Fajkos, M. L. Maddox, and J. H. Fried, 1971. Chem. Commun.: 292-293.

43. Isoe, S., Y. Hayase, and T. Sakan, 1971. Tetrahedron Letters 40: 3691-3694.

44. Plempel, M., 1957. Arch. Mikrobiol. 26: 151-174.

45. Plempel, M., 1963. Planta 59: 492-508.

46. van den Ende, H., 1968. J. Bacteriol. 96: 1298-1303.

47. van den Ende, H., A. H. C. A. Weichman, D. J. Reyngoud, and T. Hendricks, 1970. J. Bacteriol. 101: 423-428.

48. Austin, D. J., J. D. Bu'Lock, and G. W. Gooday, 1969. Nature 223: 1178-1179.

49. Sutter, R. P., D. A. Capage, T. L. Harrison, and W. A. Keen, 1973. J. Bacteriol. 114: 1074-1082.

50. Sutter, R. P., and M. E. Rafelson, 1968. J. Bacteriol. 95: 426-432.

51. Sutter, R. P., unpublished data.

52. Bu'Lock, J. D., B. E. Jones, S. A. Quarrie, and D. J. Winstanley, 1974. Arch. Microbiol. 97: 239-244.

53. Sutter, R. P., 1970. Science 168: 1590-1592.

54. Sutter, R. P., D. A. Capage, and T. L. Harrison, 1971. First Internatl. Mycological Congress, Exeter, Abstract, p. 92.

55. Sutter, R. P., T. L. Harrison, and G. Galasko, 1974. J. Biol. Chem. 249: 2282-2284.

56. Capage, D. A., 1971. M.S. Thesis, West Virginia University.

57. Austin, D. J., J. D. Bu'Lock, and D. Drake, 1970. Experientia 26: 348-349.

58. Delbrück, J., and R. P. Sutter, 1973. Phycomyces Summer Work-Shop Report, Cold Spring Harbor, N.Y., pp. 28-29.

59. Nieuwenhuis, M., and H. van den Ende, 1975. Arch. Mikrobiol. 102: 167-169.

60. Bu'Lock, J. D., B. E. Jones, and N. Winskill, 1974. J.C.S. Chem. Comm. 708-709.

61. Robison, G. A., R. W. Butcher, and E. W. Sutherland, 1968. Ann. Rev. Biochem. 37: 149-174.

62. Nachmansohn, D., 1970. Science 168: 1059-1066.

63. Wurtz, T., and H. Jockusch, 1975. Develop. Biol. 43: 213-220.

64. Mesland, D. A. M., J. G. Huisman, and H. van den Ende, 1974. J. Gen. Bicrobiol. 80: 111-117.

65. van den Ende, H., personal communication.

66. van den Ende, H., and D. Stegwee, 1971. Botan. Rev. 37: 22-36.

67. Machlis, L., 1972. Mycologia 64: 235-247.

68. Gooday, G. W., 1974. Ann. Rev. Biochem. 43: 35-49.

69. Bu'Lock, J. D., 1975. *In* Genetics of Industrial Microorganisms, Second Internatl. Symp., MacDonald (ed.), Academic Press, London, in press.

70. Whitaker, J. P., and R. P. Sutter, 1975, manuscript in preparation.

71. Sebek, O. K., and H. K. Jaeger, 1966. Bact. Proc., Abstract, p. 12.

72. Thomas, D. M., R. C. Harris, J. T. O. Kirk, and T. W. Goodwin, 1967. Phytochem. 6: 361-366.

73. Whitaker, J. P., 1974. M.S. Thesis, West Virginia University.

74. Mastrorocco, D., 1975. M.S. Thesis in preparation, West Virginia University.

75. Jelinek, B. G., 1976. M.S. Thesis, West Virginia University.

76. van den Ende, H., B. A. Werkman, and M. L. van den Briel, 1972. Arch. Mikrobiol. 86: 175-184.

77. Werkman, T. A., and H. van den Ende, 1973. Arch. Mikrobiol. 90: 365-374.

78. Werkman, T. A., and H. van den Ende, 1974. J. Gen. Microbiol. 82: 273-374.

79. Swadley, J. M., 1970. M.S. Thesis, West Virginia University.

80. Bu'Lock, J. D., B. E. Jones, S. A. Quarrie, and N. Winskill, 1973. Naturwiss. 60: 550-551.

81. Weinberg, E., 1971. Perspectives Biol. Med. 14: 565-577.

82. Pitt, G. A. J., 1971. *In* Carotenoids, O. Isler (ed.), Birkhauser Verlag, Basel, p. 717.

83. Rietz, P., O. Wiss, and F. Weber, 1974. Vitamins and Hormones 32: 237-249.

STEROID INDUCTION OF DIFFERENTIATION: *ACHLYA* AS A MODEL SYSTEM

Paul A. Horgen

Department of Botany and Erindale College
University of Toronto
Mississauga, Ontario, Canada

I. INTRODUCTION

The fungus *Achlya* belongs to a group of eucaryotic microbes commonly known as water molds. In nature, these interesting creatures form a major part of the mycological flora of ponds, lakes, and rivers, where they grow on leaves, twigs, fruits, dead insects, and various other bits of organic matter. The genus is in a class of fungi known as the Oömycetes. Phylogenetically, this class differs from the other classes of true fungi. Differences in the chemistry of the cell wall [1], differences in the organization of certain biosynthetic pathways [2,3], differences in the mol wt of the ribosomal RNAs [4], and differences in the composition of certain chromosomal proteins [5] help establish the Oömycetes' evolutionary uniqueness.

To the developmental biologist, *Achlya* (and other Oömycetes) possesses certain characteristics that are of interest, especially for studying growth and morphogenesis. *Achlya bisexualis* has the fastest known growth rate reported for a eucaryote [6]. This means it is possible to obtain large quantities for biochemical analysis quickly and easily. Furthermore, it is possible to induce *Achlya* to differentiate either sexual or asexual structures in a highly synchronous manner. Development of asexual sporangia can be induced with Ca^{2+} [7] and, as LéJohn et al. have shown in an earlier article in this volume, this mode of development provides a useful system for studies of cell differentiation.

Sexual reproduction in most species of *Achlya* involves the formation of male sex organs (antheridia) and female sex organs (oögonia) on the same thallus (homothallic forms). *Achlya bisexualis* and *Achlya ambisexualis*, however, produce sex organs only when there is cooperation between two individuals of differing mating types (heterothallic forms). In his classic experiments, the late John R. Raper showed that sexual reproduction in these species was initiated and controlled by diffusible substances (hormones) that are reciprocally secreted by the sexually reacting partners [8]. For a detailed discussion of Raper's experiments and initial hypotheses, see the review by Barksdale [9]. Through the diligent work of Alma Barksdale, Trevor McMorris, and collaborators, two hormones involved in the mating reaction of *Achlya* have been isolated and chemically characterized [9,10].

A. Steroids and Mating

The following sequence of events is involved in sexual reproduction and mating in *Achlya* (Fig. 1):

1. Female strains continually secrete a sterol, antheridiol, which induces the production of sex organ initials (antheridial branches) in male strains. The hormone is active at concentra-

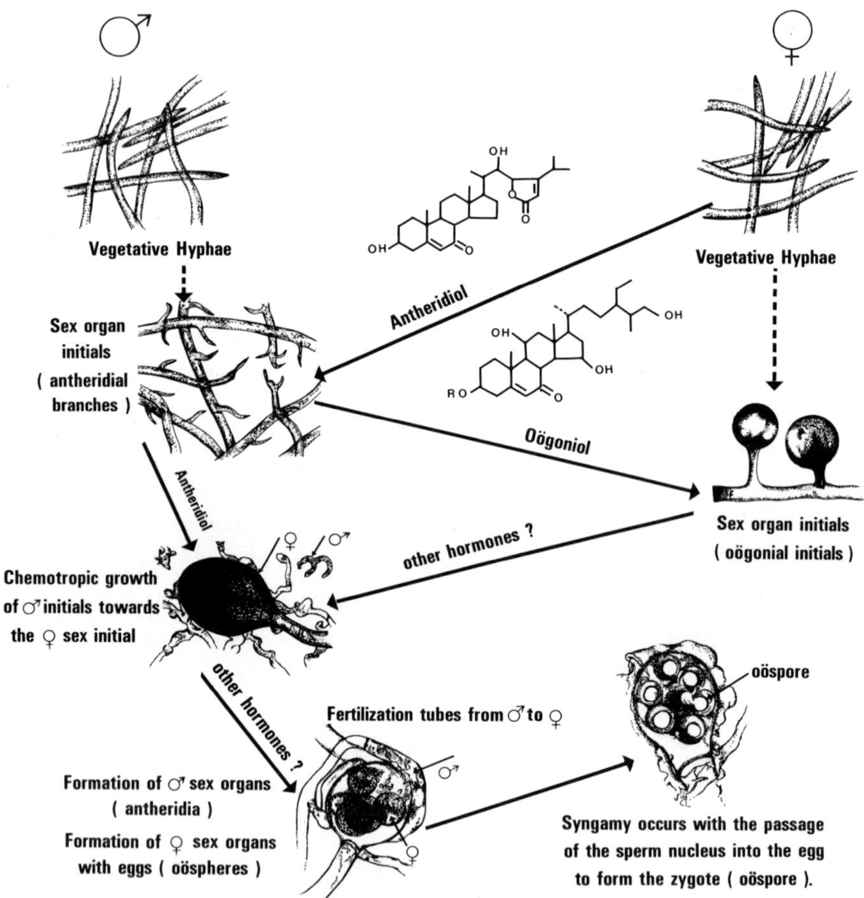

FIG. 1. Sexual hormones and mating in the water mold *Achlya*. Pen and ink drawings made from microscopic observations and photographs of Barksdale [9,13]. For an explanation of the diagram, see the text.

trations as low as 6×10^{-12} g/ml when assayed with *A. ambisexualis* E87.

2. In addition to the induction of sex organ initials, male strains of *Achlya* that have been exposed to antheridiol are stimulated to secrete a second hormone, oögoniol, which acts on the female mating partner by inducing the formation of female sex organ initials (oögonial initials).
3. As the female sex organ develops, it continually secretes antheridiol, causing the chemotropic growth of the male initials towards the female.
4. During the development of the female sex organ, a cross wall is formed, delimiting the oögonia from the remainder of the hyphae. Nuclei within the oögonia undergo meiosis [11,12] and the eggs (oöspheres) are eventually differentiated. The events involved in the differentiation of oögonia may involve only oögoniol [13] or, perhaps, other as yet uncharacterized hormones [9].
5. The male sex organs continue growing toward the oögonia and eventually become physically attached to the female sex organs. At a point during the morphogenesis of the antheridia, cross walls form which separate the sex organs from the remainder of the antheridial hyphae. Cross wall formation is another consequence of the action of antheridiol [13]. Nuclei within the antheridia undergo reduction division [14,15] to produce male gametic nuclei.
6. A fertilization tube forms between the antheridium and the oögonium; the male gametic nuclei are "injected" into the female organ [9] and the antheridial nuclei fuse with the oöspheres to produce zygotes (oöspores). The oöspores, under proper environmental conditions, will then germinate to produce new diploid individuals.

Achlya is perhaps the most primitive eucaryotic organism to produce and respond developmentally to steroid sex hormones. In more complex eucaryotic systems, one has to be concerned with discriminating between cells that produce and secrete the hormone, and cells that specifically respond both biochemically and morphologically to the hormone ("target cells"). With *Achlya*, the hormones (antheridiol and oögoniol) are produced and secreted by opposing sexual partners, so in a sense they are very similar to insect pheromones [16]. Furthermore, the entire thallus (either male or female) is the target tissue. The basic sterol backbone of both antheridiol and oögoniol is similar to mammalian sexual hormones (androgens and estrogens) and to ecdysone, an insect hormone. Thus, *Achlya* is an excellent system with which to investigate the basic action of steroid hormones.

For the last several years, we have been interested in examining the effects of antheridiol on the early stages of male sex organ morphogenesis. In the presence of 5×10^{-11} g/ml antheridiol, *A. ambisexualis* E87 will begin producing sex organ initials 3 to 4 h after the addition of hormone at 25°. We have been examining various aspects of biochemical morphogenesis during the first 8 h after the exposure of cultures to antheridiol.

II. CURRENT RESEARCH

A. Ribosomal RNA Synthesis

One of the early effects of antheridiol appears to be on the metabolism of cellular RNA. In the presence of the hormone, there is a dramatic stimulation of the incorporation of [^{3}H]uridine into total cellular RNA (Fig. 2A). This stimulation occurs almost immediately after the addition of the sterol. The antheridiol effect on total RNA is expressed as both an increase in the specific activity of the RNA and as an increase in the amount of total RNA extracted per milligram dry wt of tissue. The enhanced incorporation represents new RNA synthesis and not artifacts produced by fluctuations in the nucleotide pool size (Table 1). The specific activity and total levels of UTP are not significantly affected by antheridiol. The majority of this new RNA synthetic activity is directed toward the synthesis of ribosomal RNA (rRNA) [17]. This is manifested in the cells by the accumulation of new ribosomes in hormone-

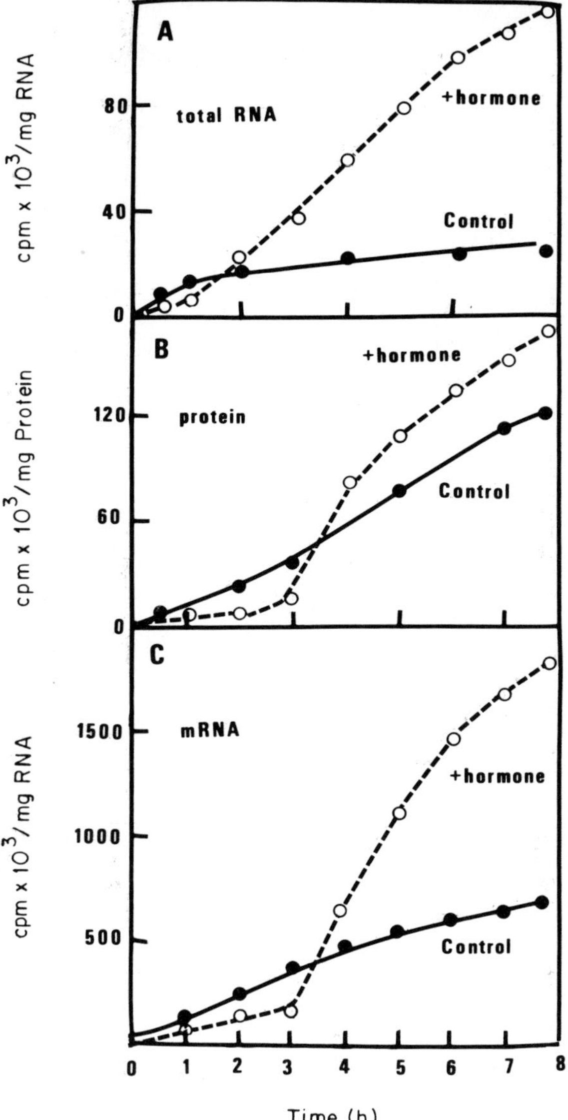

FIG. 2. The effect of antheridiol on the biosynthesis of RNA and protein during male sex organ initiation in *A. ambisexualis*. (A) Total phenol-extractable RNA. Cultures were exposed to continuous labeling conditions with [^{3}H]uridine [18] from time zero in the presence (+ hormone) and absence (control) of 5 × 10^{-11} g/ml of antheridiol. Concentration of RNA determined by A_{260}/A_{280} ratios. Radioactivity was determined by liquid scintillation. (B) Total hot TCA-precipitable protein. Cells were labeled with [^{3}H]leucine using steady state conditions [18] in the presence and absence of antheridiol. Concentration of protein determined by A_{280}/A_{260} ratios or by the Lowry procedure. (C) Poly(A)+ RNA. Cells were labeled using steady state conditions in the presence or absence of hormone [18]. Total RNA was extracted and passed through an unmodified cellulose column [18]. RNA concentration and radioactivity determined as described above.

TABLE 1

The Effect of Antheridiol on Uridine Triphosphate and Adenosine Triphosphate Cellular Pools[a]

Incubation system	UTP nMoles/ 500 mg tissue	ATP nMoles/ 500 mg tissue	UTP-Specific activity cpm/500 mg tissue	ATP-Specific activity cpm/500 mg tissue
1.5 h Control	115	445	1276	3114
1.5 h Hormone	121	479	1211	3091
3.0 h Control	127	502	1141	2993
3.0 h Hormone	116	487	1143	2910
4.5 h Control	123	479	1082	2876
4.5 h Hormone	119	486	1067	2791

[a] [^{3}H]uridine or [^{3}H]adenosine was added at time zero in the presence or absence of hormone [18]. ATP and UTP were isolated and purified using thin layer chromatography. Nucleotide spots were eluted and assayed spectrophotometrically (utilizing molar extinction coefficient) as well as with liquid scintillation. Data are expressed on the basis of 500 mg wet wt of tissue. For details of nucleotide pool measurements, see Horgen et al. [17].

treated tissue (Fig. 3A). Ribosome levels per milligram dry wt of tissue increase during the early stages of antheridiol-induced sex organ differentiation. Double-labeling experiments indicate that there is some turnover of existing ribosomes (Fig. 3B), which occurs in addition to the accumulation of newly formed ribosomes in the hormone-treated mycelia. The kinetics of [^{3}H]uridine incorporation into ribosomes (Fig. 3B) is similar to that observed for total cellular RNA (Fig. 2A). In addition, the kinetics of incorporation of label into the 26S and 18S RNA species of the ribosome (Fig. 3C) is also similar to the pattern observed for total RNA (Fig. 2A). Hence, one early effect of antheridiol is to stimulate the synthesis of rRNA and of new ribosomes.

B. Protein Synthesis

Antheridiol also stimulated the incorporation of [^{3}H]leucine into protein. Silver and Horgen [18] found that this dramatic enhancement of protein synthesis occurs 3 h after addition of hormone, at a time associated with the morphological appearance of branch initials (Fig. 2B). By contrast, a lag of no longer than 30 min (after hormone treatment) is observed for the incorporation of labeled precursors into total RNA, ribosomes, or rRNA. This temporal separation between the synthesis of total cellular RNA and the initiation of protein synthesis resulted in the asking of a series of new questions which dealt with hormonal effects on gene activation and gene product accumulation.

C. Poly(A)+ Messenger RNA Synthesis

Higher eucaryotes possess a sequence of polyadenylic acid [poly(A)] at the 3' end of the majority of cellular messenger RNAs (mRNAs). Silver and Horgen [18] were first to report the existence of such poly(A)+ presumptive mRNAs in a mycelial fungus. We have

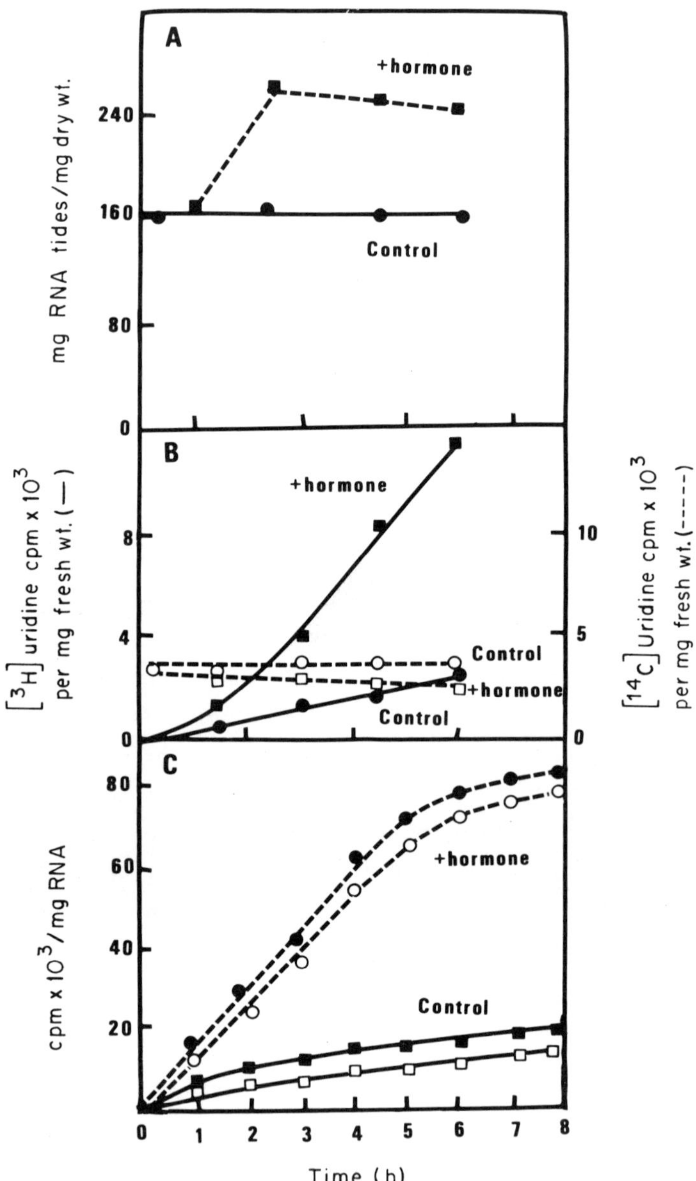
A
+hormone
Control
mg RNA tides/mg dry wt.
240
160
80
0
B
+hormone
Control
+hormone
Control
[3H] uridine cpm x 10³ per mg fresh wt. (—)
[14C] Uridine cpm x 10³ per mg fresh wt. (-----)
8
4
0
10
5
0
C
+hormone
Control
cpm x 10³/mg RNA
80
60
40
20
0 1 2 3 4 5 6 7 8
Time (h)

also demonstrated that *Achlya* mRNA is methylated at the 5' end of the molecule (data not shown). Because it is possible to isolate a poly(A)+ mRNA from *Achlya* (Table 2), it was therefore possible to examine the effects of antheridiol on the synthesis of this RNA fraction [18]. Antheridiol was found to have a stimulatory effect on the metabolism of *Achlya* poly(A)+ mRNA (Fig. 2C). The stimulation occurred at a time (3 h after the addition of hormone) immediately prior to the morphological appearance of the branched initials, and at a time associated with enhanced protein synthesis (Fig. 2B). Cordycepin (an inhibitor of posttranscriptional polyadenylation) and actinomycin D, at high concentrations (10 μg/ml), inhibited the accumulation of this messenger fraction [18]. Both of these drugs also blocked morphogenesis. Neither the UTP pool

FIG. 3 (opposite). The effect of antheridiol on the biosynthesis and turnover of cellular ribosomes. (A) The effect of hormone on ribosome levels in *A. ambisexualis*. Ribosomes were isolated from hormone-treated and control cultures according to the procedure of Lin and Key [42]. Membrane components were extracted in ethanol-ether-chloroform and the RNA was hydrolyzed [17]. Data are expressed quantitatively as mg RNAtides per mg dry wt of tissue. (B) Determination of ribosome turnover during hormone stimulation of sexual differentiation. In this double-labeling experiment, cultures were prelabeled with [^{14}C]uridine for 14 h, chased for 30 min with 2000 ml unlabeled uridine, and suspended in Mating Media [18] containing [^{3}H]uridine [17]. One-half of the culture was treated with 5×10^{-11} g/ml antheridiol (+ hormone) and the other half (control) was left untreated. Samples were taken, fresh weights determined, and ribosomes were isolated [17]. Radioactivity of RNA was determined for ^{14}C-label and for ^{3}H-label. RNA was extracted and electrophoresed [17]. The disappearance of ^{14}C-label from time zero is an indication of ribosome turnover whereas the incorporation of ^{3}H-label from time zero is an indication of new ribosome biosynthesis. (C) The effect of antheridiol on the synthesis of 26S and 18S rRNA. Cultures were continuously labeled with [^{3}H]uridine. RNA was extracted and electrophoresed as described by Horgen et al. [17]. The concentration of the 26S and 18S RNAs was determined from A_{260} nanometer measurements of the gels. The gels were sliced and radioactivity determined by liquid scintillation. The data are expressed as specific activity of the RNA taken at various time points: (●) 26S; (○) 18S hormone; (■) 26S; (□) 18S control.

TABLE 2

Base Composition Analysis
of *A. ambisexualis* RNA Rich in Adenylic Acid[a]

RNA Fraction	Base composition (%) A	U	G	C
+ Hormone (bound)	33.5	27.5	21.5	17.3
Control (bound)	34.5	27.8	20.5	16.6
Unbound (both hormone and control)	29.4	29.9	23.1	17.4

[a]Phosphorus-32-labeled RNA fractions were collected from cellulose columns [18]. The RNA was hydrolysed in 0.3 N KOH at 37° for 24 h. Base composition was determined using descending paper chromatography following the procedure of Lane [43]. The cellulose-bound fractions represent the poly(A)+ RNA; the unbound fractions were poly(A)- RNA (a mixture of 26S, 18S, 5S, and 4S RNAs [18]).

nor the ATP pool levels were appreciably altered by the presence of antheridiol (Table 1). Furthermore, the kinetics of incorporation of labeled precursors ([^{3}H]uridine or [^{3}H]adenosine) into poly(A)+ RNA are identical, suggesting that antheridiol does in fact stimulate the synthesis of mRNA, and that this occurs at a time associated with the morphogenesis of the sex organ initial.

The size distribution of the hormone-stimulated messages synthesized is heterodisperse [18], suggesting that many gene products are needed for this complex developmental process. Thus it appears that antheridiol stimulates transcription, perhaps the transcription of new genes that are required for sex organ morphogenesis.

Under somewhat different hormonal conditions, Timberlake [19] has made generally similar observations and has arrived at similar conclusions.

D. Chromosomal Protein Modification

While it is still not clear whether true fungi possess basic chromosomal proteins similar to those of higher eucaryotic cells [20-22], Horgen et al. [5] demonstrated that *Achlya* possesses histone-like chromosomal proteins with electrophoretic mobilities and amino acid compositions not unlike the histone proteins found in higher plant and animal cells [5]. Histones of higher eucaryotes are subject to posttranslational modifications [23]. These modifications involve group substitutions such as acetylation, methylation, and phosphorylation, which alter the charge of the residues in the peptide chains of the basic molecules [23]. In the case of acetylation, this chemical change results in a decrease of the net positive charge of the histones, therefore affecting the basic protein interaction with the genetic material. Histone acetylation has consequently been suggested as one mechanism associated with the activation (derepression) of new genes during cellular differentiation in general and also during differentiation that is mediated by hormonal stimuli [24-26].

Horgen and Ball [27] showed that there is a corresponding modification of basic proteins associated with antheridiol-stimulated differentiation of male sex organ initials. This modification involved the acetylation of specific *Achlya* basic nuclear proteins (Fig. 4). Reports from higher animal cells indicate that only histone fractions H4 and H3 are appreciably labeled with [^{14}C]acetate [28]. Although the exact chemical nature of the individual *Achlya* basic nuclear proteins has not been fully elucidated, it is interesting and perhaps significant that the *Achlya* protein band which incorporates the most [^{14}C]acetate (Fig. 4A) is electrophoretically similar to histone H3 of calf thymus [29,30]. The kinetics of labeling the *Achlya* nuclear proteins with [^{14}C]acetate involve a 3 h lag period prior to the time when specific acetylation of the *Achlya*

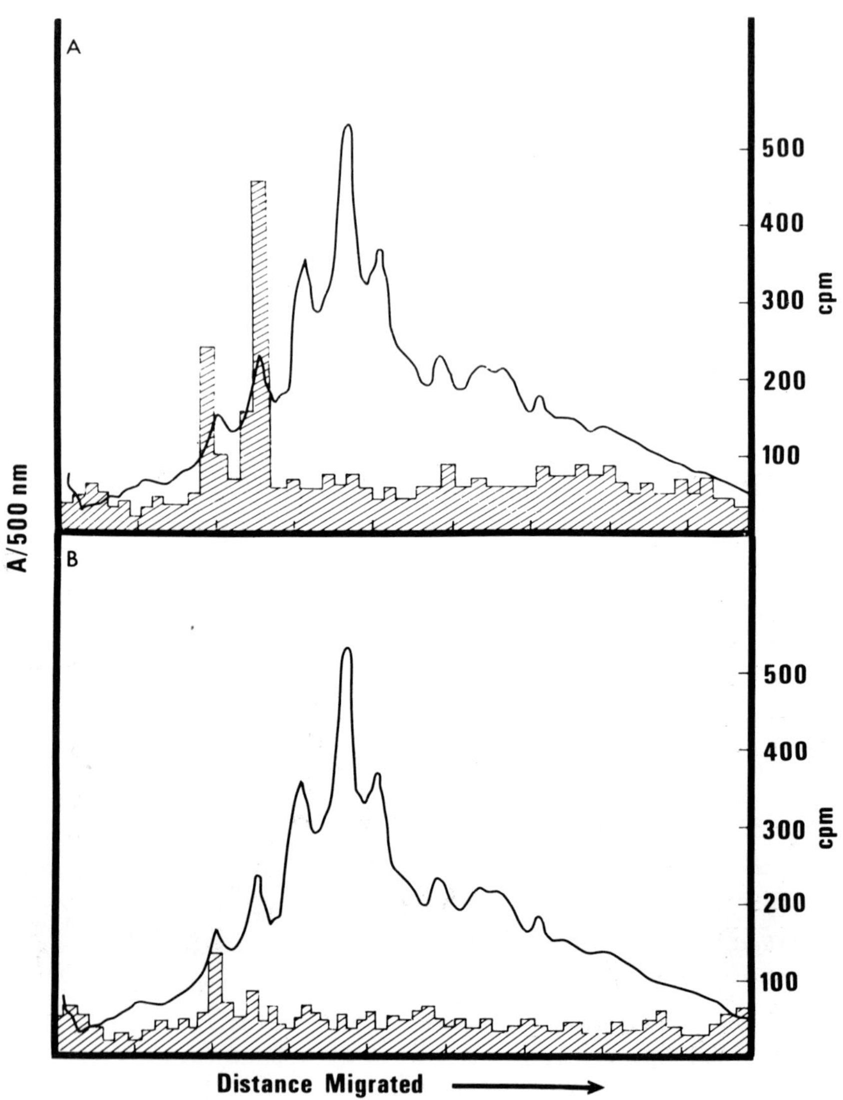

FIG. 4. Acetylation of *Achlya* basic histone-like nuclear proteins. Basic nuclear protein extracted as described by Horgen et al. [5]. Cultures were labeled with ^{14}C acetate at 24° in the presence or absence of antheridiol [27]. 50 μg basic protein was applied to the gels which were electrophoresed, sliced, and counted, as described by Horgen and Ball [27]. Zero migration is located at the positive pole. (A) 6-h antheridiol-treated (similar radioactivity profile for 3-, 4-, and 5-h hormone-treated). (B) Vegetatively growing control cultures (similar radioactivity profile for 1- and 2-h hormone-treated cultures).

histone-like proteins occurs [27]. Cycloheximide had no effect on the incorporation of [^{14}C]acetate into the histone-like proteins, suggesting that the acetylation reaction probably occurs after completion of the synthesis of the polypeptide chain [27]. Thus another way in which antheridiol appears to affect cellular metabolism is by inducing chemical modifications of the *Achlya* basic chromosomal proteins.

III. FINAL COMMENTS

Studies with the *Achlya* system so far indicate that there is a concomitant enhancement of the synthesis of both RNA and of protein associated with antheridiol stimulation of the morphogenesis of male sex organ initials. The hormone appears to affect both gene activation and gene product accumulation. The first genes that appear to be activated by antheridiol are the ribosomal genes (Jaworski and Horgen [31] have examined some of the biochemical characteristics of the ribosomal cistrons of *Achlya* and report that they are part of the reiterated portion of the *Achlya* genome). The gene products that accumulate are the 26S and 18S rRNAs as new cellular ribosomes within the hormone-stimulated tissues (Fig. 5). It is apparent that gene products in the form of ribosomal proteins (either newly synthesized or from preexisting cellular pools) would also be needed for the accumulation of new ribosomes observed in sexually developing mycelia. We have found that, whereas total protein synthesis is depressed for 3 h in the presence of hormone, [^{3}H]leucine or [^{14}C]amino acids are still incorporated into ribosomal proteins. The level of incorporation in hormone-treated tissue during the first 2 h is considerably higher than in the control. One might also predict that these newly synthesized ribosomes will be utilized later in sex organ development for translation of more new gene products.

The increase in rRNA synthesis is seen early (within 30 min) after the addition of antheridiol. Immediately prior to the morphological appearance of the antheridiol branches (at 3 h), the hormone

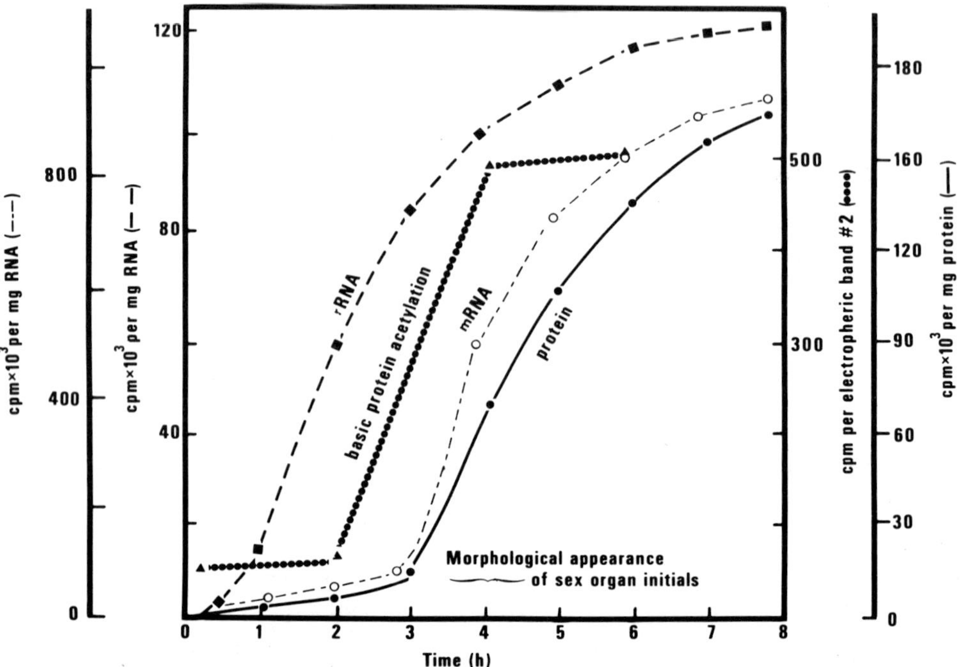

FIG. 5. The stimulation of gene activation and gene product accumulation by the steroid hormone antheridiol. Composite of information in Figs. 2, 3, and 4.

also stimulates the synthesis of mRNA (Fig. 5). However, prior to this enhancement of poly(A)+ mRNA synthesis, there is a modification of the chromosomal proteins of *Achlya*. This modification appears in the form of the acetylation of specific histone-like basic nuclear proteins (Fig. 5). The addition of acetyl groups to these histone-like proteins is thought to make them less positive in charge, decreasing their affinity for negatively charged DNA. It is, therefore, reasonable to suggest that this chromosomal protein modification may be involved in the activation of genes which are required for the differentiation of male sex organs. Indeed, shortly after the observed chromosomal protein modification, there is an enhancement of mRNA synthesis, and this does appear to be associated with the accumulation of gene products in the form of new proteins. This enhancement of apparent protein synthesis occurs immediately prior to the morphological appearance of male sex organ initials (Fig. 5). In the *Achlya* system then, one sees definite patterns emerging; patterns of biochemical changes which profoundly affect the morphogenesis of male sex organ initials.

A. Areas for Future Study

One of the primary areas for future research is to attempt to characterize the gene products that are produced in response to antheridiol. It has been suggested by Thomas and Mullins [32] that a prerequisite for antheridiol stimulation of sex organ initials is a localized softening of the cell wall by the enzyme cellulase [32]. Reports of hormone-stimulated increases in the activity of this enzyme [33,34] and ultrastructural observations of weakened areas of the cell wall [35] tend to strengthen this hypothesis. Cellulase would therefore appear to be at least one important gene product involved in sexual morphogenesis in *Achlya*. In a recent study, Groner et al. [36] reported that shortly after antheridiol induction an unidentified "induced protein" with a mol wt of 69,000 is synthesized. This induced protein is made preferentially and appears immediately prior to the formation of antheridiol branch initials [36]. Silver and Horgen [18], however, found that antheridiol stimulated the synthesis of a population of poly(A)+ mRNAs of heterodisperse size. Our data suggest that perhaps many gene products are required for this complex differentiation process. Recent studies on enzymes other than cellulase currently being conducted in my laboratory suggest that, whereas intranuclear proteinase activity increases in response to antheridiol, there does not appear to be any change in cellular ribonuclease activity during antheridiol induction of development (data not shown).

Another area for future research is to look more specifically at *Achlya* mRNAs, specifically mRNAs synthesized in response to antheridiol. Dodd et al. [37] have studied the reassociation kinetics of *A. ambisexualis* DNA. *Achlya* DNA could be separated into two components (Fig. 6): a fast-reacting component representing the repeated portion of the genome (17.6% of the nuclear DNA) and a slow-reacting component corresponding to the unique portion of the genome (80.1% of the nuclear DNA). The multiple-copy DNA

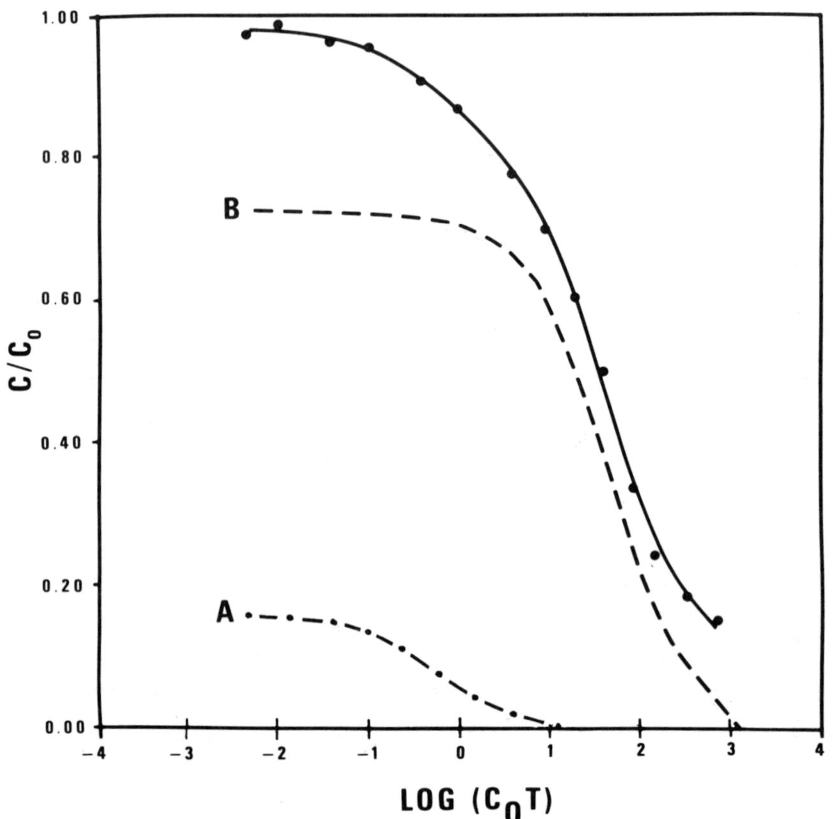

FIG. 6. Reassociation kinetic analysis of *Achlya ambisexualis* nuclear DNA. DNA isolated from *Achlya* nuclei was sheared to a single-stranded fragment size of 200 nucleotides and reassociated in 0.12 M phosphate buffer at 60° [37]. The extent of reassociation was monitored by hydroxylapatite chromatography. Component A has a $C_0t_{1/2}$ value of 0.495 and represents 17.6% of the genome (repeated DNA). Component B has a $C_0t_{1/2}$ value of 41 and represents 80.1% (unique DNA). For a detailed discussion of the methodology used in generating these data refer to Dodd et al. [37].

was 91 times repeated when compared to the single-copy DNA. We estimated the genome size of *A. ambisexualis* to be 11 times that of *Escherichia coli* [37] and about the same as that of *Dictyostelium* [38]. We are now able to isolate *Achlya* single-copy DNA and intend to utilize this DNA in hybridization studies to identify steroid-specific mRNAs. It may also be possible to synthesize DNA complementary to *Achlya* mRNA using the enzyme reverse transcriptase and do DNA-RNA hybridization studies. Once information is available on steroid-specific mRNAs, it will then be possible to attempt in vitro translation of these messages and identification of the protein products.

Essentially nothing is known about the nonhistone chromosomal proteins of *Achlya*. Whereas histone-like proteins could conceivably be involved as nonspecific effectors of gene activation [23-26], nonhistone chromosomal proteins have been implicated as regulatory effectors of specific genes, especially in developmental systems mediated by steroid hormones [39,40]. We are presently examining the acidic chromosomal proteins of *Achlya* and the possible effect of antheridiol on these proteins.

An interesting and important area of investigation is the role of steroid hormones on the in vitro transcriptional activity of chromatin preparations. Since considerable information has accumulated regarding fungal RNA polymerases (see review by Griffin et al., Ref. 41), it is possible to do reconstitution experiments with *Achlya* nuclear RNA polymerases and the other components of the chromatin. In addition to our studies on *Achlya* chromatin, we are presently involved in studying the in vitro RNA synthesizing activity of nuclei isolated from differentiating and nondifferentiating mycelia of *Achlya*.

Certainly another exciting area of investigation in the *Achlya* system will be investigation of the nature of the steroid receptor. In animal systems, receptor proteins are localized in the cytosol and are involved in transporting the hormone from the cytoplasm to the nucleus [39,40]. With the possibility of synthesizing radioactive antheridiol (or oögoniol), these studies are possible and are certainly needed. Since there is a continuum of maleness to femaleness in the various strains of *Achlya* [13], it will be interesting to eventually compare the hormone reception properties of the different strains.

We have recently done experiments which suggest that *Achlya* does possess "hormone-receptor" mechanisms. Incubation of *Achlya* chromatin with antheridiol-cytosol mixtures of cytosol from male strain E87 increases the RNA synthetic capacity of the chromatin (Table 3). Neither cytosol, nor antheridiol alone, nor hormone-cytosol mixtures of cytosol isolated from female strain 734 by themselves stimulated transcription of the chromatin. These results suggest that hormone-cytosol interactions similar to the hormone-receptor mechanisms operable in complex animal systems exist in the primitive eucaryote *Achlya*. They also demonstrate that hormone-target cell interactions can exist in eucaryotic microbes.

Another area completely open for investigation is the role of oögoniol on the biochemical morphogenesis of female sex organs.

In conclusion, it is evident that *Achlya*, which is the most primitive eucaryote known to respond developmentally to steroid hormones, seems to do so in a manner similar to higher eucaryotes (see review by O'Malley and Means: Ref. 39). Because it is a microorganism, because it is extremely fast-growing, because the entire thallus acts as the target tissue for the steroid, and because it is possible to generate mutants of *Achlya* for genetic experiments (something not so easily done in the target tissue of adult animals), we feel that *Achlya* has the potential to lend itself as a useful "model system" with which to study the way in which steroid hormones control differentiation. We, and others, have made some beginnings in realizing this potential. In the future, we hope to continue studying this interesting organism and the role sterols play in controlling its development.

TABLE 3

The Effects of Cytosol, Antheridiol, and Cytosol-Antheridiol Mixtures on Transcription of *Achlya* Chromatin

Chromatin preincubated with	Incubation temperature °C	p moles incorporated per min /50 μg DNA	% of control
[a]E87 cytosol	30	112	100
Antheridiol	30	97	86
E87 cytosol + antheridiol	30	415	370
734 cytosol + antheridiol	30	117	104
[b]E87 cytosol + antheridiol	30	271	318
E87 cytosol	30	85	100
E87 antheridiol	30	75	88

[a]Transcription with heterologous RNA polymerase. 50 μg of chromatin DNA were added along with 10 μg of *E. coli* RNA polymerase (fraction IV containing sigma factor, 4000 units/mg) to each reaction mixture. Either 1×10^{-9} g (2.1 nM) antheridiol or 1.25 to 1.50 mg of cytosol or a combination of both were added to each preincubation mixture. The final buffer concentrations in the 1 ml reaction mixture were 50 mM Tris-HCl (pH 8.0), 10 mM dithiothreitol, 10 mM $MgCl_2$, 20% glycerol, 0.1 M KCl, 200 mM ammonium sulphate, and 0.4 mM K_2HPO_4. Unless otherwise indicated, these mixtures were preincubated for 15 min at 25°. 0.4 mM UTP, 0.4 mM ATP, 0.4 mM CTP, 0.4 mM GTP, and 10 μCi [^{3}H]UTP were added to begin the RNA synthetic reaction. Each reaction mixture was then incubated at the designated temperature for 15 min. Reactions were terminated; the products were collected and the radioactivity determined as described by Horgen and Key [44].

[b]Transcription with endogenous *Achlya* RNA polymerases. Preincubation mixtures were similar to *a* except 200 μg chromatin DNA were added and no *E. coli* RNA polymerase. Chromatin was isolated by a modification of the procedure of Rizzo and Nooden [45].

ACKNOWLEDGMENTS

The work reported in this paper was supported by research grants from the National Research Council of Canada, from Brown-Hazen (Research Corporation), and from internal research funds provided by the University of Toronto, Erindale College. Individuals involved in various aspects of the studies reported include Dr. Julie C. Silver, Dr. N. Straus, Steven F. Ball, Robin Smith, Barbara Schuerch, Robert Sutherland, Janice Dodd, and Gary Craig. The technical assistance of E. Thompson is gratefully acknowledged.

REFERENCES

1. Bartnicki-Garcia, S., 1970. *In* Phytochemical Phylogeny, J. B. Harborne (ed.), Academic Press, New York, p. 81.
2. Klein, R. M., and A. Cronquist, 1967. Quart. Rev. Biol. 42: 105-296.
3. Hütter, R., and J. A. DeMoss, 1967. J. Bacteriol. 94: 1896-1907.
4. Lovett, J. S., and J. A. Haselby, 1971. Arch. Mikrobiol. 80: 191-204.
5. Horgen, P. A., R. T. Nagao, L. S. Y. Chia, and J. L. Key, 1973. Arch. Mikrobiol. 94: 249-258.
6. Griffin, D. H., W. E. Timberlake, and J. C. Cheny, 1974. J. Gen. Microbiol. 80: 381-388.
7. Griffin, D. H., 1966. Plant Physiol. 41: 1254-1256.
8. Raper, J. R., 1950. Bot. Gaz. 112: 1-24.
9. Barksdale, A. W., 1969. Science 166: 831-837.
10. McMorris, T., R. Seshardri, G. Weihe, and A. Barksdale, 1975. J. Amer. Chem. Soc. 97: 2544-2545.
11. Sansome, E., 1965. Cytologia 30: 103-117.
12. Barksdale, A. W., 1966. Mycologia 58: 802-804.
13. Barksdale, A. W., 1967. Ann. N.Y. Acad. Sci. 144: 313-319.
14. Barksdale, A. W., 1968. J. Elisha Mitchell Sci. Soc. 84: 187-193.
15. Bryant, T. R., and K. L. Howard, 1969. Amer. J. Bot. 56: 1075-1083.

16. Markl, H., and M. Lindauer, 1965. *In* The Physiology of Insecta, Vol. II, M. Rockstein (ed.), Academic Press, New York, p. 92.

17. Horgen, P. A., R. Smith, J. C. Silver, and G. Craig, 1975. Can. J. Biochem. 53: 1341-1345.

18. Silver, J. C., and P. A. Horgen, 1974. Nature 249: 252-254.

19. Timberlake, W. E., 1976. Develop. Biol. 51: 202-214.

20. Hsiang, M. W., and R. D. Cole, 1973. J. Biol. Chem. 248: 2007-2013.

21. Leighton, T. J., B. C. Dill, J. J. Stock, and C. Phillips, 1971. Proc. Nat. Acad. Sci. USA 68: 677-680.

22. Stumm, C., and J. Van Went, 1968. Experientia 24: 1112-1113.

23. Allfrey, V. G., 1970. Fed. Proc. 29: 1447-1460.

24. Pogo, B. G., V. Allfrey, and A. E. Mirsky, 1966. Proc. Nat. Acad. Sci. USA 55: 805-812.

25. Mukerjee, A. B., and M. M. Cohen, 1969. Exp. Cell Res. 54: 257-260.

26. Libby, P. R., 1972. Biochem. J. 130: 663-669.

27. Horgen, P. A., and S. F. Ball, 1974. Cytobios 10: 181-185.

28. Vidali, G., E. L. Gershey, and V. G. Allfrey, 1968. J. Biol. Chem. 243: 6361-6366.

29. Coukell, M. B., and J. O. Walker, 1973. Cell Diff. 2: 87-95.

30. Panyim, S., and R. Chalkely, 1969. Arch. Biochem. Biophys. 134: 577-589.

31. Jaworski, A. J., and P. A. Horgen, 1973. Arch. Biochem. Biophys. 157: 260-267.

32. Thomas, D., and J. T. Mullins, 1965. Science 154: 84-85.

33. Thomas, D., and J. T. Mullins, 1969. Physiol. Plant. 22: 347-353.

34. Nolan, R. A., and A. K. Bal, 1974. J. Bacteriol. 117: 840-843.

35. Mullins, J. T., and E. A. Ellis, 1974. Proc. Nat. Acad. Sci. USA 71: 1347-1350.

36. Groner, B., N. Hynes, A. Sippel, and G. Schatz, 1976. Nature 261: 599-601.

37. Dodd, J. G., P. A. Horgen, and N. A. Straus, 1975. Cytobios 13: 31-36.

38. Firtel, R. A., A. Jacobsen, and H. F. Lodish, 1972. Nature New Biol. 239: 225-228.

39. O'Malley, B., and A. Means, 1974. Science 183: 610-620.

40. Spelsberg, T. C., 1974. *In* Acidic Proteins of the Nucleus, I. L. Cameron and J. R. Jeter (eds.), Academic Press, New York, p. 248.

41. Griffin, D. H., W. Timberlake, J. Cheny, and P. A. Horgen, 1975. *In* Isozymes, Vol. I, Molecular Structure, C. L. Markert (ed.), Academic Press, New York, p. 69.

42. Lin, C. Y., and J. L. Key, 1967. J. Mol. Biol. 26: 237-247.

43. Lane, B., 1963. Biochim. Biophys. Acta 72: 110-113.

44. Horgen, P. A., and J. L. Key, 1973. Biochim. Biophys. Acta 244: 227-235.

45. Rizzo, P. J., and L. D. Nooden, 1974. Biochim. Biophys. Acta 349: 407-414.

THE CONTROL OF MORPHOGENESIS IN *DICTYOSTELIUM DISCOIDEUM*

Antony J. Durston

Hubrecht Laboratory
International Embryological Institute
Universiteitscentrum "De Uithof"
Utrecht, The Netherlands

I. INTRODUCTION

The cellular slime mold *Dictyostelium discoideum* is a mononucleate soil amoeba which feeds and divides as a unicellular organism. On a solid substratum, after the food supply is exhausted, the amoebas begin multicellular morphogenesis [1] (Fig. 1). After an 8-h interphase, the amoebas collect into hemispherical mounds fed by radial, randomly bifurcating streams of cells (aggregation stage). Each aggregate secretes an external slime sheath, containing mucopolysaccharide and protein, which is elastic at the apex and rigid at the periphery of the aggregate. Incoming cells are thus constrained to enter a vertical, approximately cylindrical

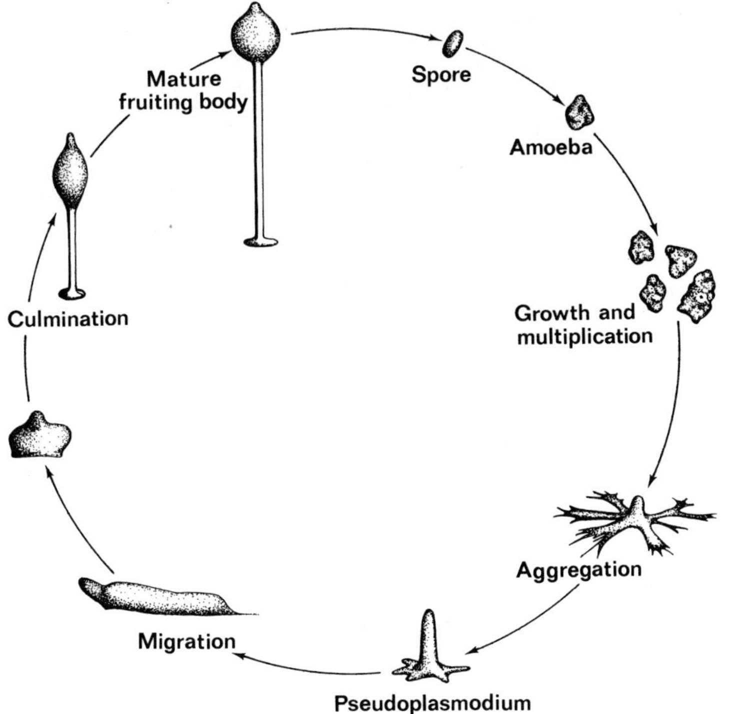

FIG. 1. The life cycle of the cellular slime mold *Dictyostelium discoideum*.

structure, producing the pseudoplasmodium or slug stage. The slug falls over and migrates for a variable time, leaving behind a trail of collapsed slime sheath, before it finally rounds up and constructs a fruiting body (culmination stage). A vertical cellulosic tube (stalk) is now secreted within the culminating slug and is extended by addition to its top end. Some cells enter the stalk, vacuolate, and die (stalk cells). Most of the others are raised on top of the elongating stalk and secrete a thick cell wall to become viable spores. A few cells which were originally at the rear of the slug are left at the base of the stalk and make a characteristic basal disk of dead, vacuolated cells. The pattern of cell types in the fruiting body is approximately proportionate over the 10^4-fold volume range of the organism (sorocarps contain 10 to 10^5 cells) [2]. It originates in a regulative axial pattern of embryonic cell types in the migrating slug [3].

The *Dictyostelium* life cycle is of great interest for developmental biologists because it poses, in elementally simple form, the problem of how an organism is assembled from component cells during embryogenesis. In the following section, I discuss investigations of the mechanisms of multicellular pattern formation in the development of *D. discoideum*.

II. CURRENT RESEARCH

A. The Control of Aggregation

Because aggregation occurs through the coordinated movement of a two-dimensional array of initially separate cells, much can be learned about its control by direct observation using time-lapse cinemicrography. Time-lapse films of aggregation reveal beautiful and complex spatio-temporal movement patterns [1,4-6]. The individual cells approach centers by making pulsatile movements. These movements are a result of directional pseudopodial extensions. The cells move about 20 μm in 100 sec. The movements are repeated rhythmically by each cell (once every 2 to 10 min), and movements of individual cells are coordinated in space and time, so that periodic waves of (centipetal) cell movements propagate centrifugally from each aggregation center at constant or increasing velocities in the order of 50 μm/min. When waves from neighboring centers collide they are mutually annihilated and thus define boundaries between domains of influence for the respective centers. Visible wave propagation continues until the end of aggregation.

The late aggregate forms an apical nipple of cells (the tip), which is the origin for any later waves. The tip persists as a distinct structure in later morphogenesis [at the front of the migrating slug and at the apex of the developing fruiting body (Fig. 1)] and it is of great importance. It has proved to act as an embryonic organizer (morphogenetic boundary region) for all stages of development in which it is present [6-8]. The above observations are evidence for a complex control system regulating movement during morphogenesis.

There is clear evidence that cells of *D. discoideum* aggregate by chemotaxis to cyclic adenosine-3',5'-monophosphate (cAMP) [9-11]. Aggregating cells are also known to secrete a soluble cAMP phosphodiesterase (soluble PDE) [12] as well as a PDE inhibitor [13] and to have a cell-bound PDE [14] as well as a surface-bound nonenzymatic cAMP binding protein [15,16]. Of the above, PDE inhibitor, cell-bound PDE, and the cAMP-binding protein (as well as cAMP secretion and chemotactic sensitivity) are aggregation stage-specific and are not seen in vegetative cells. The functions of the cAMP-related proteins are unknown, but there is reason to suspect that the nonenzymatic binding protein may be the chemotactic receptor (see below).

A number of tests have been used to measure the chemotactic sensitivity of *D. discoideum* cells to cAMP [11,17-20]. In the most sophisticated of these [18,19], calibrated pulses of cAMP are delivered locally to a monolayer of cells on an agar surface by electrophoresis using a glass microelectrode with an internal diameter of about 2 μm. The pulses diffuse symmetrically from the electrode tip and are depleted at a calculable rate by the PDE enzymes to give standard spatio-temporal cAMP profiles. The test shows, in agreement with a previous result [11], that the cells of *D. discoideum* manifest full chemotactic sensitivity at 4 h after starvation. They are then attracted from within a defined radius of the electrode tip, the radius depending on the pulse size. This suggests that there may be a threshold signal for chemotaxis (i.e., that the chemotactic response is highly nonlinear). If calculated as the maximum cAMP concentration reached at the chemotactic radius following a cAMP pulse, the threshold corresponds to between 1 and 4×10^{-9} M cAMP (based on standard pulse amplitudes of 6×10^{9} and 6×10^{10} molecules). This is in agreement with an independent result for the minimum effective cAMP concentration for chemotaxis [17]. It is very similar to the dissociation constant of the nonenzymatic cAMP binding protein, supporting the contention that this may be the chemotactic receptor [15]. The PDE enzymes have higher Km values [12,14]. At this time, we know nothing more about the quantitative dependence of the chemotactic response on the perceived signal nor anything about the nature of the perceived signal (whether cells detect a space or a time gradient of cAMP). A quantitative analysis, using appropriate test environments (e.g., stationary space gradients of cAMP), is required.

Bonner [22] made an important observation about the chemotactic response, namely, that chemotactically responsive cells have differentiated to a polar state. Vegetative (nonchemotactic) cells are isotropic and move by extruding characteristic pseudopods [23, 24] from randomly situated points around the cell periphery. Chemotactically responding cells, observed after the beginning of aggregation, make pseudopods only from about the front third of the cell surface, and the pseudopods are induced synchronously among the local cell population by chemotactic signals. This polarity is fixed, since cells switching allegiance between aggregation centers 180° apart usually preserve polarity and make "U turns" [22]. It does not account for the directional nature of chemotaxis, since cells make signal-directed pseudopods from the permitted part of the cell surface and hence can make "U turns." It does appear *necessary* for directional chemotaxis during natural aggregation signals, since I have isolated a mutant [10] which lacks normal polarity and consequently fails to aggregate. This mutant makes spatio-temporal waves of cell movements which are normal except that the responding cells fail to make a net centripetal movement with each wave. Instead, the cells make mainly forward pseudopods as each wave approaches and backward pseudopods as it passes. Thus they move forward and then return to their original positions. As discussed later, polarity appears to be required if cells are to be refractory to retrograde stimulation by the natural chemotactic waves of secreted cAMP.

Polarity appears to develop during the course of aggregation. Cells responding to the first few aggregation waves make synchronous center-directed movement steps in response to the waves and randomly timed and directed movements between waves. The latter occur by apolar pseudopod formation from any part of the cell surface. As aggregation proceeds, apolar movements gradually cease and the only movements are polar and synchronous in response to chemotactic signals. This gradual polarization appears to be initiated by perception of chemotactic signals, since cells are capable of making polar chemotactic movements as early as 4 h before the onset of aggregation if they are signaled appropriately. It may represent a second polar response of cells to cAMP other than directional pseudopod formation. The frequency of pseudopod formation decreases as polarity develops, suggesting that cells become generally more refractory for pseudopod formation while also developing an antero-posterior gradient of refractivity.

B. Signal Relaying during Aggregation

The nondecremental waves of movement seen during aggregation are not expected from a chemotactic response to cAMP signals diffusing from aggregation centers. Signal amplitude and velocity (i.e., velocity of cAMP concentration contours) should diminish rapidly with distance from the center because of diffusion and extracellular PDE activity. The waves imply a signal-relaying competence: that cells sensing a chemotactic signal are simultaneously induced to make a new one, so that a local signal initiates a propagating wave of cAMP secretion [25]. The simplest type of signal-relaying in response to a cAMP signal was indicated by the finding that cAMP pulses applied locally to an appropriate aggregation field can initiate typical aggregation waves [18]. It has now been verified by the finding that pulses of cAMP can induce secretion of pulses of tritiated cAMP ($[^3H]$cAMP) by a labeled population of *D. discoideum* cells [26,27].

Little more is known about the parameters or mechanism of the relaying response except the following:

1. Wave initiation by cAMP pulses is pulse amplitude-dependent in an all or nothing fashion, indicating that there may be a threshold signal for signal relaying. Calculated as a concentration, this is 10^{-7} M [19]. No detailed quantitative results are available from the pulse amplification ($[^3H]$cAMP) experiments, but these give a positive response with concentrations as low as 6×10^{-8} M [27].
2. The low velocity of aggregation waves indicates that they are not rate-limited by diffusion from cell to cell and that there is a delay of the order of 15 sec between signal sensing and signal secretion in each cell [28]. The pulse amplification experiments show a rise time of 90 sec between the applied cAMP pulse and the $[^3H]$cAMP peak, but a cascade mechanism (among the cell population) has not been eliminated as a contributing factor.
3. The unidirectional nature of wave propagation and annihilation of colliding waves imply that each wave front is succeeded by a zone of refractory cells which are unable to relay the signal. Cells must therefore enter a refractory period after relaying a signal. This deduction has been confirmed by use of cAMP microelectrodes. It is found that high-frequency (1/3 min) cAMP pulses may be gated, with only every second or third pulse

initiating a wave. The refractory period is a monotonically decreasing, time-dependent variable, equal to 9 min at the beginning and 2.5 min at the end of aggregation [6,19].

4. Fields of slime mold cells will not propagate waves at a density of less than 2.5×10^4 cells/cm^2. (At this density, an average cell has just enough randomly distributed, competent neighbors in range (i.e., anterior and posterior neighbors in a connected chain) [28-30]. The critical mean number of neighbors is 4.5, giving a range of 75 μm for the observed critical density (2.5×10^4 cells) [29]. This range can be used to calculate signal amplitude (given a realistic value for PDE degradation of cAMP, hemispherical diffusion with known rate of diffusion through the agar substrate, and a known threshold for a signal relaying) [19, 28]. Such a calculation has been made, assuming a maximally efficient durationless (δ function) signal. The calculation gives a value of 1×10^8 cAMP molecules/cell/signal [19].

    The amplification experiments quoted above have also been used for direct measurement of signal amplitude and give a lower value (mean = 6×10^6 molecules/cell/signal) [27]. They also give data on the form of the cAMP signal secreted by a population (a pulse ∿90 sec wide) [26,27]. At this time, the form and amplitude of the signal secreted by a single cell are in doubt. If Robertson's [19] calculations are reliable and if the amount of cAMP secreted per cell does not exceed his calculated value, as suggested by the amplication experiments [27], *then the relayed signal must be a very short pulse* (duration ≤ a few seconds) and the 90-sec peak width seen in amplification experiments must be an artifact due to variable delay times in individual cells [28]. A signal lasting 90 sec would require several orders of magnitude more cAMP/signal to reach the same range as a (maximally efficient) durationless pulse.

5. Coalescence of aggregating cells into radial streams is a consequence of signal relaying. Because all cells are local repeaters of the signal, aggregating cells are attracted towards their nearest anterior neighbors and azimuthal maxima in cell density grow [29].

Once they enter aggregation streams, cells make ethylenediamine tetraacetic acid (EDTA)-resistant polar contacts (the signal-sensitive front end of each cell adheres to the back of a preceding cell) [4,31]. These contacts probably serve a synaptic function (i.e., now mediate signal propagation between cells) because cells joining streams are attracted towards the junctions between cells [25]. It is noticeable that large contractile vacuoles (∿3% of the cell volume) which were expelled at random points on the periphery of the vegetative amoeba are now expelled at the rear of each polar cell, apparently into its junction with the next cell. It is conceivable that these vacuoles play a role in signal secretion, though their frequency (∿1/30 sec) far exceeds the signal frequency.

The formation of synaptic contacts would be expected to change the nature of the aggregation field. In early aggregation, where the separate cells communicate by (symmetrical) diffusion through extracellular space, wave propagation is isotropic (i.e., possible with any orientation). Restriction of signaling to junctions would make each aggregate locally anisotropic, i.e., signaling would be possible away from but not toward aggregation centers [32]. There is experimental evidence that this is indeed the case [24,32].

The intercellular contacts made during aggregation are long-lived. I have observed that individual junctions last for more than 20 min (the length of the observation period). They may, in fact, persist into the slug (since strings of connected cells can be recovered from slugs) [4].

C. Autonomous Signaling during Aggregation

In addition to signal relaying, we must infer that some cells make autonomous signals and so spark off new waves of signal secretion. Two possibilities are conceivable:

1. Signal relaying and autonomous signaling are two aspects of the same property (periodic secretion of cAMP pulses due to an os-

cillator which can be driven faster than its natural frequency by external signals). Individual cells with the highest natural frequencies act as pacemaker centers and entrain others.

2. Autonomous signaling and signal relaying are separate properties, not necessarily present simultaneously in each cell.

There is clear evidence for item 2, both from periodicity measurements of aggregation waves as will be described and from measurements of emergence of signal-relaying competence (X2) [33] and of autonomous signaling competence (X3) [34,35]. X2(t) is determined by measuring the time, as a function of cell density, when a cell population can just propagate a wave of signaling in response to an applied signal from a microelectrode. This gives the time when the population contains a critical density of relaying-competent cells, and hence gives X2(t), which is the fraction of relaying-competent cells as a function of time. Critical density itself varies with cell density because PDE dissipation of the signal varies with density. It is determined as a function of cell density by diluting relaying-competent, mature, wild-type cells with a relaying-deficient but otherwise normal mutant. X2 is an S-shaped function with the first few cells gaining competence at 400 min, most cells at 500 min, and all cells by 600 min after starvation. X3(t) is measured by recording times of initiation of aggregation in uniform, small populations of cells, containing about 500 cells in a high-cell density drop on an agar surface. In such populations, aggregation begins well after all cells have acquired the chemotactic and signal-relaying competences. It is initiated by emergence of autonomous cells. If we assume independent emergence of autonomy in each cell we can calculate the mean number of autonomous cells per drop as a function of time and hence X3(t). The fraction of unaggregated drops at any one time is then the zero term of a poisson series. On this assumption, X3 is approximately linear, crossing the origin at 500 min and increasing at about 10^{-3} cells/h. Virtually all aggregating cells relay signals but very few can make spontaneous signals.

D. Signal Periodicity in Aggregation Waves

Much can be learned about autonomous signaling and the roles of autonomous signaling and signal relaying in aggregation from measuring the periodicity of aggregation waves. The waves are of the two geometrical types expected in a two-dimensional, isotropic-sensitive medium. They are either expanding rings, initiated at points, or expanding spirals, with an inner free end rotating about a central core [32,36].

A ring is the wave form expected from a localized, spontaneous signal (as from an autonomous cell), which should spark off a symmetrical wave of signal relaying among neighboring cells. Rings are the most common wave form and are the first waves seen. They usually occur as sequences of concentric rings, each initiated at the same point, indicating existence of continuously or continually signaling autonomous sources. These concentric waves organize aggregates with radial streams, as shown in Fig. 2a.

The periodicity of the waves should reveal the nature of the autonomous source. If the source makes a continuous superthreshold signal, successive waves will be separated by the refractory period for signal relaying (i.e., by the minimum possible interval). If the source makes intermittent superthreshold signals, then each signal will initiate a wave only if the signal interval exceeds the

FIG. 2 (opposite). Geometries of aggregates and periodicity of aggregation waves. (a) Concentric ring waves organize aggregates with radial streams, as in this example. The long axes of major streams are approximately perpendicular to the wave fronts and each stream propagates a segment of each wave. (b) Periodicity of waves from a concentric ring wave aggregation center. The ordinate records temporal intervals (minutes) between successive waves. The abscissa shows age of the center (minutes from starvation). The time of tip formation is indicated with an arrow. (c) Spiral waves organize aggregates with spiral aggregation streams. (The streams rotate counter to the spiral wave.) This aggregate has an open center which is circumvented by the inner end of the spiral wave. (d) Periodicity of a spiral aggregation wave. The ordinate shows temporal interval (minutes) between successive coils of the spiral wave. The abscissa shows age of the spiral (minutes from starvation).

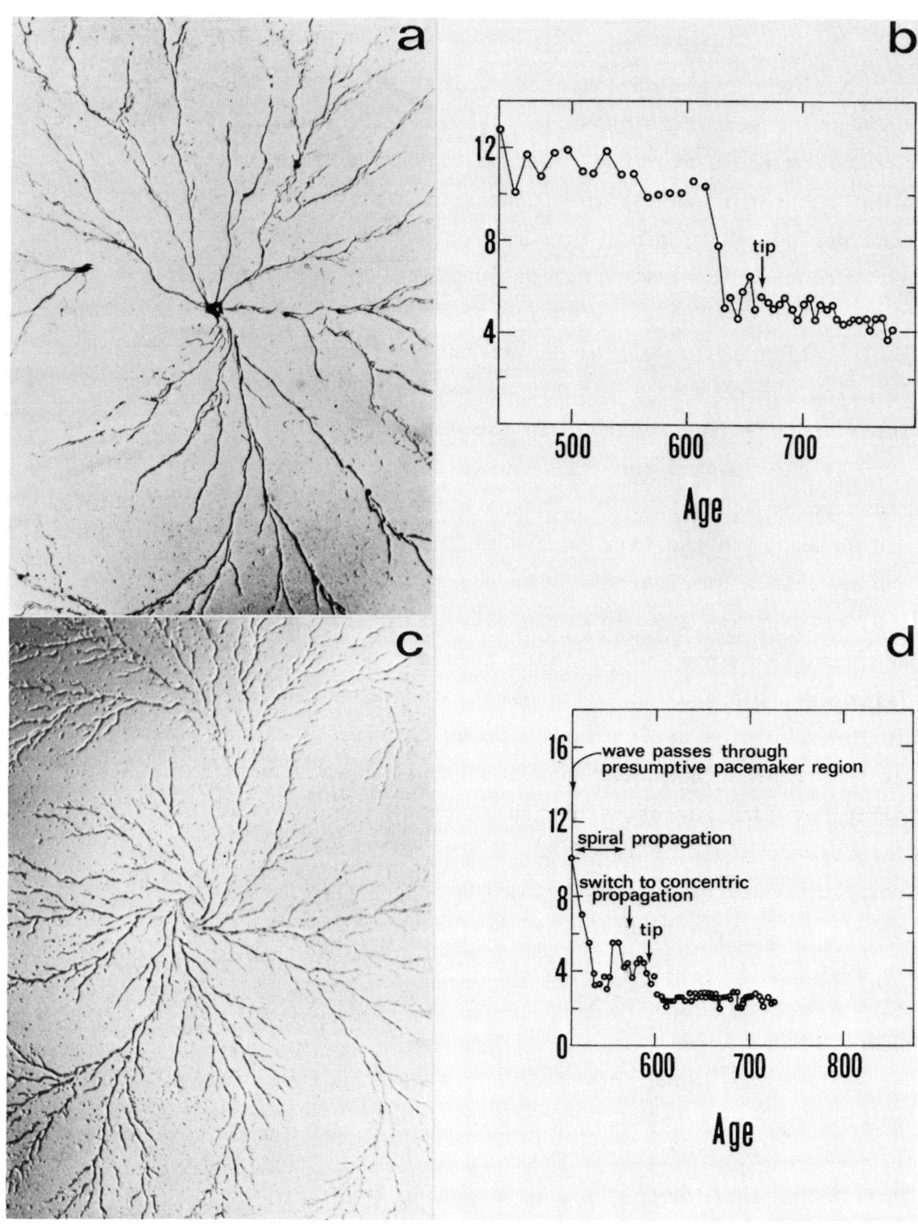
a
b
12
8
4
tip
500
600
700
Age
c
d
16
wave passes through
presumptive pacemaker region
12
spiral propagation
switch to concentric
propagation
8
tip
4
0
600
700
800
Age

refractory period. Every nth signal will initiate a new wave if the signal interval is less than the refractory period [nT > Tr > (n - 1)T; T = signal interval, Tr = refractory period].

A plot of wave periodicity against age for a concentric ring aggregation center is shown in Fig. 2b. The figure shows that early waves from the center had a periodicity of 10 min and that after about 3 h (or 17 waves) they underwent a frequency doubling to become periodic at 5 min. This result is typical and it implies that early concentric ring waves are initiated by a periodic signal (with period 5 min) which is gated for the first several waves [6]. The early gating could well be due to the refractory period for signal relaying, since this initially exceeds 5 min but not 10 min and later drops to less than 5 min (see earlier comments).

We can compare the time course for the refractory period, as measured by Robertson [19], using a continuous cAMP source to initiate waves, with the time course of the natural gating. The comparison reveals a discrepancy. The electrode-induced (refractory period) waves reach a 5-min period 120 min after the first wave, whereas the natural waves reach this period about 200 min after the first wave propagated. The electrode-induced waves were started at the same time after starvation as the first natural waves (500 min). Fig. 2b does not show the first waves to pass the reference point for measurement, since these came from other centers. The discrepancy would be resolved if the refractory period for signal relaying measures developmental time by counting aggregation waves (i.e., if it is a function of the number of times a cell has relayed a signal). Both the (high-frequency) electrode-induced and (low-frequency) natural waves reach a 5-min period after about 20 waves.

The older center develops a visible tip (Fig. 2b), which becomes the origin for all subsequent waves. In the (unusual) example shown, tip formation causes no change in wave frequency for 2 h until the center stops wave propagation (wave propagation is replaced by continuous cell movement to the center).

A spiral is a single, continuously propagating wave. It should therefore run indefinitely (and organize an aggregate with the aggregation center at the center of the spiral wave) by signal relaying alone, without any autonomous signaling.

Spirals are initiated when wave fronts are broken appropriately by encountering inhomogeneities in aggregation fields. For example, a wave may encounter an area containing a subcritical density of cells for wave propagation (a hole) [32,37]. The wave will be broken to make two free ends which will circumvent the left and right sides of the hole, respectively, and given no further disturbance, meet to make a single-wave front with a kink distal to the hole. However, if the hole is broken during passage of the wave (e.g., by migration of cells into the hole or by maturation of cells to relaying competence), the outcome can be a number of holes, two of which have one free end of the wave rotating about their circumference and will become the centers of spirals. About 50% of the spirals seen initially have open centers (Fig. 2c) and are probably formed in this way. Spirals can be initiated in other ways, e.g., by interaction of closely successive waves with areas of variable refractory period (Fig. 3) [32,38].

Spiral periodicity is predictable. Newborn spirals will have a period dependent on their origin. In a hole spiral, the initial period will equal the circumference of the hole divided by the velocity of the wave. In solid-centered spirals, the initial period is likely to be the refractory period (Fig. 3). In older spirals, the center of the spiral will invariably become a continuously excitable cell mass (due to aggregation) and the inner free end of the spiral will find a minimum trajectory equal to the refractory wavelength [32]. The spiral period thus always becomes the refractory period. Fig. 2d shows a typical time course for a solid-centered spiral. All spiral waves have temporal frequencies similar to this example. The intervals between the first coils of the spiral are undefined (e.g., in a hole spiral, they depend on the velocity of the wave and the circumference of the hole). The coil period then decreases rapidly to a defined low value (finally one coil per 2.5 to 3 min). This phase is associated with collapse of the open center in open-centered spirals. Spiral propagation eventually ceases and is

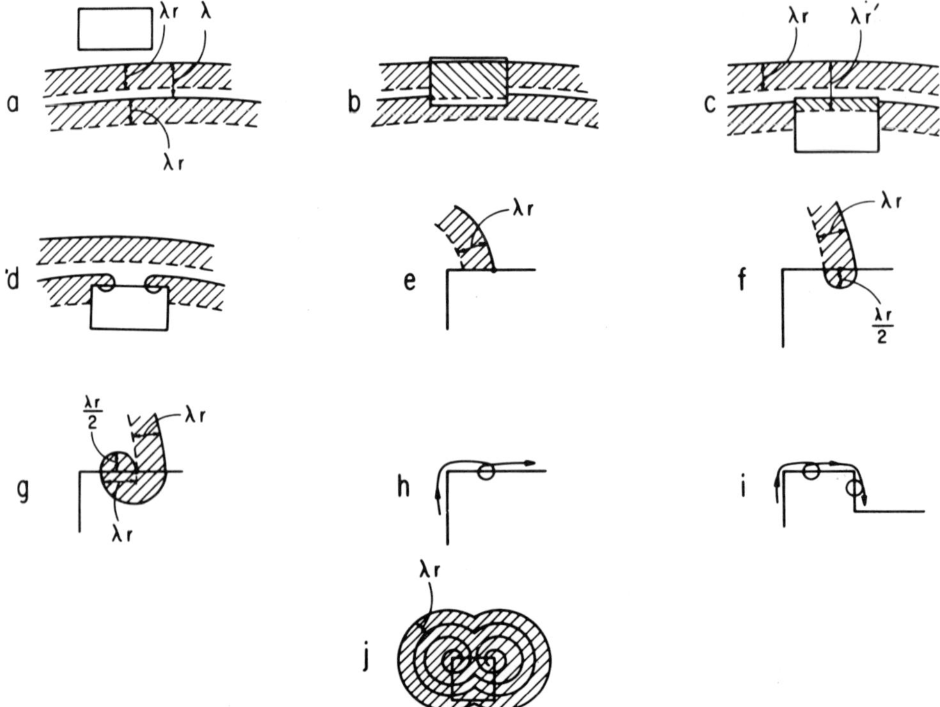

FIG. 3. Initiation of spiral waves via an interaction between two waves and an area of nonuniform refractory period in an aggregation field. (a) Two wave fronts, separated by a wavelength (λ) which slightly exceeds the refractory wavelength of the local aggregation field (λr), approach a patch of tissue with a long refractory wavelength (λr') which exceeds λ. Shaded areas are refractory. Nonshaded areas are sensitive. (b) The second wave front is broken and develops two free ends, one at each side of the patch. (c) The first wave leaves the patch, part of which remains refractory after the refractory zone of the first wave has left the surrounding tissue. (d) The free ends of the second wave propagate around the patch until they reach a part of the patch which has become nonrefractory, and then spiral into it. There is a difficulty in making a stable spiral in this way, and this is as follows: (e) Consider the point in time at which the patch of refractivity becomes sensitive and allows entry of the external wave front. (f) The wave front propagates semicircularly through the patch for a radius of λr/2, until it reaches the back of the refractory zone of the externally propagating wave. (g) It then reenters the external aggregating field and makes a second hemispherical external wave, which cannot reenter the patch as long as the wave trajectory is a straight line because it is perpetually behind the last propagating wave front in the patch and therefore is within the refractory zone (λr') of this wave. (h) The free end of the wave front therefore makes only one turn and generates a transient spiral. (i) A similar process will occur wherever the wave front encounters a convex corner to the patch. (j) Stable spirals may arise if the refractory period of the patch becomes equal to that of the field (λr' = λr). If this occurs in the whole patch before completion of the first turn

generally replaced by concentric ring wave propagation, stabilizing at 2.5 min, which is associated with formation of the tip. The spiral in this example started as a double spiral (two interlocked spiral waves) and was transiently a single spiral before it gave way to concentric propagation. The transient increase in period between the first minimum and the switch to concentric propagation is associated with loss of one of the two spiral waves. Note that the spiral period starts at about 7 min and drops rapidly to 2.5 to 3 min as would be predicted.

In view of their evolution to the minimum period (i.e., the refractory period), one would expect that spirals should be very stable wave forms. They should, within limits imposed by the onset of anisotropy and loss of physical contact between neighboring aggregates, advance the boundaries of their aggregation territories at the expense of neighboring aggregates and eventually take these over, as occurs in other excitable media [26,32]. In fact, one mutant of *D. discoideum* (80) [39] does make stable spirals as the end product of its wave propagation, but in wild-type aggregation, spirals are unstable. Their instability is mysterious, but it is often associated with formation of the tip (and mutant 80 makes no tips).

All late aggregates [spiral (Fig. 2d) and concentric (Fig. 2b)] make an apical tip as previously described. The tip is the origin for any waves made in the late aggregate and these waves are always concentric rings. This is true whether the aggregate originally made spiral or concentric ring waves. Mature tips either initiate high-frequency (one per 2.5 min) concentric ring waves or else initiate continuous cell movement towards the aggregation center. This behavior is mimicked by cAMP microelectrodes set to release cAMP continuously, and I have suggested that the mature tip is a continuous source of cAMP [6,32]. The mature state is usually reached simultaneously with or very soon after appearance of the

---

(Continued from page 308.)
of the spiral, a pair of counterrotating spirals would be expected, as shown in the figure. Other outcomes are possible if only parts of the patch change their refractivity. The periods of refractivity spirals will be equal to the local refractory period of the external medium. This is a time-dependent variable between 2 and 9 min, so their periods will be in this range and may change (decrease) with time.

tip. Occasionally concentric aggregates with tips show 5-min periodic wave propagation for an extended time (>1 h) before mature behavior is manifested. I deduce that some immature tips are either ineffective as pacemakers or periodic with a 5-min period.

E. Mutant Analysis

Most laboratory strains of *D. discoideum* are haploid. It is therefore easy to obtain mutants from them by the same methods as are used to induce mutagenesis in bacteria and yeasts [40]. Some of these mutants have been exploited by using one of two different genetic systems to begin constructing a genetic map of *D. discoideum* and to begin a genetic analysis of aggregation [41-44]. I thought it worthwhile to begin a functional analysis of defects in some morphogenetic mutants by analyzing films of their natural aggregation and of their responses to cAMP pulses delivered from a microelectrode [45,46]. I found that defects occur in each of the competences described above (Table 1). All of the mutants examined grew and moved normally as vegetative cells.

1. Chemotactic mutants were either absolutely unresponsive to cAMP pulses (G50, A5, 7) or made a quantitatively abnormal response (R46) or were apolar (10).
2. Relaying-deficient mutants were tentatively identified as chemotaxis-positive mutants, unable to make waves (or streams) in response to applied cAMP pulses over a range of amplitudes (up to 6×10^{11} molecules). All such mutants examined made some form of organized movement spontaneously, but none made spontaneous waves or streams. Two mutants (ap66, D1) made small, round aggregates (formed by movement of cells directly to the aggregation center), as would be expected if the aggregates are made by a direct chemotactic response to cAMP secreted by centers. One mutant (R46) made mobile bands of high cell density resembling water waves. The bands are often expanding rings and they can be phenocopied by placing wild-type cells on 10^{-3} M cAMP agar.

TABLE 1

Aggregation Competence Defects
in Morphogenetic Mutants of *Dictyostelium discoideum*

Mutant	Source	Defects in aggregation competences	Morphogenetic defects
A5	J. Ashworth	No chemotactic response	No aggregation
7	A. J. Durston	No chemotactic response	No aggregation
50	G. Gerisch	No chemotactic response	No aggregation; final stage: independent amoebas
10	A. J. Durston	Apolar chemotaxis	No aggregation; final stage: independent amoebas
R46	E. Rossomando [47]	Altered time for chemotaxis. Altered chemotactic radius	Final stage: motile bands of amoeba
ap66	G. Gerisch [4,46]	No signal relaying response	Round aggregates; final stage: normal fruiting bodies
1	A. J. Durston	No signal relaying response	Round aggregates; no further morphogenesis
R46	E. Rossomando	No signal relaying response	Final stage: motile bands of amoeba
91	A. J. Durston	Low X3	Many spirals; final stage: slugs
FR17	M. Sussman [48]	Aperiodic autonomous signals from early centers and tips	Final stage: abnormal fruiting bodies

I speculate that, in the wild type, these bands may occur because of cAMP degradation by the slime mold PDE enzymes. Due to PDE, each cell should act as a cAMP sink. Areas of higher cell density should therefore generate depressions in the cAMP landscape, each with a peripheral cAMP gradient. Cells in a high-density area should respond chemotactically to their peripheral gradient to generate a ring of high cell density. The ring will generate a new annular gradient at its outside edge and will respond to this by continuing to expand.

3. A number of mutants (91, 12, and also Gerisch's mutant 80) make too many spirals (and correspondingly few concentric ring centers). For example, under standard conditions, 91 makes 60% of spirals and wild-type (NC-4) makes 14%. This kind of defect is likely to be due to one of two kinds of problem.
   (a) A low rate of emergence of autonomous cells (low X3) or deficient functioning of autonomous cells, since concentric wave initiation depends directly on autonomous signaling and spiral initiation does not.
   (b) A deficiency in the tip, since demise of spirals is associated with appearance of the tip.

   In fact, the mutant 91A proves to have an X3 which is 20 times lower than that of the wild type [34]. Conversely, Gerisch's mutant 80 makes no tips [39].
4. The mutant FR17 shows defects in early autonomous signaling and in the functioning of its tip. Early FR17 concentric centers initiate aperiodic instead of periodic waves and this phenotype continues after appearance of the tip (which appears at the point of origin of the aperiodic waves). FR17 spiral waves are qualitatively normal (Fig. 4). The concentric ring waves are aperiodic in aggregates with and without tips and are occasionally initiated at points other than the tip in aggregates with tips (this is never seen in wild type). Early FR17 aggregates (synchronous aggregates) make abnormal aperiodic synchronous cell movements which are deduced to be early homologues of concentric ring waves. It is not clear whether the FR17 tip is nonfunctional and merely forms as a morphologically distinct structure in the usual place, or if it indeed makes intermittent signals.

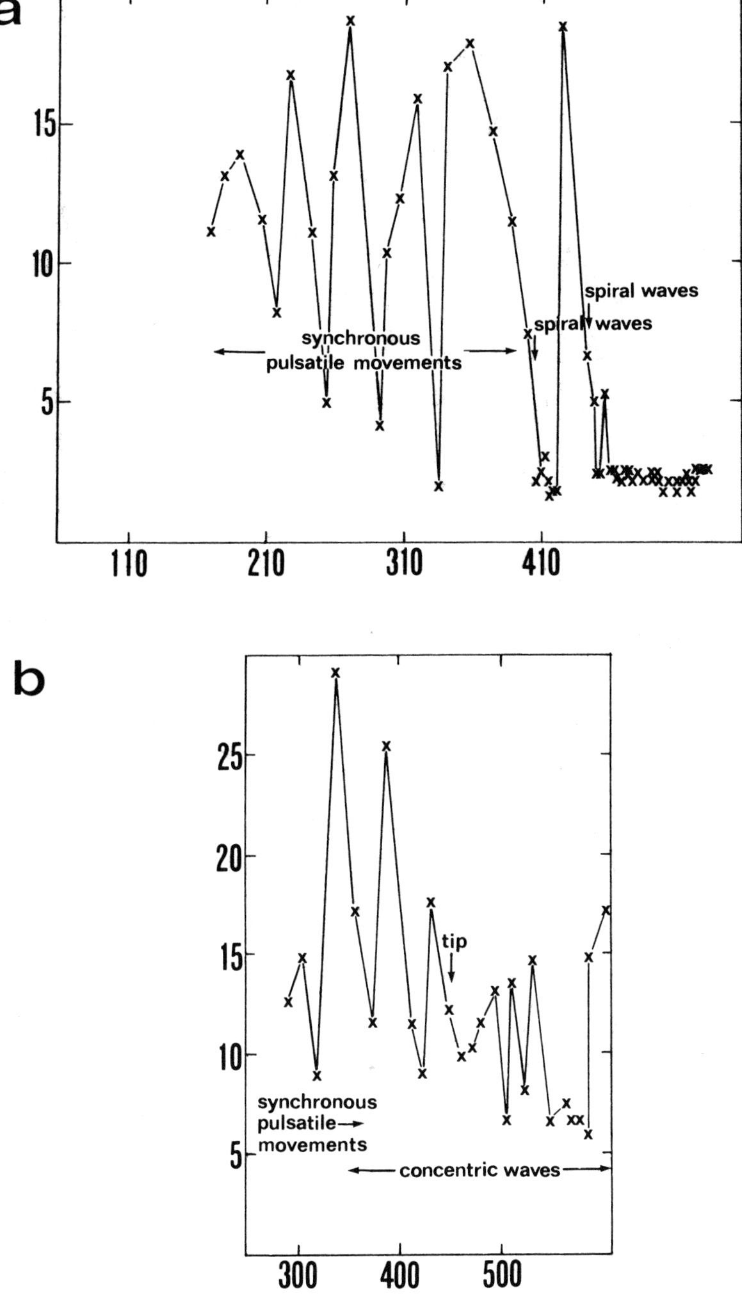

FIG. 4. The mutant FR17 has aperiodic concentric ring waves and normal spiral waves. (a) Shows temporal intervals (ordinate) against developmental age (abscissa) for a synchronous aggregate that develops to a spiral center. (b) Similarly for a synchronous aggregate that develops to a concentric center.

F. Control of Later Development

Three lines of evidence point to involvement of elements of the aggregation control system in later development in *D. discoideum*.

1. We have observed periodic pulsatile movements similar to those seen in aggregation, both in migrating slugs and in erecting fruiting bodies [45,49]. We have analyzed the movement of erecting fruiting bodies [49] and find that these elongate by making precisely periodic movement steps (period 6-½ min) superimposed on continuous movement (Fig. 5). The periodic component of fruiting body movement occurs only during the time when cells are entering the fruiting body stalk (and stops before the end of culmination). The continuous component lasts until the end of culmination and coincides in time with stalk cell vacuolation (which also plays a role in elongating the stalk since vacuolating cells increase their volume) [49,50]. We suspect that fruiting body erection is periodic because of periodic cell movements into the stalk. The periodicity of the fruiting body movements is significantly different from either the mean refractory period or the mean autonomous period during aggregation. Its correspondence with either is uncertain.
2. The polar organization of slugs and of erecting sorocarps can be disrupted by exposure to high cAMP concentrations [51]. The cause of this effect is not known. Cells from mechanically disrupted slugs and sorocarps lack overt chemotactic sensitivity to cAMP for about 50 min after disruption, and then manifest it [50,51]. The cells also reaggregate spontaneously and show normal waves and streams from about 50 min after disruption, suggesting that they have signal relaying and autonomous signaling competences at this time [52].

We must consider the possibilities that slug and sorocarp cells lack the aggregation competences and dedifferentiate to a competent state within 50 min after disruption of the multicellular structure; that they possess these competences but are

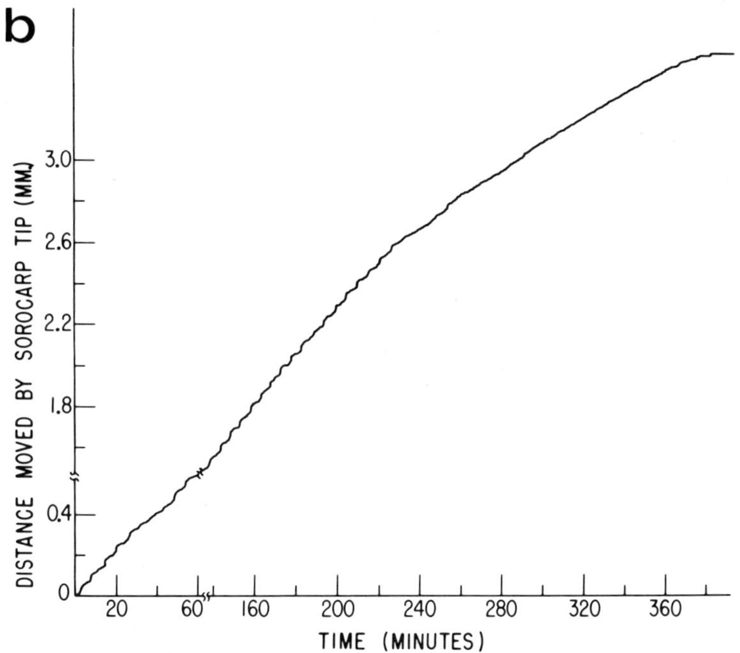

FIG. 5. (a) The later stages of the life cycle of *D. discoideum* show periodic morphogenetic movements. These are most obvious in developing sorocarps at about the stage shown. (b) A distance vs. time plot for elongation of such a sorocarp.

packaged or connected in a way that prevents an immediate, visible chemotactic response under the test conditions; that they possess the competences but make an impaired response due to damage imposed during mechanical disruption.

3. The tip of the embryonic slime mold (late aggregate, slug, and fruiting body) acts as an embryonic organizer [7,8,45]; that is, it acts as a reference point for the slime mold embryonic axis [54]. Excision of the tip delays development until a new tip is made and addition of a second tip initiates a second axis. The new tip takes over a part of the existing structure [7,8, 45].

The basis of this effect is understood in aggregation: it is that the tip is a continuous source of cAMP and therefore controls the movement of the aggregating cells, either by initiating refractory period waves of signal relaying or by inducing continuous centripetal chemotaxis [6,32]. It has now been shown that tips from late aggregates, slugs, and early sorocarps are all equivalent as organizers (i.e., that a tip from any of these stages will organize a new axis in any other stage) [8,45]. It has also been shown that tips from each of these stages continuously secrete cAMP, whereas other parts of slugs and fruiting bodies do not [7]. These results are consonant with independent demonstrations of gradients of bound and secreted cAMP in slugs and fruiting bodies [55,56]. They suggest that the organizing ability of the tip may occur due to its action as a cAMP source.

The results described above justify further enquiry into the roles of chemotaxis, signal relaying, and autonomous signaling in later morphogenesis of *D. discoideum*.

III. FINAL COMMENTS

In the preceding section, I evaluated our understanding, at the cellular level, of the control of aggregation in *Dictyostelium discoideum*. As I hope became clear from the various subsections, this understanding is very incomplete. We know some of the parameters of the chemotactic, signal relaying, and autonomous competences. We also know the roles of these competences. We know almost nothing about the parameters or importance of cell polarization and contact formation. We should seek to make a more complete description of the aggregation competences so that it will be possible to ask sophisticated and quantitative molecular questions. Experimental [47] and theoretical [57] approaches are already in progress toward understanding the biochemical oscillations which underlie the signal relaying and autonomous periodic competences.

TABLE 2

Some Examples of Periodic Events
in Development of Various Organisms

Organism	Periodic event	Period	Ref.
Acetabularia	Regeneration: contractions of the regenerate and action potentials	minutes	58
Dictyostelium discoideum	Aggregation: cell movement into aggregates	2-10 min	4,6
Dictyostelium discoideum	Culmination: fruiting body elongation	6-½ min	49
Some other cellular slime molds	Aggregation: cell movement into aggregates	minutes	59
Hydra and *Tubularia*	Head regeneration: contractions of the regenerate	minutes	60,61
Campanularia	Stolon growth	10-15 min	62
Locust	Chitin deposition	15-30 min	59
Chick	Waves of optical density, changes in the blastoderm	minutes	63
Chick	Migration of heart mesoderm cells	10 min	59

The real value of the investigations described above is that they make aggregation in *D. discoideum* the only developmental event for which any details of a morphogenetic field are known. We should use the insights obtained with this system to help us understand other morphogenetic fields. There are two approaches:

1. Investigate the role of the aggregation competences in later development of *D. discoideum*. It is especially important to know if they have any relation to the pattern of differentiation. The consensus of the slime mold literature is that this pattern originates as a regulative axial pattern in the slug. Understanding such patterns is perhaps the most difficult problem in pattern formation.
2. Investigate whether morphogenetic fields resembling the aggregation control system in *D. discoideum* are important during embryogenesis in other organisms.

In fact, periodic morphogenetic events and propagating waves of cellular behavior are common during embryogenesis and related phenomena (e.g., regeneration) in organisms as diverse as colonial bacteria, hydroids, and vertebrate embryos. Some examples are listed in Table 2. These periodic events are being investigated in a number of systems [58-63].

ACKNOWLEDGMENTS

I thank Dorothy Parsons for typing the manuscript, and Eva Bartova and Ilse Aleven for the illustrations.

REFERENCES

1. Bonner, J. T., 1967. The Cellular Slime Molds, Princeton University Press, Princeton, New Jersey.
2. Bonner, J. T., and M. Slifkin, 1949. Amer. J. Bot. 36: 727-734.
3. Gregg, J., 1965. Develop. Biol. 12: 377-393.

4. Gerisch, G., 1968. Current Topics in Developmental Biology 3: 157-197.

5. Bonner, J. T., 1944. Amer. J. Bot. 31: 175-182.

6. Durston, A. J., 1974. Develop. Biol. 37: 225-235.

7. Raper, K., 1940. J. Elisha Mitch. Sci. Soc. 56: 241-282.

8. Rubin, J., and A. Robertson, 1975. J. Embryol. Exp. Morphol. 33: 227-241.

9. Bonner, J. T., 1947. J. Exp. Zool. 106: 1-26.

10. Konijn, T. M., D. Barkley, Y. Chang, and J. T. Bonner, 1968. Amer. Nat. 102: 225-233.

11. Bonner, J. T., D. S. Barkley, E. M. Hall, T. M. Konijn, J. W. Mason, G. O'Keefe, and P. B. Wolfe, 1969. Develop. Biol. 20: 72-87.

12. Chang, Y., 1968. Science 160: 57-59.

13. Riedel, V., and G. Gerisch, 1971. Biochem. Biophys. Res. Comm. 42: 119-124.

14. Malchow, D., B. Nagele, H. Schwartz, and G. Gerisch, 1972. Eur. J. Biochem. 28: 136-141.

15. Mato, J., and T. M. Konijn, 1975. Personal communication.

16. Malchow, D., and G. Gerisch, 1974. Proc. Nat. Acad. Sci. USA 71: 2423-2427.

17. Konijn, T. M., 1974. Antibiotics and Chemotherapy 19: 96-110.

18. Robertson, A., D. Drage, and M. Cohen, 1972. Science 175: 333-334.

19. Robertson, A., and D. Drage, 1975. Biophys. J. 15: 765-775.

20. Bonner, J. T., A. Kelso, and R. Gillmor, 1966. Biol. Bull. 130: 28-42.

21. Gerisch, G., D. Malchow, and B. Hess, 1974. *In* Biochemistry of Sensory Functions, L. Jaenicke (ed.), Springer Verlag, Berlin, p. 279.

22. Bonner, J. T., 1950. Biol. Bull. 99: 143-151.

23. Cohen, M., and A. Robertson, 1971. J. Theoret. Biol. 31: 101-118.

24. Cohen, M., and A. Robertson, 1972. *In* Statistical Mechanics: New Concepts, New Problems, S. Rice, K. Freed, and J. Light (eds.), p. 131.

25. Shaffer, B., 1962. Advan. Morphogen. 2: 109-182.

26. Shaffer, B., 1975. Nature 255: 545-547.

27. Roos, W., V. Nanjundiah, D. Malchow, and G. Gerisch, 1975. F E B S Letters 53: 139-142.

28. Cohen, M., and A. Robertson, 1971. J. Theoret. Biol. 31: 119-130.

29. Cohen, M., and A. Robertson, 1972. *In* Cell Differentiation, R. Harris, P. Allin, D. Viza (eds.), Munksgaard, Copenhagen, p. 36.

30. Shante, V., and S. Kirkpatrick, 1971. Adv. in Physics 20: 325-357.

31. De Haan, 1959. J. Embryol. Exp. Morphol. 7: 335.

32. Durston, A. J., 1973. J. Theoret. Biol. 42: 483-504.

33. Gingle, A., 1975. J. Cell Sci. 20: 21-27.

34. Durston, A. J., 1974. Develop. Biol. 38: 308-319.

35. Hashimoto, Y., A. Robertson, and M. Cohen, 1976. J. Cell Sci. 21: 243-259.

36. Winfree, A., 1972. Science N.Y. 175: 634-636.

37. Wiener, N., and A. Rosenblueth, 1946. Arch. Inst. Cardiol. Mexico 16: 105.

38. Krinskii, V. I., 1971. *In* Systems Theory Research, Vol. 20, A. A. Lyupunov (ed.), Plenora Press, New York, p. 1.

39. Gerisch, G., 1972. Personal communication.

40. Yanagisawa, K., W. F. Loomis, and M. Sussman, 1967. Exp. Cell Res. 46: 328-334.

41. Katz, E., and M. Sussman, 1972. Proc. Nat. Acad. Sci. USA 69: 495-498.

42. Williams, K., R. Kessin, and P. Newell, 1974. Nature 247: 142-143.

43. MacInnes, M., and D. Francis, 1974. Nature 251: 321-324.

44. Newell, P., and M. B. Coukell, 1975. Personal communication.

45. Robertson, A., M. Cohen, D. Drage, A. Durston, and D. Wonio, 1972. *In* 3rd Lepetit Colloquium, L. Sylveshi (ed.), North Holland, Amsterdam, p. 299.

46. Gerisch, G. 1971. Naturwissenschaften 58: 430-438.

47. Rossomando, E., and M. Sussman, 1973. Proc. Nat. Acad. Sci. USA 70: 1254-1257.

48. Sonneborn, D., G. White, and M. Sussman, 1962. Develop. Biol. 7: 79-93.

49. Durston, A. J., M. H. Cohen, D. Drage, M. Potel, A. Robertson, and D. Wonio, 1976. Develop. Biol. 52: 173-180.

50. Raper, K., and D. Fennell, 1952. Bull. Torrey Botan. Club 79: 25-51.

51. Nestle, M., and M. Sussman, 1972. Develop. Biol. 28: 545-554.

52. Garrod, D., 1974. J. Embryol. Exp. Morphol. 32: 57-68.

53. Durston, A. J., unpublished results.

54. Wolpert, L., 1969. J. Theoret. Biol. 25: 1-48.

55. Bonner, J., 1949. J. Exp. Zool. 110: 259-271.

56. Pan, P., J. Bonner, H. Wedner, and C. Parker, 1974. Proc. Nat. Acad. Sci. USA 71: 1623-1625.

57. Goldbeter, A., 1975. Nature 253: 540-542.

58. Goodwin, B., 1974. Personal communication.

59. Robertson, A., and M. Cohen, 1972. Ann. Rev. Biophys. Bioeng. 1: 409-464.

60. Durston, A. J., and S. Newman, unpublished results.

61. Cooke, J., and G. Webster, 1972. Personal communication.

62. Wyttentach, C., S. Crowell, and R. Suddith, 1973. J. Morphol. 139: 363-375.

63. Robertson, A., personal communication.

PART III

Dormancy and Germination

EDITORS' INTRODUCTION

Most eucaryotic microbes are capable of entering a phase of dormancy, which is generally characterized by the appearance of a thickened cell wall, reduced metabolic and synthetic activities, and heightened resistance to adverse environmental conditions. The breaking of dormancy (germination) is typically characterized by dramatic increases in oxygen consumption and the synthesis of various macromolecules. The use of unicellular microbial spores and cysts in the study of cellular differentiation provides some unique advantages over other developing systems. One of the most important of these includes the ability to acquire a large, homogeneous population of cells which can be induced to undergo a reversible differentiation in a very precise manner. Thus by altering specific environmental parameters, vegetative cells can be made to encyst or sporulate and, subsequently, the dormant cells (cysts or spores) can be induced to germinate. As a result, microbial dormancy-germination systems make themselves amenable to many kinds of studies of the events and controls in cellular differentiation. In this chapter such systems have been exploited for the study of membrane behavior, the function of extracellular enzymes, and the regulation and requirements of macromolecular synthesis during dormancy and germination.

MEMBRANE BEHAVIOR

There is no doubt that cellular membranes are dynamic structures in a constant state of flux [1], and current models of membrane structure, such as the "fluid mosaic model" [2], attempt to compensate for this. Still, many aspects of membrane synthesis, construction, and repair remain to be elucidated. John E. Thompson has utilized the encystment process in *Acanthamoeba* in an analysis of membrane behavior during cellular differentiation. In his article, he demonstrates the existence of active exchange of membrane components between the cytosol and the plasma membrane.

EXTRACELLULAR ENZYMES

Many eucaryotic cells excrete enzymes into the extracellular environment. Of particular interest to developmental biologists is the role of excreted enzymes in modification of the cell surface. For example, the extracellular release of protease appears to effect certain membrane glycoprotein changes that dictate the behavior of transformed cells [3]. Changes in surface carbohydrates also appear concomitant with morphogenesis and differentiation [4]. These changes may be effected by extracellular glycosidases. In his article, Danton H. O'Day describes the patterns of excretion of several hydrolytic enzymes and provides evidence that an acid protease and β-glucosidase may be critical enzymes for cell wall removal during germination of microcysts of *Polysphondylium pallidum*.

MACROMOLECULAR SYNTHESIS

Because the breaking of dormancy (germination) involves the sequential reinitiation of macromolecular syntheses, it has appealed to investigators interested in the means which eucaryotic cells have evolved to regulate cellular transcriptive and translative mechanisms. Although the diversity of microbial dormant structures discourages generalizations, it can be said that coincident protein synthesis is usually essential for germination. On the other hand, DNA synthesis generally is not essential, while the requirement for RNA synthesis varies. The patterns and requirements of macromolecular synthesis in various eucaryotic microbes have been recently reviewed [5,6].

In this part, James L. Van Etten and his co-workers discuss the germination of *Rhizopus stolonifer* sporangiospores in which both RNA and protein synthesis initiate simultaneously and appear to be essential for germination. In their studies they reveal significant changes in the protein and RNA synthetic apparatus during germination in *Rhizopus*. Of particular interest is the early initiation of mRNA synthesis during germination, before the activity of RNA polymerase II can be detected.

James S. Lovett and his co-workers elucidate a different sequence of macromolecular synthetic events during zoospore germination in *Blastocladiella emersonii*. Protein and RNA synthesis begin early during rhizoid outgrowth, but only protein synthesis is essential at this time. They demonstrate the presence of a completely functional preprogrammed system for protein synthesis in dormant zoospores that appears to be suppressed by the presence of a ribosome inhibitor.

REFERENCES

1. Siekevitz, P., 1972. Ann. Rev. Physiol. 34: 117-140.
2. Singer, S. J., and G. L. Nicolson, 1972. Science 175: 720-731.
3. Hynes, N. O., 1974. Cell 1: 145-156.
4. Kleinschuster, S. J., and A. A. Moscona, 1972. Exp. Cell Res. 70: 397-400.
5. Sussman, A. S., and H. A. Douthit, 1973. Ann. Rev. Plant Physiol. 24: 311-352.
6. Van Etten, J. L., L. D. Dunkle, and R. H. Knight, 1975. The Fungal Spore: Form and Function, D. J. Weber and W. M. Hess (eds.), Wiley-Interscience, New York, p. 243.

ENCYSTMENT AND EXCYSTMENT IN *ACANTHAMOEBA CASTELLANII*: A MODEL SYSTEM FOR EXAMINING THE BEHAVIOR OF MEMBRANES DURING DEVELOPMENT

John E. Thompson

Department of Biology
University of Waterloo
Waterloo, Ontario, Canada

INTRODUCTION

Acanthamoeba castellanii is a small, free-living soil amoeba that can exist in either of two states of differentiation--a vegetative stage (trophozoite) and a resting stage (cyst) [1]. The trophozoite possesses a full complement of subcellular organelles and, unlike some of the larger amoebas, is surrounded by a naked plasma membrane [2]. The vegetative cells can be cultured axenically in an enriched growth medium containing yeast extract and proteose peptone. They grow exponentially in this medium with a mean generation time of 7 h to a population density of about 10^6 cells/ml. At this point the growth rate declines and eventually reaches a plateau as the cells enter stationary phase [3].

Like many protozoa, *Acanthamoeba castellanii* adapts to alternating favorable and unfavorable environmental conditions by undergoing reversible differentiation. When the vegetative cells encounter environmental stress, they react by transforming into cysts [4]. The specific nature of the environmental stimulus which induces encystment is not completely clear. However, for laboratory cultures the transformation can be induced by depriving the cells of nutrients [4]. In response to the encystment stimulus, the trophozoite undergoes metabolic changes that direct the progressive synthesis of a cyst wall, which in the mature cyst comprises about 37% of the cell dry wt. The cyst wall is composed to a large extent of protein (33%) and cellulose (35%), and at maturity comprises two layers, an exocyst and an endocyst [5,6]. Encystment culminates in the formation of a partially dehydrated cell that can withstand prolonged starvation and desiccation. Metabolic activity is very low [7].

Excystment occurs once propitious environmental conditions return [4], and can be induced in the laboratory by transferring cysts from the salt medium (encystment medium) back into growth medium. Mattar and Byers [8] have followed excystment of *Acanthamoeba castellanii* induced in this manner by phase-contrast microscopy, and were able to observe actual emergence as well as formulate a sequence of morphological changes leading up to the event. As might be expected, excystment is dependent upon macromolecular synthesis. Emergence is inhibited by both actinomycin D and cycloheximide, implying that RNA and protein synthesis are both required [8]. In contrast, emergence is essentially unaffected by the presence of hydroxyurea [8], suggesting that DNA synthesis is not required until postemergence. Thus the patterns of macromolecular synthesis during germination in *Acanthamoeba castellanii* may be similar to *Rhizopus stolonifer*, which is described in a subsequent article in this volume by James L. Van Etten.

II. CURRENT RESEARCH

Inasmuch as *Acanthamoeba castellanii* can be mass-cultured axenically and induced to undergo reversible differentiation, it is well suited for a study of membrane behavior during cell differentiation. However, in order to realize this objective it has first been necessary to develop methods for synchronizing differentiation. There has also been a need to establish dependable morphological criteria reflecting the sequence of each differentiation process, so that temporal correlations between changes in membrane properties and specific stages of germination or cyst formation can be made. These prerequisites have been met, at least to the point where meaningful investigations of the membrane systems in *Acanthamoeba* are possible. Studies designed to characterize the behavior of its membranes and to recognize the mechanistic basis for this behavior have been carried out, particularly in respect of the plasmalemma.

A. Synchronous Encystment and Excystment

Synchronous encystment of *Acanthamoeba castellanii* (strain Neff) can be induced by transferring the amoebas from optimal growth medium into a buffered salt solution [7,9]. The first morphological indications of cyst initiation visible by phase-contrast microscopy are a general rounding of the organism and the disappearance of filopodia. A measure of the synchrony of transformation is illustrated in Fig. 1. Initiation of encystment is complete within 20 h, and by 48 h the cysts are all mature.

Excystment occurs when the mature cysts are placed back in the enriched growth medium. However, the usefulness of this procedure in experimental investigations of excystment is limited by the fact that it gives rise to incomplete and highly asynchronous germination [8,10]. During an investigation of ways to surmount this problem we found that capability for excystment is related to cyst age.

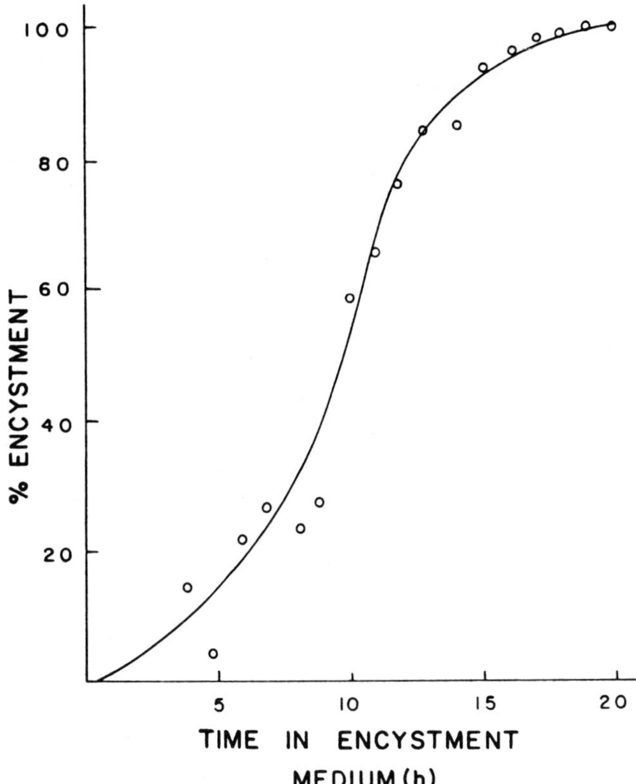

FIG. 1. Synchrony of encystment for *Acanthamoeba castellanii*. Trophozoites were transferred from growth medium to encystment medium and the progress of encystment was monitored by phase-contrast microscopy.

Rates of excystment for cultures aged in the encystment salt medium for varying periods of time are illustrated in Fig. 2. These data were obtained by counting numbers of empty cyst walls in excysting cultures. The cyst walls persist as recognizable structures for at least 75 h following emergence of the trophozoite [8], and are readily distinguishable from intact cysts and amoebas by phase-contrast microscopy. In general, with increased length of the aging period there is a corresponding increase in both the synchrony and the extent of excystment. Cultures 30 to 50 h of age are

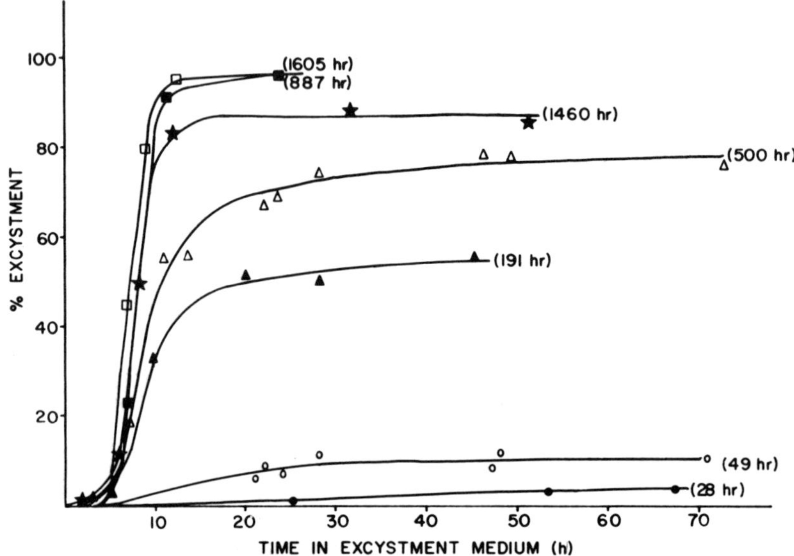

FIG. 2. Synchrony of excystment for *Acanthamoeba castellanii*. Cysts were aged in encystment medium for periods of time indicated in parentheses and then transferred to growth medium. Excystment in the growth medium was scored by counting empty cyst walls with a phase-contrast microscope.

capable of only about 10% excystment, and this occurs essentially asynchronously (Fig. 2). By 500 h a much higher proportion of the organisms is able to excyst and some degree of synchrony is evident. After 800 h, 85 to 95% of the organisms excyst within about 12 h of the initial transfer of cysts into the growth medium.

It is clear, therefore, that mere exposure of newly matured cysts to a renewed source of nutrient is not sufficient to spark excystment. Cysts which look mature are formed after only 24 h in encystment medium yet organisms which have been in this medium for as long as 50 h are still for the most part incapable of excysting when transferred to the enriched growth medium. This suggests that some metabolic or physiological process, initiated when encystment begins, must run its full course irrespective of the presence of nutrients before excystment will occur. This process may constitute a final phase of cyst maturation that is not manifested morphologically. The poor degree of excystment evident in asynchronous cultures may be attributable to an excystment inhibitor produced by the few trophozoites that emerge and begin to replicate. Such an inhibitor does accumulate in actively growing cultures of *Acanthamoeba* trophozoites in concentrations directly proportional to the cell population density [11]. There is sufficient time in asynchronous cultures with low excystment rates for trophozoites to significantly increase their numbers by replication, but this does not hold true for synchronous cultures. Thus, the inhibitor would be expected to have a pronounced effect on young cultures and virtually no effect on synchronously excysting cultures.

B. Surface Topology During Encystment and Excystment

The most conspicuous morphological changes accompanying differentiation of this organism occur at the cell surface. These alterations in surface structure have been documented by scanning electron microscopy to enable recognition of specific stages during encystment and excystment. The organisms were processed as described previously [9,10].

Vegetative cells are characterized by smooth, filamentous filopodia arising from a convoluted and folded surface (Fig. 3A). Trophozoites prey upon bacteria in their native soil habitat, engulfing them by phagocytosis [1]. Amoebas in suspension ingest latex beads in a manner that simulates naturally occurring phagocytosis [12,13], and bead ingestion as seen by scanning electron microscopy is illustrated in Fig. 3B. The bead becomes attached to the external surface of the plasma membrane. Thereafter the surface membrane creeps up around the edge of the bead giving rise to a gullet-like tube of membrane which serves as a food cup and ultimately engulfs the bead. Large, mouth-like openings corresponding to the open external ends of the tubes are evident (Fig. 3B). These open ends apparently close once the bead has been engulfed. Expenditure of plasma membrane is clearly a major consequence of phagocytosis, and there is

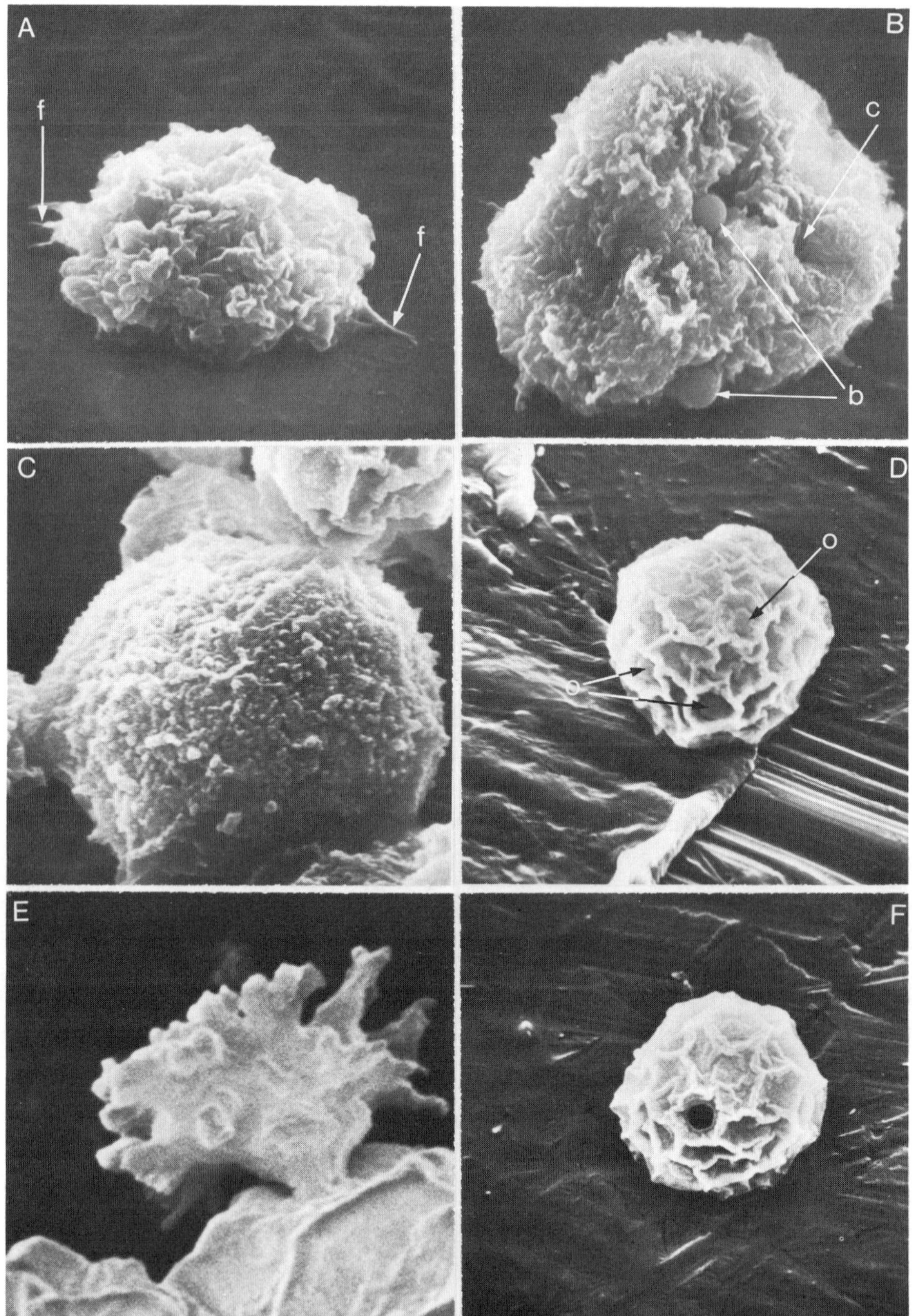

FIG. 3. Scanning electron micrographs of *Acanthamoeba castellanii*. (A) An amoeba prior to the addition of latex beads (× 3515). (B) An amoeba ingesting latex beads (× 2982). (C) Immature cyst (× 2821). (D) Mature cyst (× 3293). (E) An amoeba emerging from its cyst (× 13083). (F) An empty cyst wall (× 3462). b, bead; c, food cup; f, filopodia; o, ostiole. (From Refs. 9, 10, and 13.)

evidence to suggest that the filopodia and extensive surface convolutions serve as a reserve source of plasma membrane for use during active ingestion [13].

The first morphological signs of encystment visible by scanning electron microscopy are a shortening and thickening of filopodia. As the filopodia recede the surface convolutions become less severe. Ultimately, thin ridges become quite distinct over the cell surface (Fig. 3C). These ridges gradually become thicker and interconnected as the cyst matures. The fully mature cyst features prominent interconnecting ridges which delimit shallow, irregularly shaped craters (Fig. 3D). Several ostioles are present and can be recognized by a distinct circular plug (operculum), which is also delimited by a ridge.

Germination is first apparent by the appearance of a cytoplasmic bud protruding from the surface of the cyst (Fig. 3E). The cytoplasmic bud is delimited by a circular ridge, which implies that the amoeba is emerging through an ostiole from which the operculum has been removed. The remainder of the cyst surface still features the characteristic ridges and craters of a mature cyst. Postemergence is characterized by the presence of free trophozoites and empty cyst walls. The distinguishing feature of the empty cyst wall is the irregularly shaped hole through which the amoeba has emerged (Fig. 3F). In every other respect the empty cyst wall has the same surface topology as a mature cyst.

C. Pinocytosis and Phagocytosis by Amoebas

Perhaps the most important property of the plasma membrane in this organism is its role in mediating endocytosis. The amoebas are clearly able to phagocytize particles [12,13], and there is also evidence to indicate that they take in carbohydrates and amino acids by pinocytosis [14]. Comparisons of phagocytic activity in logarithmic and stationary phases of growth have revealed that exponentially growing cells ingest latex beads progressively as a

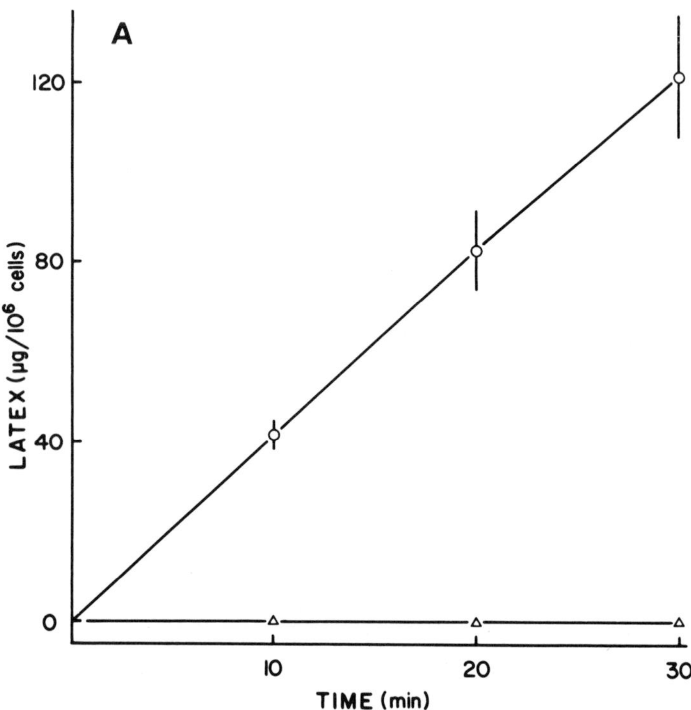

FIG. 4. Endocytotic activity of exponential and stationary phase amoebas. (A) Phagocytosis of polystyrene latex beads. The amoebas were resuspended at a population density of 3.4×10^6 cells/ml in the growth medium from which they had been harvested. Polystyrene latex beads (1.01 μm diam) were added at a concentration of 3.6 mg/ml. O, exponential phase cells; Δ, stationary phase cells. Standard errors of the means are indicated; n = 4.

function of time, but that this capability is essentially lost once the cells achieve stationary phase (Fig. 4A). Levels of ingested latex were determined by extracting the cells with dioxane and measuring dissolved latex in the extract spectrophotometrically [15]. Dioxane extracts of stationary phase cells incubated in the presence of latex beads consistently yielded undetectable levels of solubilized latex. The significance of this observation was confirmed by examining wet mounts of the cells with a phase-contrast microscope at the end of the ingestion period; exponential phase cells were

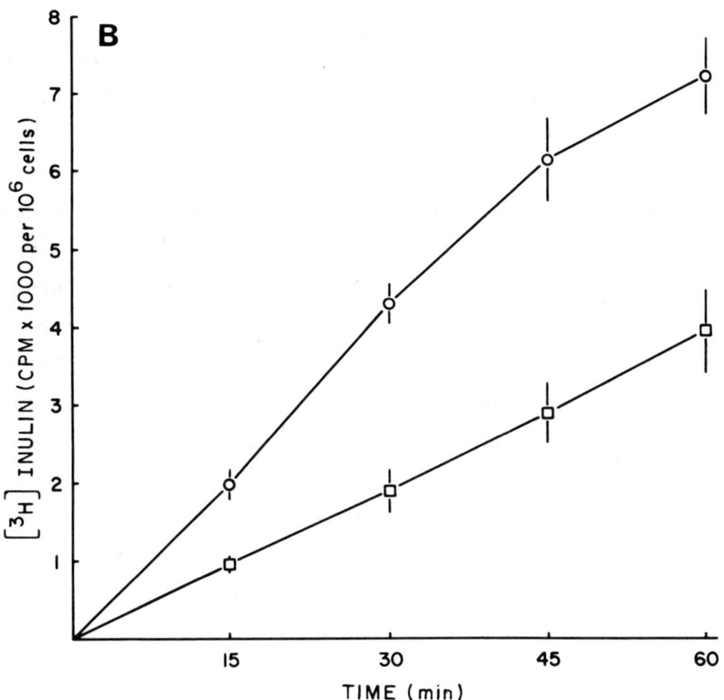

(B) Pinocytosis of [^{3}H]inulin. The amoebas were resuspended at a population density of 1.2×10^6 cells/ml in the growth medium from which they had been harvested. [^{3}H]inulin (1.5×10^{15} dpm/mole) was added at a concentration of 4.6×10^6 dpm/ml. ○, exponential phase cells; □, stationary phase cells. Standard errors of the means are indicated; n = 3 or 4.

packed with internalized beads, containing at least 100 per cell, whereas stationary phase cells regularly contained only 2 to 8 beads per cell. The dioxane extraction procedure for quantifying internalized latex is presumably not sufficiently sensitive to pick up such a low concentration.

Comparative studies of pinocytosis for logarithmic and stationary phase cells revealed that ingestion by this mode is also reduced for cells in plateau phase (Fig. 4B). Pinocytotic activity was measured by monitoring the uptake of [^{3}H]inulin [14]. The alteration in pinocytosis is not as drastic as for phagocytosis in that stationary phase cells are still capable of ingesting inulin. However, they do so at a rate only about half that characteristic of exponential cells.

The waning endocytotic activity of stationary phase cells would appear to be more a reflection of altered conditions within the cells themselves than of the growth medium. This is supported by the observation that simply replenishing the nutrient status of stationary phase cells had no detectable refurbishing influence on phagocytosis, and only marginally improved pinocytosis. This was true despite the fact that in experiments designed to test this, the stationary phase cells were preincubated in fresh growth medium for up to 4 h prior to initiating ingestion. Had the impairment of endocytosis been due only to an impoverished growth medium, such deficiencies should logically have been surmounted after a period of this duration in fresh medium, particularly since stationary phase cells retain upwards of 50% of log phase pinocytotic activity.

Attenuation of endocytosis may well explain previous observations of a reduced need for stationary phase cells to discharge cytoplasmic water [16]. Pinocytosis is doubtless the mechanism by which the trophozoites of *Acanthamoeba* ingest liquid growth medium. In fact were it not for its ability to pinocytose this organism, which feeds predominantly by phagocytosis in its native environment, could probably not be cultured axenically in chemically defined liquid growth medium. The ingestion of fluid arising from pinocytosis is counterbalanced by excretion of cellular water mediated by the contractile vacuole [17]. Inasmuch as phagocytosis by these amoebas occurs in such a manner as to virtually exclude entry of liquid medium [15], it is probably the alteration in pinocytosis rather than phagocytosis that lessens the requirement of stationary phase cells for stringent osmotic control. Indeed, the reduction in contractile vacuole activity of stationary phase cells as compared with those in exponential growth [16] is of much the same magnitude as the decrease in pinocytosis.

D. Properties of Isolated Plasma Membrane from Amoebas

The observation that the curtailed endocytotic activity of stationary phase cells is not directly attributable to nutrient depletion suggests that the amoeboid plasma membrane sustains modifications coincident with this change in growth behavior, which impede the ingestion mechanism. To investigate this the properties of isolated plasma membranes from log and stationary phase amoebas were compared. The membranes were isolated as described previously [18]. This procedure has since been shown to compare favorably with an alternative method developed by Ulsamer et al. [19]. The membrane preparations were essentially free of enzymatic markers for cytoplasmic membranes, and electron micrographs of the isolated fractions showed no evidence of contamination by cell organelles or actin (Fig. 5). Gel electrophoresis has confirmed that actin is not a major component of these isolated membrane fractions [3]. For both exponential and stationary phase cells the isolated membranes were in the form of vesicles and strands, and displayed a distinct trilamellar feature measuring 85 to 100 Å (Fig. 5).

The isolated membranes possess ATPase (EC 3.6.1.3), 5'-nucleotidase (EC 3.1.3.5), alkaline phosphatase (EC 3.1.3.1), and various enzymes of lipid metabolism [18-20]. Cytochemical tests have shown that alkaline phosphatase is also present on the contractile vacuole membrane [21]. Indeed the presence of this enzyme in preparations of purified plasma membrane is thought to partially reflect the fact that the contractile vacuole becomes associated with the plasmalemma during discharge [3]. Since *Acanthamoeba castellanii* eliminates a volume of water approximately equal to its weight in 15 to 30 min [16], a substantial proportion of the cells in a population would be engaged in water expulsion at any one time. The vacuolar membrane associated with the cell surface during discharge would in essence be part of the plasmalemma, albeit transiently, and would isolate with it during fractionation, thus accounting at least in part for the presence of alkaline phosphatase.

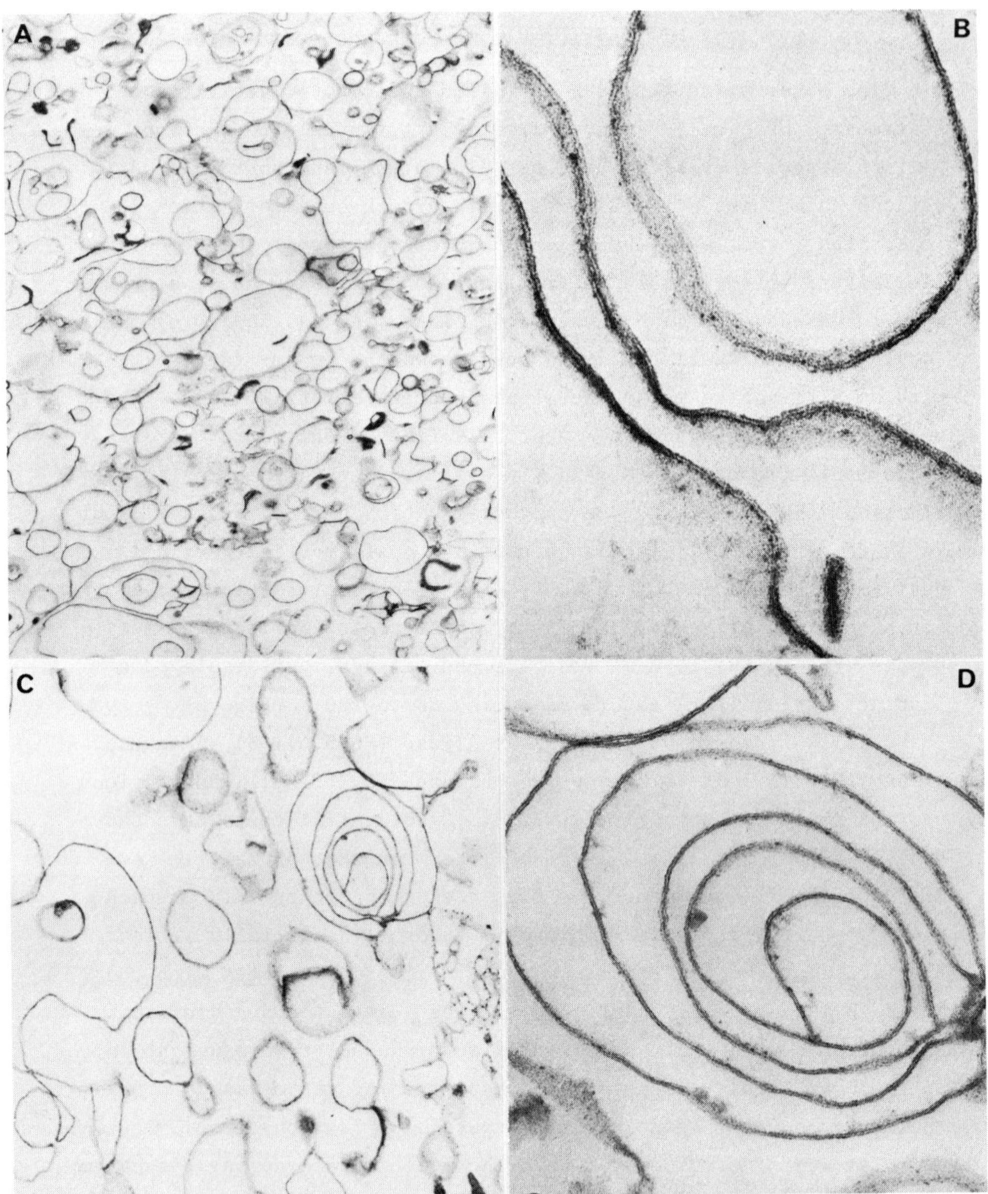

FIG. 5. Transmission electron micrographs of isolated plasma membranes from trophozoites of *Acanthamoeba castellanii*. (A) From cells in stationary phase (× 13612). (B) From cells in stationary phase (× 65587). (C) From cells in exponential phase (× 22275). (D) From cells in exponential phase (× 55687).

A comparison of alkaline phosphatase activities for isolated plasma membrane from exponential and stationary phase cells revealed a marked difference. A specific activity of 84.4 μM substrate hydrolyzed/mg protein/h ±17.4 (S.E. for n = 4) was obtained for purified membranes from exponential cells, yet the corresponding value for plasma membranes from stationary phase cells was only 17.6 ± 11.3 (S.E. for n = 7). This difference is also reflected in the enrichments of alkaline phosphatase in the plasma membrane fractions, relative to homogenates, on a specific activity basis. Preparations from exponentially growing cells showed enrichments of from 10.8 to 23.6, whereas corresponding values for stationary phase cells ranged from 1.9 to 7.6.

The reduced specific activity of alkaline phosphatase in surface membrane fractions from stationary phase cells was not due to higher levels of contamination by mitochondrial and microsomal membranes; nor does the reduction reflect a decrease in the cellular level of the enzyme, for its activity per cell remains essentially constant during the transition to stationary phase growth. Rather the diminished plasma membrane activity appears to be due to a marked shift in enzyme localization from the cell surface to cytoplasmic membranes. Three fractions, a 10,000*g* pellet, which includes the major proportion of fragmented plasma membrane, mitochondria, and lysosomes as well as a microsomal fraction and a soluble fraction, were assayed (Fig. 6). For both exponential and stationary phase cells very little of the alkaline phosphatase activity (5 to 6%) was found to be soluble. The majority of the activity (72%) sedimented in the 10,000*g* pellet for exponential cells, leaving only 22% in the microsomal fraction. Yet, for cells in stationary phase a substantially higher proportion of the enzyme (40%) sedimented in the microsomal fraction and the proportion in the 10,000*g* pellet was correspondingly lower (55%). Recoveries of enzyme activity in these experiments ranged from 79 to 98%.

It would appear, therefore, that in stationary phase cells a greater proportion of the total membrane bearing alkaline phosphatase is internalized, as opposed to being associated with the

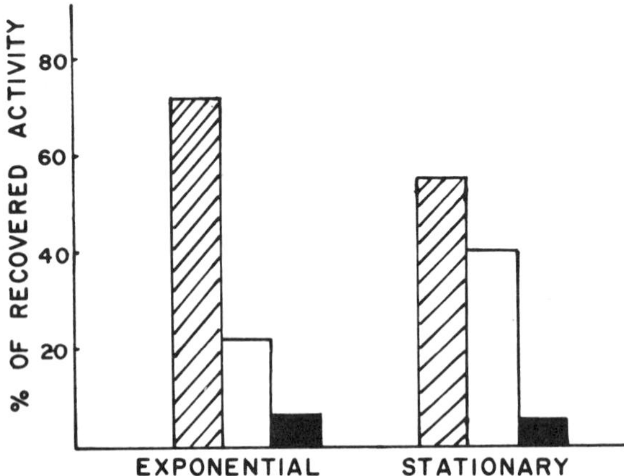

FIG. 6. The subcellular distributions of alkaline phosphatase for exponential and stationary phase amoebas. Recovered activity is the sum of the total activities in each of the isolated fractions. Hatched bars, 10,000*g* pellet; clear bars, microsomal fraction; solid bars, soluble fraction.

plasmalemma, than is the case for exponentially growing cells. While it is true that a decreased specific activity of alkaline phosphatase on the plasma membrane could also reflect an increase in the abundance of protein per unit membrane, this appears not to be the case since the electrophoresis patterns of plasma membrane from exponential and stationary phase cells are virtually identical (Fig. 7). Any increase in protein of sufficient magnitude to account for the change in alkaline phosphatase specific activity would certainly have registered on the gels, and since the patterns are identical, could only have been in the form of a proportional increase in all membrane proteins per unit plasmalemma. This is a highly implausible prospect. The gel patterns are much more consistent with the conclusion that there is an effective loss of alkaline phosphatase from the plasma membrane coincident with the attainment of stationary phase, and indeed the increased levels of microsomal alkaline phosphatase in stationary as compared with exponential phase cells support this contention. Since alkaline phosphatase is known to be present on the contractile vacuole, its diminished pre-

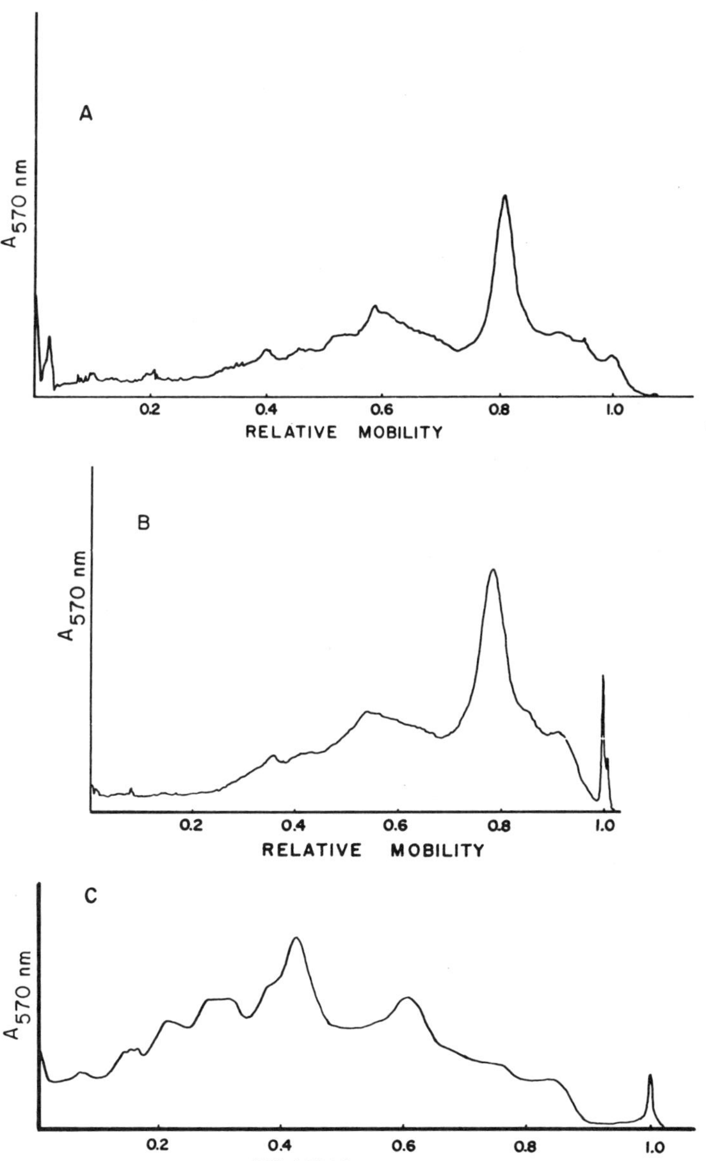

FIG. 7. Polyacrylamide gel electrophoresis scans of isolated plasma membrane from trophozoites of *Acanthamoeba castellanii*. The gels were stained for protein with Coomassie blue. India ink was used to mark the location of the tracking dye and accounts for the peaks at an electrophoretic mobility of 1. (A) Untreated membrane from exponential phase cells run on a 10% gel. (B) Untreated membrane from stationary phase cells run on a 10% gel. (C) Membrane from stationary phase cells treated with 0.5 M KCl and electrophoresed on a 7.5% gel.

sence on the plasmalemma and increased presence in microsomal fractions can in part be interpreted as reflecting the decreased incidence of water expulsion in stationary as opposed to exponential phase cells. Thus the redistribution of this enzyme within the cell is consistent with the attenuation of endocytotic activity in stationary phase cells.

Isolated plasma membranes were also examined electrophoretically to determine whether changes in macromolecular composition are incurred as the cells reach a growth plateau. Solubilization of the isolated membranes in sodium dodecyl sulfate and polyacrylamide gel electrophoresis were carried out as described previously [3,22]. Coomassie blue-stained gels were essentially similar for the two types of membrane (Fig. 7A and B). Each featured a major band at a relative mobility (R_f) of about 0.8, which corresponded to a polypeptide with a mol wt of about 15,000 and accounted for 40 to 50% of the total protein. Gels for membrane from both types of cells also featured three additional protein bands of high enough intensity to register in spectrophotometric scans, which ran at R_f values ranging from 0.5 to 0.7. These bands were more discernible in our gels than in those of Korn and Wright [22] for plasma membrane from exponential cells of *Acanthamoeba* presumably because we applied 50 to 60 μg of protein to the gels whereas they applied only 3 to 12 μg. In addition there were 6 to 8 minor bands of lower intensity corresponding to proteins ranging in mol wt from 35,000 to about 70,000. These bands were not of sufficient intensity to register distinctly in spectrophotometric scans, but were clearly visible by eye and could therefore be compared for actively growing and stationary phase cells. Small differences in both the number of such bands and their relative mobilities were observed, but they were inconsistent and are thought to reflect varying amounts of minor contaminants rather than discrete changes in the plasma membrane. For both types of membrane the two phosphonoglycan bands reported previously [3,22] were evident at R_f values of about 0.85 and 0.95 in gels stained with periodic acid-Schiff reagent.

The essential homology of the protein-banding patterns for the two types of membrane suggests that there is no change in protein composition coincident with the transformation to stationary phase. However, when plasma membrane isolated from stationary phase cells is treated with salt (0.5 or 1 M KCl) or a mixture of 5 mM EDTA and 1 mM β-mercaptoethanol in the manner known to remove spectrin from erythrocytes [23], the major 15,000 mol wt protein characteristically running at an R_f of about 0.8 is removed (Fig. 7C). By contrast such treatment has no effect on plasma membrane from exponential phase cells. In gels for untreated membrane (Fig. 7A and B) the bands at R_f values of 0.55 to 0.65 are of higher intensity than those at R_f values of 0.30 to 0.40, while in gels for KCl-treated membrane (Fig. 7C) the reverse is true. This presumably reflects less complete destaining in the low-mol wt region of the gels for untreated membrane. Coomassie blue readily diffuses out into the gel from regions of high concentration such as that corresponding to the 15,000 mol wt protein, and in the event such stain lodges in the vicinity of other protein bands, it is much more difficult to remove than simple background stain in the region of the gel where there are no protein bands. This phenomenon would tend to artificially accentuate the intensity of protein bands in close proximity to one which is in extreme quantitative preponderance, such as those adjacent to the 15,000 mol wt protein in gels for untreated membrane.

It would appear therefore that the transition to stationary phase is marked by a molecular reorganization of the plasmalemma whereby the major low-mol wt protein becomes more loosely associated with the membrane. This could well account for the reduced endocytotic activity of stationary phase cells; it could for example render sites on the cell surface, to which particles being ingested normally attach, cryptic. The question of whether the 15,000 mol wt protein is integral or peripheral has not yet been resolved.

E. Molecular Exchange Between Plasma Membrane and Soluble Cytoplasm

There is now substantial evidence to support the concept that membranes are dynamic structures whose macromolecular components are continuously being degraded and synthesized [24]. In fact turnover of macromolecules appears to be a characteristic feature of virtually all membranes in the eucaryotic cell. This phenomenon could well play a role in mediating membrane repair, membrane modification, or membrane differentiation. In our studies with *Acanthamoeba castellanii* we have attempted to illuminate further the question of membrane turnover by examining the extent to which renewal of plasma membrane protein constituents is mediated by interaction at a molecular level between plasmalemma and soluble cytoplasm or cytosol, as distinct from a turnover of membrane fragments.

An in vitro experimental approach was used. Basically, isolated plasma membranes were incubated for 15- to 30-min intervals in a corresponding cytosol fraction (microsomal supernatant) which had been obtained from cells cultured in the presence of ^{14}C-labeled amino acids, and the incorporation of radioactivity into the membrane was monitored. A preponderance of the label (60 to 92%) associating with the membrane under these conditions is in the form of protein; the remainder is extractable with lipid solvents. Moreover, as the ratio of cytosol protein:membrane protein in the incubation mixture is made higher, there is a corresponding increase in the specific radioactivity of the incubated membrane (Fig. 8). Ultimately, however, this increase in membrane radioactivity reaches a plateau and remains constant with further increase in the cytosol:membrane ratio (Fig. 8). This saturation consistently occurred at values for the ratio ranging from 10 to 12, but the specific radioactivities of the treated membrane at saturation varied among experiments. Such variation doubtless reflects differences in the specific radioactivity of the isolated cytosol fraction from experiment to experiment. In addition, when cytosol was present in saturating amounts in the incubation mixture, the

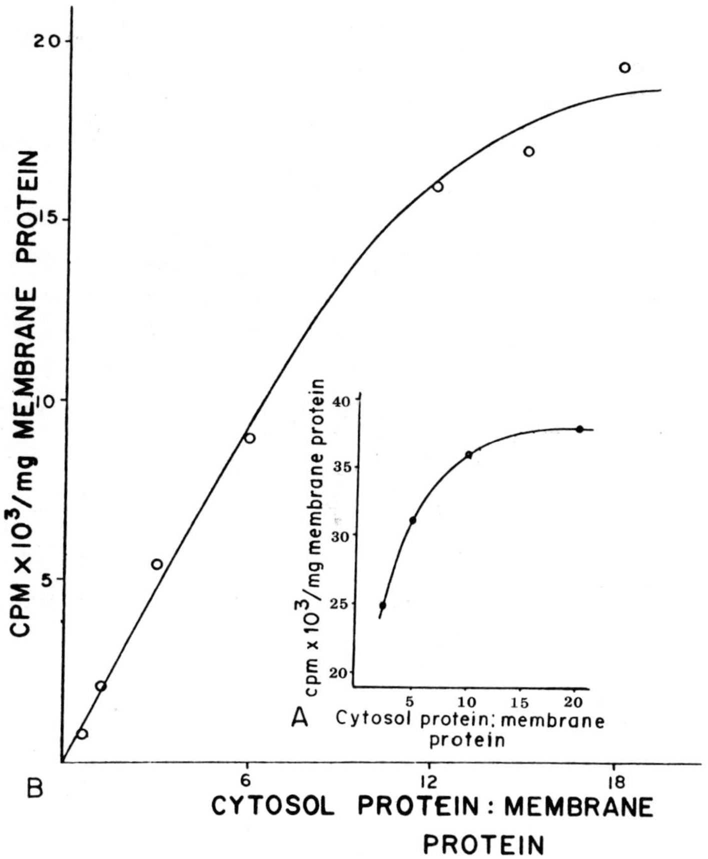

FIG. 8. Changes in the specific radioactivity of incubated plasma membrane from *Acanthamoeba* trophozoites with increasing ratios of cytosol protein to membrane protein in the incubation mixture. The incubation mixture consisted of 0.05 M Tris, pH 7.5, plasma membrane suspended in 1 mM $NaHCO_3$, pH 7.5, and cytosol fraction in a total volume of 5 ml. The incubations were for 15 min. Each point is an average of duplicates. (A) and (B) illustrate separate experiments.

incorporation of radioactivity into the membrane increased linearly as a function of time (data not shown).

As a further parameter by which to judge the specificity of interaction between plasma membrane and cytosol under these conditions, treated membranes were electrophoresed and the patterns of radioactivity and protein distribution along the length of the gels compared. For some of these experiments the cytosol fraction to be used for the incubation was labeled by culturing the cells in the presence of ^{14}C-labeled amino acids as referred to above. However, in order to obtain higher radioactivity levels in the treated membrane additional experiments were carried out in which the proteins of the isolated cytosol fractions were labeled with ^{125}I by using lactoperoxidase [25]. The iodination reactions were stopped with cysteine before these cytosol fractions were used for membrane incubations. In addition controls were run in which membrane was incubated in cytosol treated with ^{125}I in exactly the same manner except that cysteine was present right from the start to inhibit any iodination of protein.

The electrophoresis patterns for membranes incubated in cytosol proved to be closely similar to those for corresponding untreated membrane (Fig. 9A and B). Moreover, comparisons of protein-banding patterns and corresponding radioactivity profiles for gels of incubated membrane revealed that virtually all of the bands visible in the spectrophotometric scans are also labeled. This was true both for plasma membrane from stationary phase cells which had been treated with KCl and for membrane from exponential phase cells which had not been treated (Fig. 9B and C). By contrast, in the controls for the iodination experiments only background counts were obtained for the gel slices. The spectrophotometric scans in Fig. 9 show better resolution of the protein-banding patterns than do those in Fig. 7, because more protein (100 μg/gel) was applied and a superior scanner was used.

There are a number of indications that the incorporation of radioactivity into plasma membrane during its incubation in the cytosol fraction manifests a highly specific process. For one thing, labeled cytosol protein was always present in excess even at cytosol:membrane ratios which did not result in saturation of the membrane with radioactivity during the incubation period. This indicates that only a limited number of cytosol proteins are capable of being incorporated into the membrane and that these proteins comprise a very small proportion of the entire protein complement in the cytosol. Moreover, the implication is very strong that such proteins in the soluble fraction are membrane proteins, somehow rendered soluble, which are exchanging with their counterparts in the membrane. That the presence of radioactivity in the incubated membrane reflected molecular turnover rather than net incorporation of material is supported by the observation that the protein-banding pattern on gels for experimental membrane was closely similar to that on gels for control membrane which had not been incubated in soluble fraction. The specific nature of the incorporation is also reflected in the fact that as the cytosol:membrane ratio in the incubation mixture was increased, the specific radioactivity of the membrane reached a plateau.

Comparisons of protein patterns and corresponding radioactivity profiles of polyacrylamide gels for incubated membrane also provide support for a selective incorporation in that the radioactivity peaks correlate with Coomassie blue positive bands of membrane protein. There is evidence from other systems to indicate that virtually all proteins of a membrane are susceptible to turnover in vivo, but have widely divergent half-lives [24,26-28]. The electrophoresis data obtained in this study also suggest that most, if not all, of the proteins of *Acanthamoeba* plasma membrane turn over in the in vitro system. Moreover, the superimposed profiles of protein bands and radioactivity counts (Fig. 9) suggest that the specific radioactivities of the proteins are dissimilar, which is to be expected if they have different rates of turnover.

It seems reasonable to conclude, therefore, that proteins native to the plasma membrane are present in the cytosol and are capable of being incorporated into preexisting plasmalemma. These proteins may well be in the form of macromolecular complexes, perhaps with phospholipid, but they clearly are not aggregates of sufficient size to be sedimentable by centrifugation that would pellet even the smallest of membrane fragments.

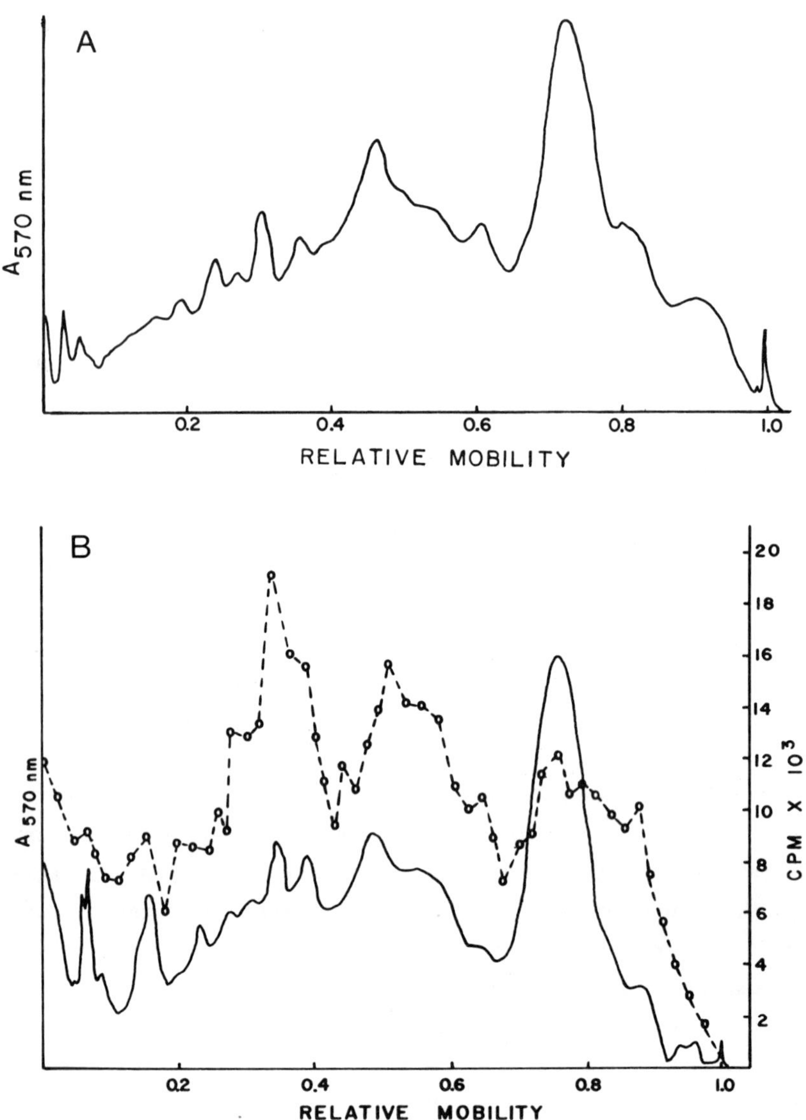

FIG. 9. Polyacrylamide gel electrophoresis scans and corresponding radioactivity profiles of isolated plasma membrane from *Acanthamoeba* trophozoites. Incubation conditions were as in Fig. 8 except that the cytosol:membrane protein ratio was 18:1. Solid lines show spectrophotometric scans of the protein-banding pattern after Coomassie blue staining. Broken lines show the pattern of radioactivity obtained by slicing the gels and counting the slices. (A) Unincubated membrane from exponential phase cells run on a 10% gel. (B) Membrane from exponential phase cells incubated in ^{125}I-labeled cytosol and electrophoresed on a 10% gel.

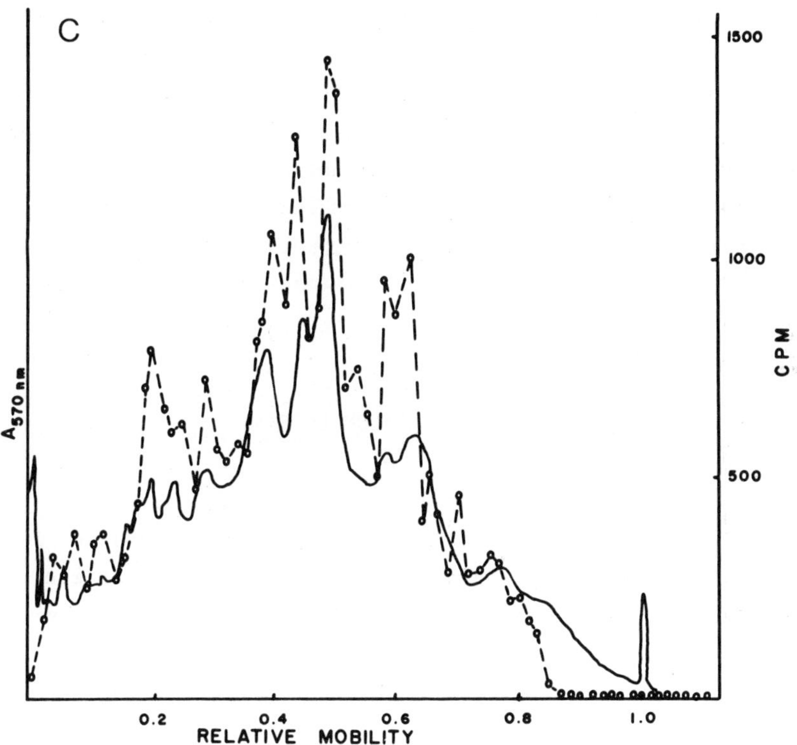

(C) Membrane from stationary phase cells incubated in ^{14}C-labeled cytosol, treated with 0.5 M KCl and electrophoresed on a 7.5% gel.

There is increasing evidence to suggest that various species of membrane in the cell constitute a dynamic system within which there is a flow of membrane or perhaps of membrane components [29-33]. In this context pools of macromolecules in the cytosol could serve as mediators of membrane flow and perhaps even of membrane modification. Indeed this prospect is highly consistent with some previous observations on membrane differentiation and turnover. Leskes et al. [34] in an examination of glucose-6-phosphatase synthesis in hepatocytes of developing rats found that the enzyme is introduced throughout preexisting endoplasmic reticulum. The old membrane thus acts as a structural matrix into which new molecules are inserted. Moreover, the lipids and proteins of membranes do not turn over at the same rate, thus implying that membrane fragments are not the means of incorporation [24]. Relatively little is in fact known of the actual state in which proteins and lipids are incorporated into membranes of eucaryotes. However, in a discussion of membrane assembly Whaley et al. [30] suggest that there must be considerable movement of membrane components at a microscopically undetectable level. The results obtained in the present study indicate that, at least in vitro, uptake into membranes at the level of macromolecules or possibly macromolecular complexes does indeed occur.

III. FINAL COMMENTS

Our examination of membranes in *Acanthamoeba castellanii* has so far been restricted to a consideration of the plasmalemma in exponential and stationary phase trophozoites. However, in the sense that we can now control the onset of differentiation in the organism and recognize its various stages of development, a framework has been established for extending these membrane studies to encysting and excysting cells as well as the mature cyst. We are now attempting to prepare purified plasma membranes from cysts. This will permit a comparison of surface membranes from actively growing and essentially dormant cells, and should further illuminate the question of whether membrane properties are radically altered in quiescent cells. Knowledge of the timing of membrane modifications during encystment and excystment may give some indication of whether changes at the level of membranes contribute to the onset of physiological quiescence or germination.

The in vitro system for monitoring membrane turnover may well prove useful as a means of learning more about membrane repair, modification, and even synthesis. At the moment we are attempting to confirm the correspondence of radioactivity and protein peaks on the polyacrylamide gels of treated membrane by using double-labeling procedures in which the membrane is tagged with one isotope and the cytosol proteins with another. Once the system is perfected it may then be possible to clarify such important questions as the form in which macromolecules are incorporated into the membrane, as well as the role of the membrane matrix in regulating such incorporation. An in vitro system offers versatility for investigating problems of this nature in that it is comparatively easy to modify the various components of the system. In the present case, for example, it should be possible, utilizing radioactive tracer techniques, to alter selectively either the cytosol or the membranes by treatment with such agents as digestive enzymes. In addition the extent to which membrane modifications incurred during differentiation of *Acanthamoeba* are mediated by turnover could be investigated by treating membranes from one stage of development with cytosol from another in an effort to simulate naturally occurring membrane alterations. Such information, if unequivocal, could contribute significantly to our knowledge and understanding of how membranes behave in developing systems.

REFERENCES

1. Neff, R. J., 1957. J. Protozool. 4: 176-182.
2. Bowers, B., and E. D. Korn, 1968. J. Cell Biol. 39: 95-111.
3. Wilkins, J. A., and J. E. Thompson, 1974. Exp. Cell Res. 89: 143-153.
4. Neff, R. J., S. A. Ray, W. F. Benton, and M. Wilborn, 1964. *In* Methods in Cell Physiology, Vol. 1, D. M. Prescott (ed.), Academic Press, New York, p. 55.
5. Neff, R. J., W. F. Benton, and R. H. Neff, 1964. J. Cell Biol. 23: 66a.
6. Bowers, B., and E. D. Korn, 1969. J. Cell Biol. 41: 786-805.
7. Neff, R. J., and R. H. Neff, 1969. *In* Symposium of the Society for Experimental Biology, No. 23, Cambridge University Press, Cambridge, p. 51.
8. Mattar, F. E., and T. J. Byers, 1971. J. Cell Biol. 49: 507-519.
9. Pasternak, J. J., J. E. Thompson, T. M. G. Schultz, and K. Zachariah, 1970. Exp. Cell Res. 60: 290-298.

10. Chambers, J. A., and J. E. Thompson, 1972. Exp. Cell Res. 73: 415-421.

11. Dubes, G. R., and T. Jensen, 1964. J. Parasitol. 50: 380-385.

12. Korn, E. D., and R. A. Weisman, 1967. J. Cell Biol. 34: 219-227.

13. Goodall, R. J., and J. E. Thompson, 1971. Exp. Cell Res. 64: 1-8.

14. Bowers, B., and T. E. Olszewski, 1972. J. Cell Biol. 53: 681-694.

15. Weisman, R. A., and E. D. Korn, 1967. Biochemistry 6: 485-497.

16. Pal, R. A., 1972. J. Exp. Biol. 57: 55-76.

17. Vickerman, K., 1962. Exp. Cell Res. 26: 497-519.

18. Schultz, T. M. G., and J. E. Thompson, 1969. Biochim. Biophys. Acta 193: 203-211.

19. Ulsamer, A. G., P. L. Wright, M. G. Wetzel, and E. D. Korn, 1971. J. Cell Biol. 51: 193-215.

20. Victoria, E. J., and E. D. Korn, 1975. J. Lipid Res. 16: 54-60.

21. Bowers, B., and E. D. Korn, 1973. J. Cell Biol. 59: 784-791.

22. Korn, E. D., and P. L. Wright, 1973. J. Biol. Chem. 248: 439-447.

23. Nicolson, G. L., V. T. Marchesi, and S. J. Singer, 1971. J. Cell Biol. 51: 265-272.

24. Siekevitz, P., 1972. Ann. Rev. Physiol. 34: 117-140.

25. Marchalonis, J. J., 1969. Biochem. J. 113: 299-305.

26. Schimke, R. T., R. Granschow, D. Doyle, and I. M. Arias, 1968. Fed. Proc. 27: 1223-1230.

27. Arias, I. M., D. Doyle, and R. T. Schimke, 1969. J. Biol. Chem. 244: 3303-3315.

28. Kiehn, E. D., and J. J. Holland, 1970. Biochem. 9: 1716-1728.

29. Wise, G. E., and C. J. Flickinger, 1970. Exp. Cell Res. 61: 13-23.

30. Whaley, W. G., M. Douwalder, and J. E. Kephart, 1971. *In* Origin and Continuity of Cell Organelles, J. Reinert and H. Ursprung (eds.), Springer-Verlag, New York, p. 1.

31. Meldolesi, J., and D. Cova, 1972. J. Cell Biol. 55: 1-18.

32. Meldolesi, J., and D. Cova, 1972. *In* Role of Membranes in Secretory Processes, L. Bolis, R. D. Keynes, and W. Wilbrandt (eds.), American Elsevier, New York, p. 62.

33. Meldolesi, J., 1974. J. Cell Biol. 61: 1-13.

34. Leskes, A., P. Siekevitz, and G. E. Palade, 1971. J. Cell Biol. 49: 264-287.

MICROCYST GERMINATION IN THE CELLULAR SLIME MOLD *POLYSPHONDYLIUM PALLIDUM*: REQUIREMENTS FOR MACROMOLECULAR SYNTHESIS AND SPECIFIC ENZYME ACCUMULATION

Danton H. O'Day

Department of Zoology and Erindale College
University of Toronto
Mississauga, Ontario, Canada

I. INTRODUCTION

One of the remarkable attributes of the cellular slime molds is that various developmental phenomena can be studied as separate events free of the confusion that reigns when such events occur simultaneously. For example, growth and cell division are temporally separated from cellular differentiation and morphogenesis; differentiation proceeds once growth has ceased. By choice of a specific developmental pathway or part of that pathway, the following events may be studied essentially as sole phenomena: growth and cell division, cell communication, morphogenesis, cytodifferentiation, and induction. The developmental alternatives open to cellular slime mold cells are summarized in Fig. 1 of David W. Francis' article in this volume, and details of each of these alternatives are included in this paper and the articles by Barbara E. Wright, William F. Loomis, and Antony J. Durston in this volume.

A. Microcyst Formation

Of the developmental alternatives open to cellular slime mold amoebas, only microcyst formation provides a homogeneous population of differentiating cells. In *Polysphondylium pallidum*, microcyst formation occurs when amoebas are subjected to increased osmotic conditions such as 0.1 M sucrose or KCl [1-3], but also occurs naturally in axenic liquid culture after the stationary phase of growth has been reached [4]. There appears to be specific ionic requirements for microcyst formation [1]. During microcyst formation, individual amoebas round up and differentiate a bilayered cell wall [2] containing cellulose, lipid, protein, and some undefined glucose polymers [5,6], while intracellularly the cytoplasm becomes condensed [2]. Biochemically, microcyst formation appears to require continuous protein synthesis and is characterized by specific intracellular and extracellular increases in several lysosomal enzymes [7-9]. The end product of this developmental pathway is a homogeneous population of unicellular microcysts. These cells can be used for the analysis of the controls regulating the accumulation of specific enzyme activities and the reinitiation of specific metabolic events as well as for understanding the precise role of these changes in the morphologic expression of the germination process.

B. Morphology of Microcyst Germination

Using phase contrast microscopy, a series of discrete morphological stages is evident during the germination of *P. pallidum* microcysts (Fig. 1) [11]. The dormant microcyst is spherical with a birefringent cell wall, a uniformly dense, silvery cytoplasm, and

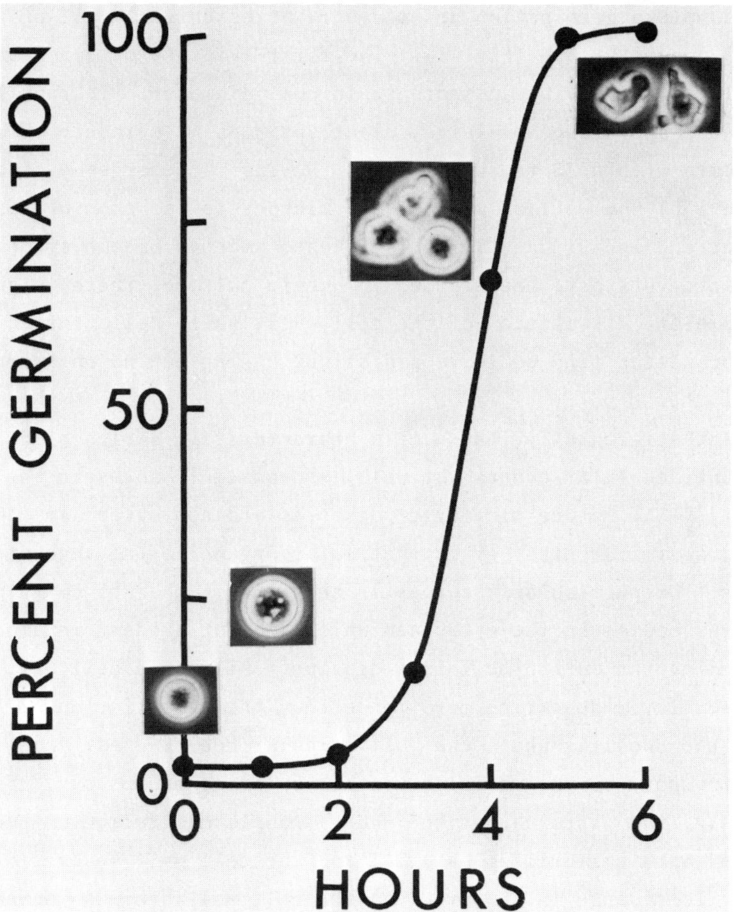

FIG. 1. Time course of germination (emergence) in *Polysphondylium pallidum*. The inserted photographs show changes in the cellular morphology during the germination process.

a diameter of 8 to 12 μm. In contrast to spores which are constitutively dormant, being held in a dormant state by specific autoinhibitors (see Ref. 11 for a review of spore germination in cellular slime molds), microcyst dormancy is environmentally enforced (exogenous dormancy; Ref. 12) and autoinhibitors are not produced. When dormant microcysts, which have been formed in axenic culture, are placed in sterile, nonnutrient phosphate buffer (0.01 M; pH 6.5), complete germination or emergence of essentially all of the amoebas occurs by 5 h (Fig. 1). The first evidence of germination is swelling, involving an increase in cell diameter. Intracellularly, the appearance of large, clear vesicles is evident. Swelling occurs within 30 to 40 min [13]. During the emergence stage (2.5 to 5 h) the wall of many of the microcysts is becoming less distinct and individual, irregular-shaped amoebas become evident. When microcysts have been formed in axenic culture, there is generally complete dissolution of the cell wall, while cells formed on dry agar plates tend to leave a distinct casing behind on emergence [2,10].

Ultrastructurally, there is a characteristic series of distinct intracellular events, as well as dramatic changes in the cell wall [2]. The mature microcyst possesses a dense cytoplasm containing a similar array of intracellular organelles as vegetative amoebas. Of particular interest is the large number of small polyvesicular bodies in the cytoplasm which are not evident in amoebas. Surrounding the entire cell is a distinct, bilayered cell wall with the inner layer appearing more dense than the fibrillar outer layer. During the swelling phase the polyvesicular bodies greatly enlarge and correspond to the vesicles visible by phase microscopy. Simultaneously, the cisternae of the rough endoplasmic reticulum become engorged with a fibrillar material and the cell wall becomes distinctly less dense in texture. As the cell wall becomes increasingly less dense, amoebas begin to emerge from the digested cell walls. At this time, the endoplasmic reticulum shrinks and the polyvesicular bodies begin to disappear.

II. CURRENT RESEARCH

A. Requirements for Macromolecular Synthesis

Early experiments demonstrated that cycloheximide added at the start of germination allowed swelling but prevented emergence [14]. This suggested that protein synthesis was essential for germination. Ultrastructural analysis of microcyst germination provided morphological evidence for the occurrence of protein synthesis [2]. By 1.5 h germination, the rough endoplasmic reticulum is greatly distended and filled with amorphous material. Studies on bulk protein synthesis reveal that dramatic [^{3}H]leucine incorporation into protein begins within 0.5 h and continues throughout germination (Fig. 2) [13]. Addition of cycloheximide at selected intervals demonstrates that this drug is effective only when added before 1.5 h,

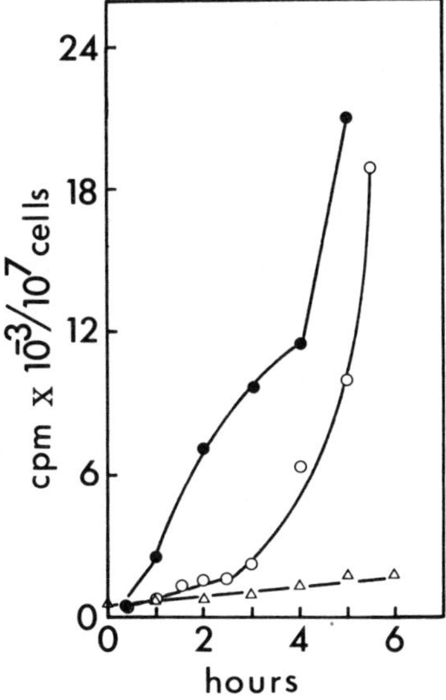

FIG. 2. Patterns of macromolecular synthesis during germination. Labeled precursors were added at 0 h and bulk incorporation into protein, RNA, and DNA was determined at subsequent intervals as detailed previously [13]. [^{3}H]leucine (●), [^{3}H]uridine (○), [^{3}H]-thymidine (△).

indicating that a very discrete period of protein synthesis is essential for germination [10,13]. Cycloheximide is completely effective in stopping protein synthesis during germination (Table 1). Taken together, these data unequivocally demonstrate the requirement for a short, specific phase of protein synthesis during germination. This requirement is common in fungal spores [12,15].

Addition of actinomycin D at the start of germination does not prevent complete germination even when used at extremely high concentrations (Table 1). α-Amanitin, at very high doses, and certain nucleic acid analogs (5-fluorouacil, iodouracil), which are effective at preventing RNA synthesis in other systems, are also ineffective during germination [10,16]. These data suggest two possibilities: microcysts are not permeable to all of these compounds or RNA synthesis is not essential for complete microcyst germination. Accumulated data support the latter possibility. Patterns of uridine incorporation suggest that RNA synthesis is initiated after protein synthesis and that significant RNA synthesis does not begin until after 2 h, when the inhibition of protein synthesis by cycloheximide no longer has any effect on germination (Fig. 2). The large increase after 2 h is mainly due to the initiation of ribosomal RNA (rRNA) synthesis which will undoubtedly be used in the reinitiation of the growth phase [17]. Furthermore, actinomycin D dramatically and immediately inhibits the incorporation of [^{3}H]uridine into RNA (Table 1), revealing that the microcysts are permeable to actinomycin and that this drug is effective at inhibiting RNA synthesis.

Recent experiments show that poly(A)+ RNA synthesis also begins after 2 h and this synthesis is actinomycin D-sensitive [17]. Related experiments indicate there are no changes in the total nucleotide pool during germination, thus ruling out the possibility that the detected changes in labeling were due to fluctuations in internal nucleotide pools.

TABLE 1

Effects of Actinomycin and Cycloheximide on Emergence, Macromolecular Synthesis, and Enzyme Accumulation During Microcyst Germination in *P. pallidum* [13]

Drug	% Amoebas[a]	$[^3H]$leucine[b] incorp. (cpm)	$[^3H]$uridine[b] incorp.	Maximal[c] alkaline phosphatase	Maximal[c] β-glucosidase
None (control)	100	14,000	19,000	42.0	18.8
Cycloheximide (400 μg/ml)	4	1,200	4,150	14.0	7.5
Actinomycin D (125 μg/ml)	100	13,400	2,900	42.5	16.4

[a]% amoebas (germination) determined after 6 h.

[b]Incorporation per 10^7 cells/ml after 6 h.

[c]Maximum developmental activity observed in total (cells plus supernatant) samples.

The developmental increases in the activities of two enzymes (alkaline phosphatase, β-glucosidase) which are synthesized during germination are completely inhibited by cycloheximide but unaffected by actinomycin D (Table 1). Together, these data strongly suggest that RNA synthesis is not essential for germination and also indicate that stable messenger RNA (mRNA) may exist in dormant microcysts for the synthesis of alkaline phosphatase and β-glucosidase as well as certain other proteins. This possibility is under investigation.

In regard to DNA synthesis, it is evident that little incorporation of thymidine occurs during germination (Fig. 2). In addition, the facts that the cell number does not increase in germinating cultures and that 5-bromodeoxyuridine has no effect on germination also support the idea that DNA synthesis is not essential for germination [10]. This is also typical of germination in fungal spores [12,15].

In summary, then, the microcysts of *P. pallidum* appear to require protein synthesis but not nucleic acid synthesis for complete germination and the sequence of initiation of macromolecular synthesis is translation → transcription → DNA replication. Thus these dormant structures differ from the spores of *D. discoideum* where RNA synthesis is not required for swelling but is essential for emergence [18]. Sussman and Douthit [12] and Van Etten et al. [15] have reviewed the patterns of macromolecular synthesis in all fungal spores which have been adequately studied to date, and it is evident that *P. pallidum* microcysts would appear to be similar to *Allomyces neomoniliformis*, *Botrydiplodia theobromae*, and *Saccharomyces cerevisiae*. In subsequent papers in this volume, James L. Van Etten et al. provide details on the macromolecular synthetic patterns during germination of *Rhizopus stolonifer* and James S. Lovett et al. cover aspects of macromolecular synthesis in *Blastocladiella emersonii*.

B. Requirements for Intracellular and Extracellular Enzyme Accumulation

Many studies focus on the changing patterns of specific enzymes during development because certain of these proteins will be the biochemical effectors of the morphological events we observe as differentiation. During germination and other developmental events, dramatic increases in certain enzyme activities can be detected intracellularly, extracellularly, or both. The question to be answered is: "which of these enzyme patterns is essential to the differentiation process?" There are several methods which can be used to answer this question. In earlier papers in this volume, William F. Loomis has invoked the use of developmental mutants while Barbara E. Wright has described the value of using kinetic models. Two other methods have been employed in the present study: use of specific enzyme inhibitors and, for extracellular enzymes, augmentation of enzyme activity.

Alkaline phosphatase

When microcysts form, the high levels of alkaline phosphatase activity present during growth decrease about three-fold [4]. When germination is induced, the intracellular level of alkaline phosphatase increases approximately three- to four-fold in 4 h (Fig. 3) [10]. Four isozymes can be detected by electrophoresis in acrylamide gels but the relative amounts of each of these forms do not appear to change during development. These isozymes are also detected during growth [4]. The majority of the alkaline phosphatase activity of *P. pallidum* is membrane-bound and is not excreted during germination [10,19]. The intracellular increase in activity is prevented by cycloheximide but is unaffected by actinomycin D, suggesting that it may be translated from stable mRNA (Table 1).

Beryllium, a potent inhibitor of alkaline phosphatase activity [20], inhibits the level of alkaline phosphatase activity in dormant microcysts and completely stops its developmental increase

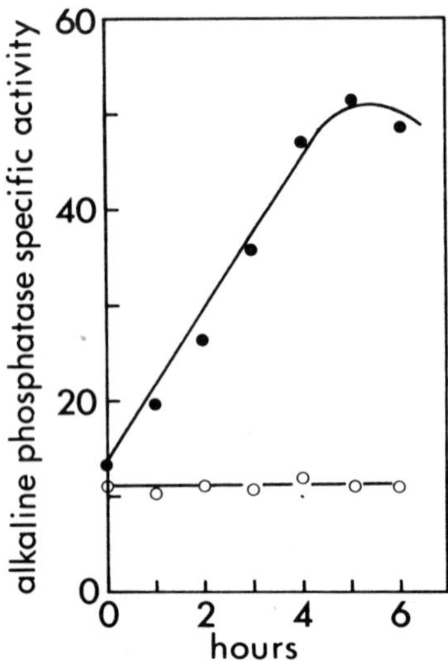

FIG. 3. The effect of beryllium on the intracellular changes in alkaline phosphatase activity during germination. Beryllium, as $BeCl_2$, was added at the start of germination. Samples were removed from control and treated cultures at hourly intervals and the intracellular activity of alkaline phosphatase was determined as detailed previously [4]. Control (●) and 100 μM beryllium (O).

(Fig. 3). Germination in the presence of beryllium is normal and temporally identical to control cultures. It appears, then, that the coincidental increase in alkaline phosphatase activity is not essential for germination. Similar experiments have revealed that the normal developmental pattern for alkaline phosphatase is not essential for fruiting body formation in *P. pallidum* [4] and *Dictyostelium discoideum* [21]. It seems reasonable to propose then that alkaline phosphatase may be a growth enzyme and that the increase in alkaline phosphatase during germination is in preparation for the subsequent reinitiation of the growth phase.

β-Glucosidase

The developmental aspects of β-glucosidase activity are especially interesting. The intracellular activity of this enzyme changes little during germination but the extracellular β-glucosidase activity increases dramatically (Fig. 4). The cells appear to excrete at least twice as much enzyme activity as they originally contained and the pattern of excretion correlates well with the emergent phase of germination [10]. This proposed function is supported by the presence of large amounts of cellulose and other glucose polymers in the microcyst wall [5,6]. If the total activity (cells plus supernatant) is calculated or assayed directly, an almost three-fold increase in β-glucosidase activity occurs by 4 h and then declines. This developmental pattern is cycloheximide-sensitive but is unaffected by actinomycin D (Table 1). Thus β-glucosidase may also be synthesized on a stable mRNA during germination.

Augmentation of the extracellular activity of β-glucosidase with purified enzyme from almonds (Sigma) markedly enhances germination (Table 2). Together with protease and/or cellulase, even greater enhancement of emergence is observed. Heat-inactivated glucosidase is ineffective at increasing the rate of emergence, revealing that the enhancement is due to the enzyme activity of the

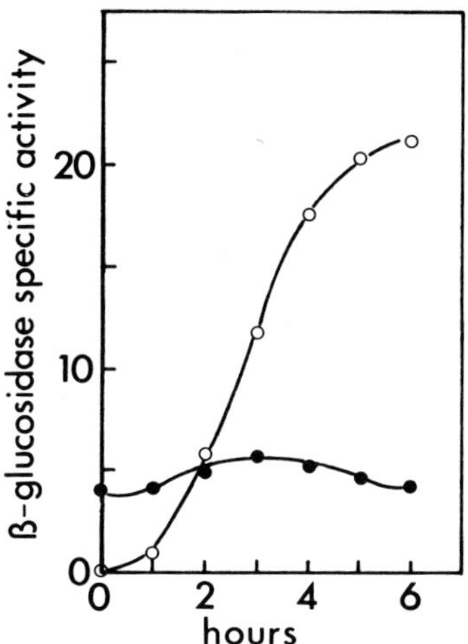

FIG. 4. Changes in intracellular (●) and extracellular (O) β-glucosidase specific activity during germination. Samples were removed from germinating cultures and separated into cell and extracellular fractions. The β-glucosidase specific activity of these fractions was determined as explained previously [10].

TABLE 2

Effect of β-Glucosidase, Cellulase, and Protease on the Rate of Germination in *P. pallidum* Microcysts[a]

Enzyme added	% Amoebas after 3 h		
	Exp. A	Exp. B	Exp. C
Control	26	30	26
β-Glucosidase	--	--	38
Inactive β-glucosidase[b]	--	--	24
Protease	38	42	--
Inactive protease[b]	23	--	--
Cellulase	56	55	--
Inactive cellulase[b]	--	17	--
Protease + cellulase	64	67	--
Protease + cellulase + β-glucosidase	--	--	75
Cycloheximide	5	3	3
Cycloheximide + enzymes[c]	--	4	4

[a]The enzymes (1 mg/ml) were dissolved in the germination medium before the microcysts were added. The source of the enzymes is noted in the text.

[b]The enzymes were inactivated by boiling.

[c]Includes β-glucosidase, protease, and cellulase.

added glucosidase. Inhibition of microcyst germination is effected by D-gluconic acid lactone, an inhibitor of *P. pallidum* β-glucosidase activity [22]. These data suggest that β-glucosidase may be a critical enzyme in cell wall removal. The isolation and characterization of the various glucose polymers in the cell wall and their ability to act as substrates for purified extracellular glucosidase should verify this function.

Acid protease

As defined by pH profiles, at least two acid protease activities exist in dormant microcysts [23]. These enzymes, called protease A and B, have pH optima of 3.5 and 6.0, respectively. Intracellularly, protease B undergoes a three- to four-fold increase in enzyme activity during development while the pH 3.5 enzyme activity changes very little (Fig. 5). Each of the enzyme activities appears in the extracellular medium during germination (Fig. 5). Leucine aminopeptidase activity also exists in microcysts, but this activity decreases during germination and is not excreted [10]. Intracellular protease activity may have several functions, and a possible role for these activities in swelling is suggested in the last part of this article.

The microcyst wall of *P. pallidum* consists of approximately 30% protein with the protein being restricted to the inner layer of the bilayered cell wall [5,6]. This large protein content in the wall suggests that extracellular protease activity would be necessary for germination. Hohl [23] has proposed a role for protease as a critical enzyme in spore germination of *D. discoideum*, but the existence of this protease has not yet been shown. Studies with purified, nonspecific protease from *Streptomyces griseus* (Sigma,

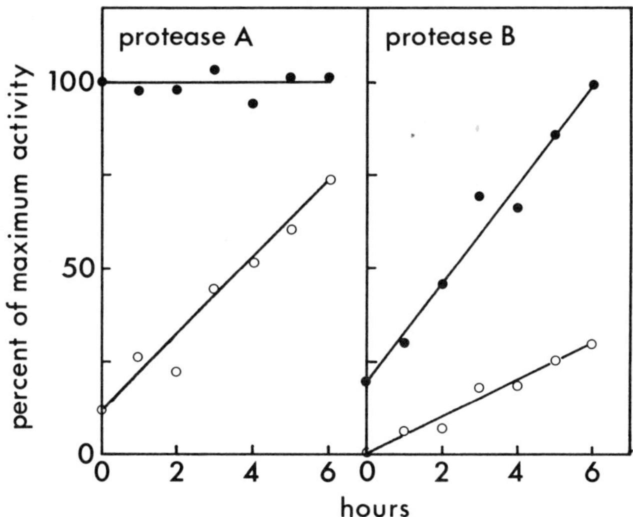

FIG. 5. Developmental changes of two acid protease activities during germination. The activity of cellular (●) and extracellular (○) fractions on casein at pH 3.5 (protease A) and pH 6.0 (protease B) was determined as described previously [23].

type I) reveal that augmentation of the extracellular protease levels significantly increases the rate of emergence (Table 2). Cellulase from *Aspergillus niger* (Sigma, type I) also enhances the rate of microcyst germination. The effect of these enzymes is due to their hydrolase activity, since heat-inactivated samples do not enhance germination. Thus it seems reasonable to postulate that acid protease as well as cellulase and β-glucosidase are involved in cell wall removal during germination. The fact that these enzymes together cannot effect emergence in the presence of cycloheximide (Table 2) indicates that there are additional enzymes involved in cell wall removal. As an essential enzyme, protease B might function during both swelling and emergence. A significant portion of the protease B activity increase occurs in the presence of cycloheximide while the remainder of the increase is cycloheximide-sensitive [24]. Thus posttranslational controls may play a role in regulating part of the protease B developmental increase.

Other enzymes

Since the microcyst cell wall lies outside of the cell proper, enzymes for its removal must be excreted from the cell. Roles for excreted acid protease and β-glucosidase activities in cell wall removal have already been discussed. It seems reasonable to assume that the excreted enzymes digest the cell wall as they diffuse through it and ultimately end up in the extracellular medium. The fact that the cell wall usually completely disappears during germination in liquid culture is probably due to the uniform distribution and general availability of enzymes in this medium after they have escaped from the cell wall.

There are three factors which will affect the rate of enzyme accumulation in the extracellular medium: rate of excretion by the cell, extracellular degradation or modification of these enzymes by extracellular enzymes, and molecular sieving by the cell wall. Since the exclusion characteristics of the microcyst cell wall are unknown and the molecular weights of the excreted enzymes have not been determined, it is difficult to draw meaningful conclusions about the rates of extracellular enzyme accumulation. Chang and

Trevithick [25] have devised a model for enzyme excretion through the walls of hyphal fungi but this model does not consider the intracellular levels of the excreted enzymes, active control of rates of excretion by the cell itself, or the effects of extracellular proteolysis.

Regardless of the determination of the exact rate of accumulation of specific enzymes, the fact is that certain enzymes do appear in appreciable quantities in the extracellular medium during germination while others do not [10]. Figure 6 presents the patterns of four enzymes that accumulate to significant levels extracellularly

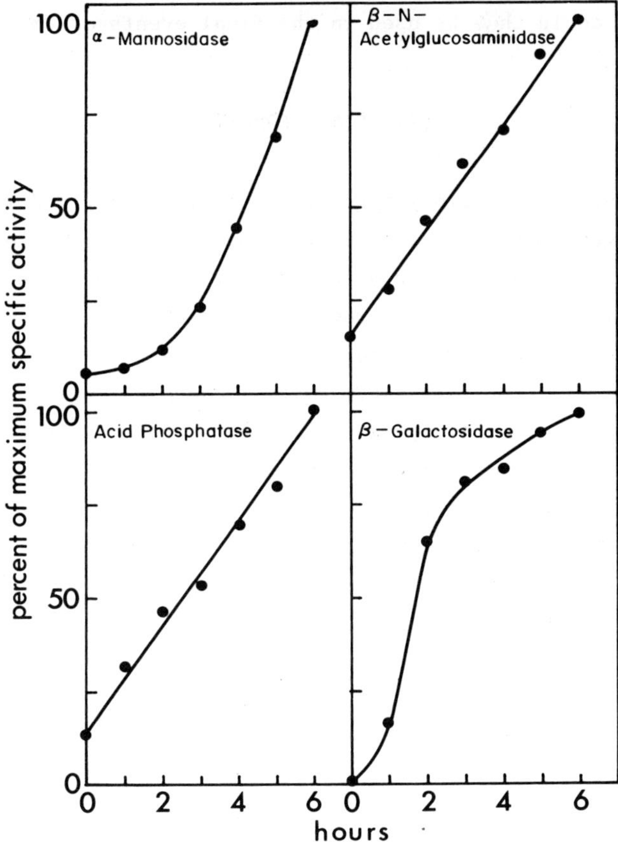

FIG. 6. Extracellular patterns of accumulation of four lysosomal hydrolase activities during germination. The activities of the enzymes in the extracellular medium were determined as detailed previously [10].

during germination. Earlier discussion in this paper revealed that coincidental changes do not mean function significance. However, some comments can be made about these patterns. The early and marked accumulation of β-galactosidase, acid phosphatase, and acetylglucosaminidase indicates that these enzymes should be further investigated as potential components of the cell wall degrading systems. The pattern of α-mannosidase suggests that if it does act in cell wall removal, it might function during the final stages of emergence or may penetrate the cell wall very slowly. The accumulations of α-mannosidase and β-N-acetylglucosaminidase are stopped by cycloheximide, suggesting that they require protein synthesis and could thus be used in the final events of emergence [22].

III. FINAL COMMENTS

The events of microcyst germination are summarized in Fig. 7. Microcyst germination is a continuous process but can clearly be divided into two distinct phases: swelling and emergence. Swelling can occur in the absence of protein or nucleic acid synthesis, while emergence requires protein but, apparently, not RNA or DNA synthesis. In microcysts, dormancy is generally environmentally enforced by high osmotic pressure and it is reasonable to assume that the reduction of osmotic pressure which permits germination allows the influx of water which induces and contributes to swelling. The increase in internal water may activate certain intracellular swelling enzymes such as protease or carbohydrate-digesting enzymes to increase the internal level of solutes by degrading polymers to monomers and thus cause maximum swelling by encouraging further influx of water. This is a modified version of a model suggested by Hohl [23] for spore swelling in *D. discoideum*. If the majority of polymer → monomer conversions were limited to the polyvesicular bodies, this would explain the dramatic increase in their size during the swelling phase. The characterization of the enzyme content of the polyvesicular bodies would be very enlightening in this regard.

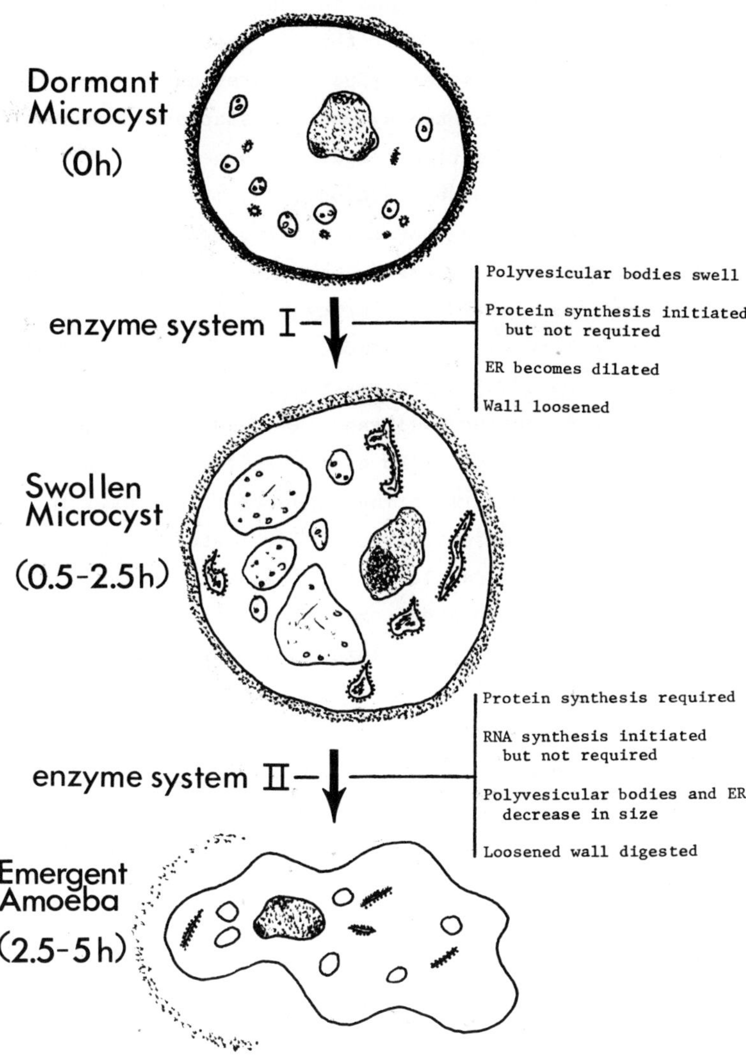

FIG. 7. Diagrammatic representation of the significant events of microcyst germination in *Polysphondylium pallidum*.

During swelling the microcyst wall also undergoes a significant loosening. This loosening appears to be due to enzymatic digestion by a complex of enzymes referred to as system I [13]. The accumulation of system I activities is apparently regulated by post-translational controls. The exact components of this system are unknown, but there are some likely candidates. The high levels of cellulose in the microcyst wall and the enhancement of germination by purified cellulase suggest that this enzyme is involved in the initial wall loosening. Rosness [26] has previously shown that cellulase activity is regulated by posttranslational controls in *D. discoideum*. Since oxidation appears to be a critical enzyme reaction in fungal cellulose degradation [27], it is possible that certain cellulose-oxidizing enzymes may also occur in system I. Thus it will be essential to verify the existence of a cellulose-degrading system in *P. pallidum* microcysts. A reexamination of the micrographs of Hohl et al. [2] indicates that the inner microcyst wall, which is composed of extremely large amounts of protein [5,6], is markedly loosened during swelling. This suggests a role for acid protease B in swelling. The cycloheximide-insensitive protease B increase during germination supports this idea. The means by which this acid protease activity is activated is not known. In regard to system I, then, at least the following questions can be asked. Which of the postulated enzymes are present in this complex? How is their activity regulated? And are they activated as a group or in sequence?

Emergence involves the removal of the loosened microcyst wall and requires the concerted action of several enzyme activities referred to as system II. The accumulation of these enzyme activities is regulated at the translational level. The means by which translation is regulated in microcysts is at present unknown. At least two enzymes appear to be components of this system: β-glucosidase and the cycloheximide-sensitive acid protease B activity. It is also likely that cellulase continues to function during emergence. The fact that these three enzymes cannot cause emergence in the presence of cycloheximide indicates that there are also other enzymes in system II. Further experiments will reveal if the extracellular accumulation of β-galactosidase, acid phosphatase, chitinase, and α-mannosidase are indicative of their part in system II. Purification of these enzymes followed by their sequential addition alone or in combinations will aid in determining which are important. The fact that all of these enzymes would appear to be synthesized using stable mRNA is especially intriguing and indicates a critical need to unequivocally demonstrate the existence of mRNA in dormant microcysts. To shed light on this problem, we are attempting to isolate polysomes and poly(A)+ RNA from dormant microcysts.

ACKNOWLEDGMENTS

I would like to thank David Gwynne, Keith Lewis, and Gale MacNab for their contributions to various aspects of this work. This work was supported by a grant from the National Research Council of Canada.

REFERENCES

1. Toama, M. A., and K. B. Raper, 1967. J. Bacteriol. 94: 1143-1149.
2. Hohl, H. R., L. Y. Miura-Santo, and D. A. Cotter, 1970. J. Cell Sci. 7: 285-306.
3. O'Day, D. H., 1973. J. Bacteriol. 113: 192-197.
4. O'Day, D. H., and D. W. Francis, 1973. Can. J. Zool. 51: 301-310.
5. Toama, M. A., and K. B. Raper, 1967. J. Bacteriol. 94: 1150-1153.
6. Buhlmann, M., 1971. Diplomarbeit, Universitaet Zuerich.
7. O'Day, D. H., 1973. Exp. Cell Res. 79: 186-190.
8. O'Day, D. H., and L. J. Riley, 1973. Exp. Cell Res. 80: 245-249.
9. Githens, III, S., and M. L. Karnovsky, 1973. J. Cell Biol. 58: 522-535.
10. O'Day, D. H., 1974. Develop. Biol. 36: 400-410.
11. Cotter, D. A., 1975. *In* Spores, Vol. 5, P. Gerhardt, H. L. Sadoff, and R. N. Costilow (eds.), American Society for Microbiology, Washington, D.C., p. 61.
12. Sussman, A. S., and H. A. Douthit, 1973. Ann. Rev. Plant Physiol. 24: 311-352.
13. O'Day, D. H., D. I. Gwynne, and D. H. Blakey, 1976. Exp. Cell Res. 97: 358-364.

14. Cotter, D. A., and K. B. Raper, 1968. J. Bacteriol. 96: 1690-1695.

15. Van Etten, J. L., L. D. Dunkle, and R. H. Knight, 1975. *In* The Fungal Spore, Form and Function, D. J. Weber and W. M. Hess (eds.), Wiley-Interscience, New York, p. 243.

16. Cotter, D. A., 1975. Unpublished results.

17. Gwynne, D. I., and D. H. O'Day, 1975. Unpublished results.

18. Bacon, C. W., and A. S. Sussman, 1973. J. Gen. Microbiol. 76: 331-344.

19. O'Day, D. H., 1973. Cytobios 7: 223-232.

20. Witschi, H. P., 1970. Biochem. J. 120: 623.

21. Loomis, W. F., 1975. *In* Isozymes, Vol. 3, C. L. Markert (ed.), Academic Press, New York, p. 177.

22. O'Day, D. H., unpublished results.

23. Hohl, H. R., 1975. *In* The Fungal Spore, Form and Function, D. J. Weber and W. M. Hess (eds.), Wiley-Interscience, New York, p. 463.

24. O'Day, D. H., 1976. J. Bacteriol. 125: 8-13.

25. Chang, P. L., and J. R. Trevithick, 1975. Arch. Microbiol. 101: 281-293.

26. Rosness, P. A., 1968. J. Bacteriol. 96: 639-645.

27. Eriksson, K. E., B. Pettersson, and U. Westermark, 1974. F E B S Letters 49: 282-285.

GERMINATION OF *RHIZOPUS STOLONIFER* SPORANGIOSPORES*

James L. Van Etten, Larry D. Dunkle, and Shelby N. Freer

Department of Plant Pathology
University of Nebraska
Lincoln, Nebraska

*Published with the approval of the Director as paper No. 4049, Journal Series, Nebraska Agricultural Experiment Station. The work was conducted under Nebraska Agricultural Experiment Station Project No. 21-17.

I. INTRODUCTION

Numerous criteria dictate selection of an organism for studying the physiological and biochemical mechanisms of cellular differentiation. Advantages are apparent for an organism or cell type that: (1) is genetically well defined; (2) undergoes an easily distinguished and measured cell differentiation (phenotypic change) in response to precise nutritional, physical, or physiological factors; (3) completes each of a series of developmental events synchronously in a short and reproducible time period; and (4) can be obtained in sufficient quantities to allow specific biochemical analyses of a cell population. Not all biological materials are amenable to the same approaches, nor do they all fulfill to the same extent the criteria outlined above. Fungal spore germination is a morphologically simple process of differentiation in which a nongrowing, metabolically quiescent cell responds to changes in environmental conditions by reactivating its physiological processes and genetic functions to produce a rapidly growing or elongating cell (usually a germ tube or hypha). The following discussion summarizes some of the physiological and biochemical processes which are associated with germination of the asexual spores (sporangiospores) of *Rhizopus stolonifer* (Ehr. ex Fr.) Lind. (syn. *R. nigricans*).

A. Life Cycle

Rhizopus stolonifer, a facultative saprophyte in the Mucorales of the class Zygomycetes (Phycomycetes), is a soil-borne fungus which causes postharvest diseases of many fruits and vegetables [1]. The fungus or taxonomically related mucoraceous forms can be a human pathogen causing deep or systemic mycoses of the lungs, sinuses, meninges, cornea, and central nervous system in patients suffering from debilitating diseases or in patients subjected to intense therapy with corticosteroids, antineoplastic agents, or irradiation [2-4]. Many fungi that are human pathogens are dimorphic, i.e., have mycelial and yeast-like growth phases that are regulated by physical or nutritional factors. Several members of the Mucorales (e.g., *Mucor rouxii*) are dimorphic, but this phenomenon has not been demonstrated in *R. stolonifer*.

Figure 1 diagrams the life cycle of *R. stolonifer* [5]. The fungus produces both sexual spores (zygospores) (H) and asexual spores (sporangiospores) (C and C') which can be harvested in gram quantities with water or dry-harvested with a vacuum pump and a series of screens [6]. The multinucleate haploid (N = 16 [7]) sporangiospores germinate in ca. 6 h to form one or more germ tubes (D and D') that branch and develop into a coenocytic mycelium (A and A'). The mycelium produces aerial hyphae or stolons which form rhizoids and sporangiophores (B and B') at the point of contact with the substrate. During the development of the sporangium (B and B') at the tip of the sporangiophore, cytoplasm and nuclei flow into the region and concentrate along the periphery of the sporangium where the sporangiospores (C and C') form. The spores become darkly pigmented as they mature and are released from the sporangium when the sporangial wall breaks. The time required to complete this portion of the life cycle (A through D) is usually 3 to 7 days.

Sexual reproduction (stages E through J) requires two compatible mycelia (designated mating types + and -). When the hyphae of two compatible mating types contact each other, the hyphal tips swell forming progametangia (E) in which cytoplasm and nuclei are

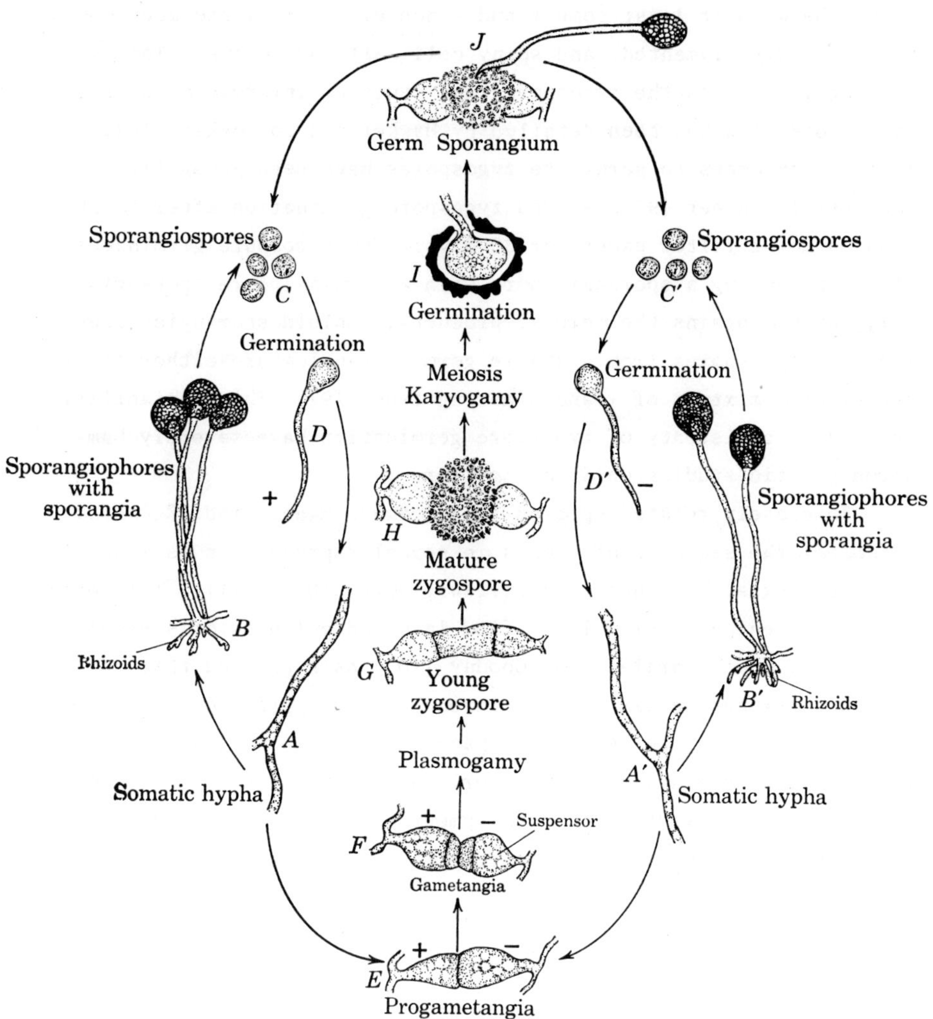

FIG. 1. Life cycle of *R. stolonifer* [5]. Reproduced with permission from John Wiley and Sons, Inc.

concentrated. A septum forms and delimits a terminal gametangium and a suspensor cell (F). The zygote (G) is formed when the protoplasts of each mating type fuse (plasmogamy) following dissolution of the gametangial walls. Although the fate of the parental nuclei within the zygote is not completely known, karyogamy apparently occurs between at least some + and - nuclei. The zygote develops a thick, darkly pigmented, and spiny cell wall and becomes clearly distinguishable as the zygospore (H). Further information on zygospore formation has been detailed by Hawker and co-workers [8]. Although attempts to germinate zygospores have been generally unsuccessful, Gauger [9] observed zygospore germination after incubation on moist filter paper for 30 days. The zygospore germinates (I) by producing a sporangiophore with a terminal germ sporangium (J), which contains the meiotic products, haploid sporangiospores. The sporangiospores from a single germ sporangium are either all +, all -, or a mixture of + and - mating types [9]. This variability and the inconsistency of zygospore germination have severely hampered genetic studies on *R. stolonifer*.

In closely related species (e.g., *Mucor mucedo* and *Blakeslea trispora*) the sequence of events in sexual reproduction is regulated by sexual hormones (see reviews, Refs. 10 and 11). Trisporic acid has been chemically identified in mated cultures of several species in the Mucorales, and Gooday [10] has suggested that this compound may be the sex hormone throughout this group of fungi. Richard P. Sutter, in an earlier paper in this volume, speculates that specific precursors of trisporic acid are the true regulators of sexual development. The involvement and identity of sex hormones in *R. stolonifer* have not been firmly established.

II. CURRENT RESEARCH

A. Ultrastructural Aspects of Germination

Several electron microscopical studies of spore germination have been conducted on *Rhizopus* spp. The spores of the three species studied--*R. stolonifer*, *R. arrhizus*, and *R. sexualis*--all have the same general intracellular organization and exhibit similar ultrastructural changes during germination. Transmission electron micrographs of nongerminated spores indicate that mitochondria are large and irregularly convoluted [12-14]. During the early stages of germination, the cristae become more numerous and the mitochondria become smaller, more regular in shape, then elongate during germ tube emergence [15]. The mitochondria apparently divide during germination [12,13], perhaps by splitting off from the lobes of the larger mitochondria in the nongerminated spore [15]. The division of mitochondria in germinating spores is prevented by sodium azide [12].

Intracellular membrane systems (e.g., endoplasmic reticulum) are sparse in nongerminated spores but increase considerably during germination [16]. Buckley et al. [16] suggested that spherosomes, small lipid- and protein-containing organelles associated with the nuclear envelope and vacuolar membranes, are involved in membrane synthesis during germination.

The cell wall of the ungerminated spore consists of two distinct layers: an outer, electron-dense layer and a thicker, inner layer that is less electron-dense. During germination a new inner wall forms inside the spore wall, and this third layer is contiguous with the wall of the germ tube [12-14,16]. Thus, one might expect that the chemical composition of the spore wall is different from that of the mycelial walls; such differences have been reported for the closely related species *Mucor rouxii* [17]. Transmission electron micrographs indicate that the outer wall layers fracture as the germ tube emerges. Figure 2 shows a sequence of scanning electron micrographs of germinating spores and confirms the breaking of the outer wall as the germ tube emerges.

B. Nutritional Requirements and Physiological Changes During Germination

Figure 3 shows oxygen uptake rates and dry wt changes associated with sporangiospore germination on a simple synthetic medium (glucose, asparagine, KH_2PO_4, and $MgSO_4$; medium A). The spores begin to swell during the first hour and continue to swell for up to 8 h [18] even though germ tubes first appear at 3.5 to 4 h. By 6 h

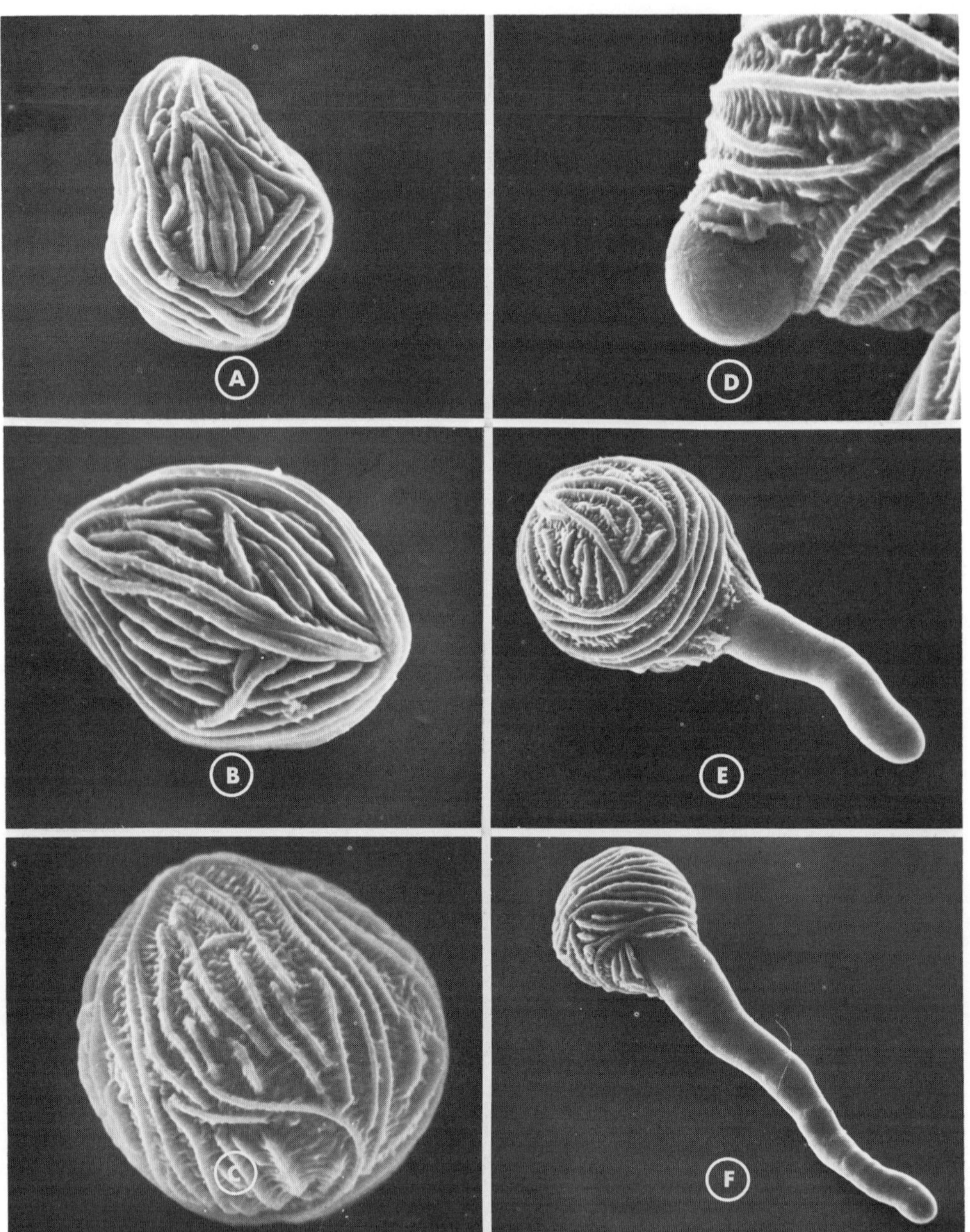

FIG. 2. Scanning electron micrographs of *R. stolonifer* at various stages of germination: (A) Ungerminated spore (× 6000); (B) swollen spore, 2 h (× 6000); (C) swollen spore, 4 h (× 6000); (D) germ tube emergence, 4 h (× 7400); (E) elongating germ tube, 5 h (× 3300); and (F) elongating germ tube, 6 h (× 2300). The spores were fixed in 4% glutaraldehyde, dehydrated in an acetone series, critical point dried, and examined in a Cambridge Stereoscan S4-10 electron microscope. Note: the spores were *not* coated with metal and details are much sharper than usually observed for fungi.

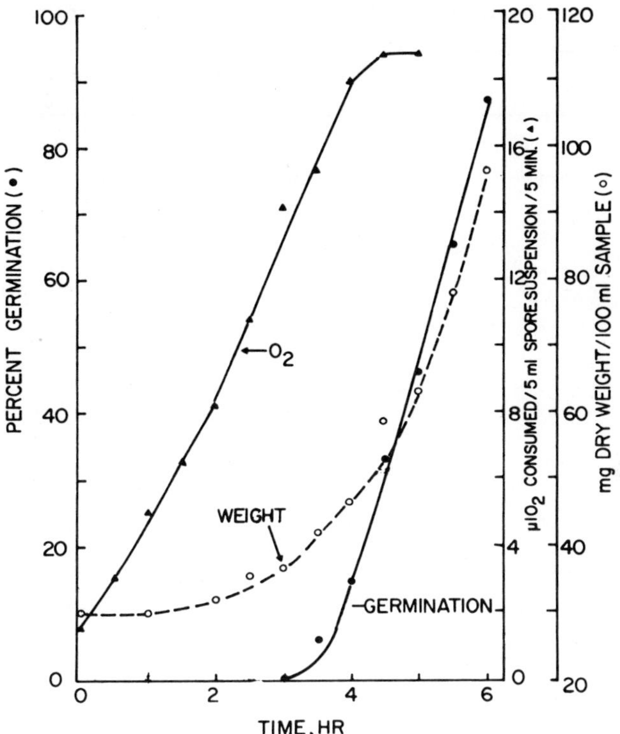

FIG. 3. Percentage of germination (●), rate of oxygen uptake (▲), and dry weight of spores (O) at various time periods during germination of *R. stolonifer* sporangiospores [19]. Reproduced with permission from the American Society of Microbiology.

about 90% of the spores contain germ tubes. During germination the dry weight of the cultures increases about three- to four-fold. Oxygen uptake increases about 10- to 12-fold during the first 4 h of germination and then remains constant for the remainder of the germination period. The Q_{O_2} of the spore is about 13 and increases to about 86 at 4 h after which it slowly decreases as the dry wt increases [19]. Respiratory inhibitors such as sodium azide, sodium malonate, and antimycin A inhibit germination (Table 1).

During the swelling process on medium A, the spore diameter increases from about 8.8 μm to 15 μm [19]. Assuming the spores are spherical, their volume increases about five-fold during swelling. Swelling is an active process and does not occur in water or in

TABLE 1

Effect of Inhibitors of Respiration and Macromolecular Synthesis on the Germination of *R. stolonifer* Sporangiospores[a]

Inhibitor	Concentration (μg/ml)	% Germination after 6 h
None	--	93
Potassium fluoride	10	96
	100	88
Sodium azide	10	0
Sodium malonate	50	45
	1000	1
Antimycin A	1	13
	10	1
Cycloheximide	10	0
Blastocidin S	50	0
Puromycin	10	84
	100	16
Chloramphenicol	100	80
	3500	41
Hydroxyurea	760	70
Ethidium bromide	10	93
	100	84
Proflavin sulfate	10	78
	100	3
Acridine orange	10	85
	100	9
Actinomycin D	20	93
Lomofungin	5	1
Daunorubicin	2.5	2
Cordycepin	12.5	90
α-Amanitin	20	90
5-Fluorouracil	10	10
	100	2
Rifampin	200	85

[a]In many cases the lack of inhibition may reflect the inability of the drug to enter the cells.

medium A containing sodium azide. Even though the spores do not swell in water, the water in the spores freely exchanges with external deuterium oxide [12]. However, permeability changes are associated with germination, since spores do not stain with methylene blue until they have begun to swell [18].

Spore swelling is not coupled directly with germ tube formation. Bussel et al. [20] noted that sporangiospores of *R. stolonifer* incubated on potato dextrose broth swelled at a faster rate and to a greater extent than those incubated on a defined medium. However, the spores required a longer time period to form a germ tube on potato dextrose broth. Some spores swelled to such an extent on potato dextrose broth that the outer wall ruptured in several places. Although a new inner wall was formed, no germ tubes developed. Consequently, fracturing of the outer spore wall and pressure within the spore are not in themselves sufficient for germ tube formation.

If any component of medium A, except for $MgSO_4$, is omitted, the spores will not swell or form germ tubes. In the absence of $MgSO_4$ the spores swell and eventually a few produce morphologically altered germ tubes. Separate experiments indicate that SO_4^{2-} rather than Mg^{2+} is the required ingredient [21].

Although the spores do not swell or form germ tubes in the absence of glucose, asparagine, or KH_2PO_4, some of the metabolic steps leading to germ tube formation can apparently take place in their absence. Hypothetically, if glucose is required only after the first hour of germination, spores incubated on medium to which glucose is added at 1 h should germinate at the same rate as spores incubated on complete medium. If exogenous glucose is added at 2 h, the spores should form germ tubes 1 h later than control spores. Alternatively, if all steps leading to spore germination are prevented by the absence of glucose, the addition of glucose to cultures at hourly intervals would result in a series of identical germination rate curves which are delayed by 1 h. Table 2 summarizes the results of a series of experiments in which a missing component was added to cultures at hourly intervals and germ tube formation monitored. The extent to which the spores advanced into

TABLE 2

Time Required for *R. stolonifer* Spores to Reach 50% Germination After the Addition of a Component of Medium A[a]

Time of addition	Missing component							
	Glucose		Asparagine		KH_2PO_4		$MgSO_4$	
	[b]Time	[c]Difference	Time	Difference	Time	Difference	Time	Difference
0	4.3	--	4.5	--	4.4	--	4.6	--
1	4.0	0.3	4.1	0.4	3.7	0.7	3.6	1.0
2	3.9	0.4	4.2	0.3	3.0	1.4	2.6	2.0
3	3.8	0.5	4.0	0.5	2.8	1.6	2.0	2.6
4	3.8	0.5	3.9	0.6	2.6	1.8	1.6	3.0
5	3.7	0.6	3.7	0.8	2.6	1.8	1.1	3.5

[a]Medium A consists of 20 g of glucose, 2 g of asparagine, 0.5 g of KH_2PO_4, 0.26 g of $MgSO_4$, and water to 1 liter.

[b]The time required for the spores to reach 50% germination after the addition of the component.

[c]The difference between the time (h) required for 50% germination of the control spores and the spores to which the component was added.

the germination sequence was estimated by subtracting the time required to reach 50% germination after the addition of the missing component from the time required for control spores to reach 50% germination. If glucose was present at 0 time, the spores reached 50% germination at 4.3 h. If glucose was added at 1, 2, 3, 4, or 5 h, the spores reached 50% germination at about 4, 3.9, 3.8, 3.8, and 3.7 h after the addition of glucose. Therefore, the spores were able to advance through about 0.6 h of the germination process (4.3 h - 3.7 h = 0.6 h) in the absence of glucose. After this time, glucose was required to complete the developmental process. Similar results were obtained when asparagine was omitted. In contrast, the spores required about 2.6 h after the addition of KH_2PO_4 to reach 50% germination. Thus, the spores completed approximately 1.8 h of the germination process in the absence of KH_2PO_4. The results of these experiments indicate, therefore, that germination of *R. stolonifer* spores may be divided into physiological stages which are regulated by the components of the culture medium. Potentially, it should also be possible to separate germination into a series of steps by obtaining temperature-sensitive germination mutants. However, our preliminary attempts to isolate such mutants have been unsuccessful.

C. Macromolecular Syntheses

Figure 4 shows the rate of incorporation of leucine into protein and of uracil or adenine into RNA during *R. stolonifer* spore germination. Protein and RNA syntheses begin within the first few minutes and rapidly increase during germination. The decreased incorporation rates during germ tube emergence probably result from increased protein and RNA precursor pools rather than from decreases in synthetic rates, since total protein and total RNA continue to increase during germination (Fig. 4). DNA synthesis begins between the first and second hours of germination and increases during the remainder of the germination process [22].

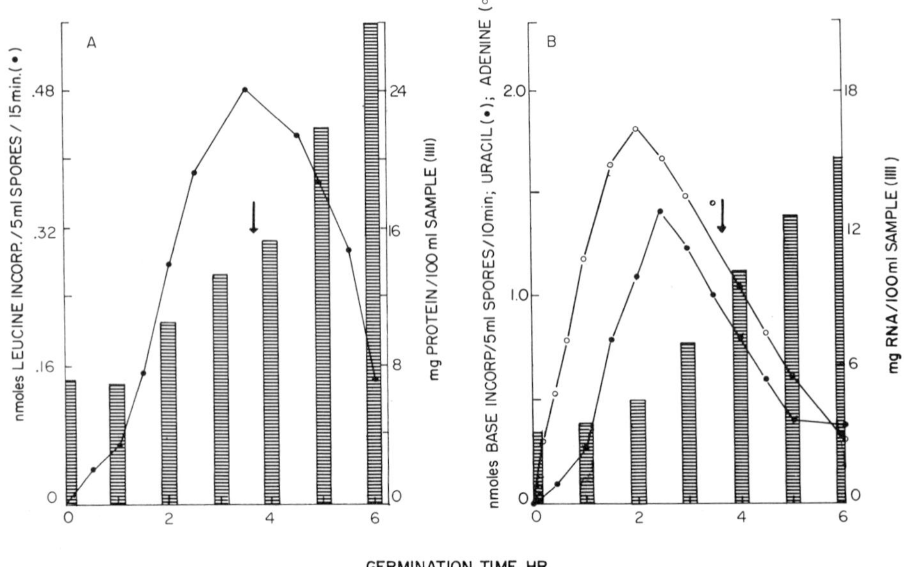

FIG. 4. (A) Incorporation of leucine into protein during 15-min pulse periods (●) and total protein content (bar); (B) incorporation of uracil (●) or adenine (○) into RNA during 15-min pulse periods and total RNA content (bar) at various stages of *R. stolonifer* spore germination. Modified slightly from Van Etten, Bulla, and St. Julian [19].

Protein synthesis is essential for germ tube formation since cycloheximide and blastocidin S inhibit both germination and leucine incorporation (Table 1 [23]). Although chloramphenicol (at 3.5 mg/ml) reduces the extent of germination, the involvement and essentiality of mitochondrial protein synthesis during germination have not been established. RNA synthesis may also be essential for germination, since some RNA synthesis inhibitors such as lomofungin and daunorubicin inhibit uracil incorporation and germ tube formation (Table 1 [24]). Other inhibitors of RNA synthesis such as actinomycin D, cordycepin, and α-amanitin have no effect on germination (Table 1), possibly because the spores are impermeable to them. At present it is not known if the synthesis of DNA is required for germination.

Protein synthesis

Many investigators have isolated components involved in protein synthesis from ungerminated and germinated spores of several fungi, and as a result several *apparent* unifying concepts have emerged. The lack of protein synthesis in spores and, hence, the maintenance of dormancy do not appear to be due to defective ribosomes, transfer enzymes, aminoacyl-tRNA synthetases, tRNAs, or mRNAs [25-27]. Only a few components of the protein synthetic apparatus have been investigated in *Rhizopus*. Polyuridylic acid-directed in vitro phenylalanine incorporation systems prepared from mycelium of *R. arrhizus* [28] or from ungerminated and germinated sporangiospores of *R. stolonifer* [29] exhibited very low activities, although the general characteristics of the systems were similar to those of other organisms. Experiments in which active ribosomes or transfer factors from the fungus *Botryodiplodia theobromae* were assayed with ribosomes and transfer factors from *R. stolonifer* revealed that the ribosomal fraction was the source of the low activity. The reason for this low ribosomal activity is not known; ungerminated and germinated spores contain 80S ribosomes from which undegraded 25-S and 18S rRNAs can be isolated [30].

Both ungerminated and germinated sporangiospores of *R. stolonifer* contain tRNAs which have acceptor activity for all 20 amino acids commonly found in protein when assayed with an aminoacyl-tRNA synthetase fraction from germinated spores [31]. However, there are differences in some of the isoaccepting species of tRNAs in ungerminated and germinated spores. Ten aminoacyl-tRNAs, prepared with tRNA and aminoacyl-tRNA synthetases isolated from ungerminated and germinated spores, were analyzed on benzoylated diethylaminoethyl cellulose columns [6]. No significant differences were detected for seven of the aminoacyl-tRNAs (alanine, arginine, aspartic acid, histidine, leucine, methionine, and phenylalanine). The two isoaccepting species of lysyl-tRNA and valyl-tRNA changed quantitatively and isoleucyl-tRNA changed qualitatively during germination. Whether these differences result from increased synthesis of the particular isoaccepting species of tRNA during germination or modification of preexisting isoaccepting species is not known. However, de novo synthesis of tRNA occurs almost immediately after the spores are placed on a medium conducive for germination [32].

R. stolonifer spores harvested in water have polysomes and thus contain mRNA [30]. In contrast, polyribosomes are not detected in dry-harvested spores. Exposure of dry-harvested spores to water for 5 to 15 min results in the appearance of polysomes. This process is inhibited by cyanide, azide, or cycloheximide. It is not known, however, if polysome formation is due to de novo synthesis of mRNA or if the water causes an association of preformed mRNA with the ribosomes.

DNA replication

Although the composition of DNA usually does not change during cellular development, there are a few reports of qualitative alterations in DNA composition during certain phases of the life cycle of some procaryotic and eucaryotic organisms [27]. Thus, potential modifications, such as amplification of DNA during spore formation or germination, were investigated by determining some of the physical and chemical properties of DNA isolated from ungerminated and germinated spores. DNA isolated from either germinated or ungerminated spores of *R. stolonifer* had identical guanine + cytosine (GC) contents of about 38 mol % as determined by thermal denaturation, buoyant density in CsCl, and spectral properties at pH 3.0 [22]. Thus, spore formation apparently does not involve detectable amplification of genes which are then deleted or fail to replicate upon germination, and germination does not involve detectable amplification of specific genes.

R. stolonifer DNA sediments in CsCl as a single band with a buoyant density of 1.696 gm/cm^3. The buoyant densities of mitochondrial and nuclear DNAs are identical in three other species of the Mucorales [33], and they are probably identical in *R. stolonifer* [22].

A DNA-directed DNA polymerase was purified several 100-fold from germinated and ungerminated spores of *R. stolonifer* [34]. The partially purified enzyme from both spore states exhibited identical characteristics; incorporation of $[^3H]$deoxythymidine monophosphate into DNA required magnesium ion, template DNA, a reducing agent, and the simultaneous presence of deoxyguanosine, deoxycytidine, and deoxyadenosine triphosphates. Heat-denatured and DNAase-activated DNAs were better templates for the enzyme than were native DNAs. The buoyant density of the reaction product was similar to that of the template DNA added in the assay mixture. The enzyme is apparently composed of a single polypeptide chain with an S value of about 5.1 and a mol wt of 70,000 to 75,000 daltons. During early stages of purification the enzyme fraction from ungerminated spores was almost completely dependent upon the addition of exogenous DNA for maximum activity, whereas the corresponding enzyme fraction from germinated spores did not require exogenous DNA. Furthermore, in the early stages of purification, the enzyme activity from germinated spores sedimented rapidly and in a heterogeneous manner in glycerol density gradients. If the enzyme preparation was treated with DNAase prior to centrifugation, the enzyme activity sedimented at about 5.1 S. Apparently, DNA polymerase from germinated spores is more tightly bound to endogenous DNA than the enzyme from ungerminated spores. This might be expected since only germinating spores synthesize DNA.

RNA synthesis

DNA-directed RNA polymerases were examined in germinated and ungerminated spores of *R. stolonifer* [35,36]. Germinated spore extracts chromatographed on DEAE-cellulose columns contained three peaks of RNA polymerase activity (designated I, II, and III in Fig. 5D). The general properties of these enzymes were similar to those reported for other organisms [37]. Enzyme preparations from ungerminated spores, however, only contained active fractions cor-

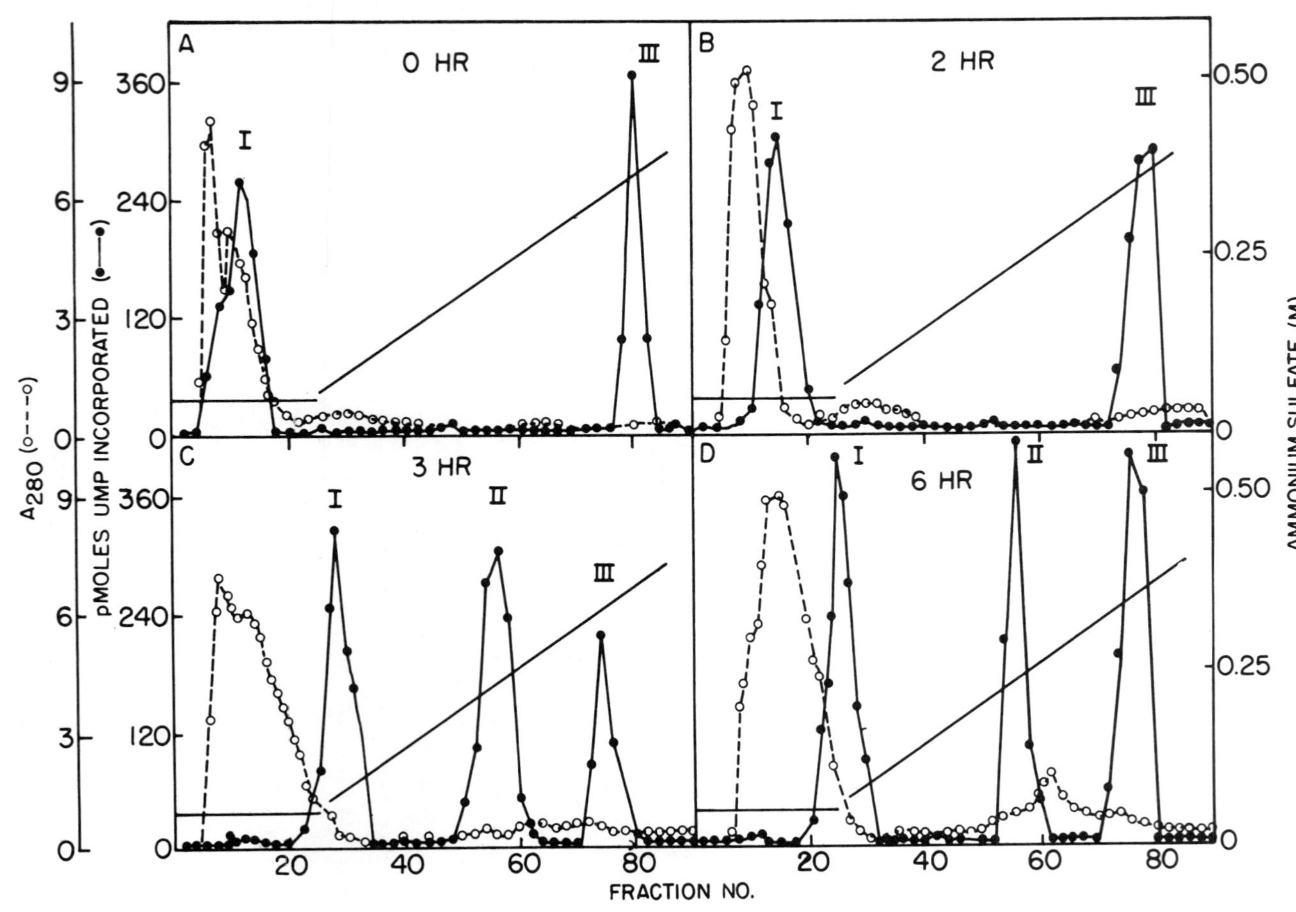
A
0 HR
B
2 HR
C
3 HR
D
6 HR
I
II
III
A_{280} (○--○)
pMOLES UMP INCORPORATED (●—●)
AMMONIUM SULFATE (M)
FRACTION NO.

FIG. 5. DEAE-cellulose column chromatography of RNA polymerases obtained from spores of *R. stolonifer* at different stages of germination. (A) Zero h, ungerminated spores; (B) 2 h, swollen spores; (C) 3 h, just prior to germ tube emergence; (D) 6 h, germinated spores; (●), RNA polymerase activity; (○), $A_{280\ nm}$ [35]. Reproduced with permission from Elsevier Scientific Publishing Co.

responding to RNA polymerases I and III. Several obvious explanations for the apparent absence of RNA polymerase II in the ungerminated spore were tested experimentally: (1) extracts from ungerminated spores were added to the RNA polymerase II fraction from germinated spores to test for an inhibitor in the spores, but no decrease in RNA polymerase II activity was observed; (2) attempts to solubilize RNA polymerase II by a variety of salt concentrations were unsuccessful; (3) RNA polymerase II from germinated spores exhibited very high activity with poly[d(A-T)] as a template and, hence, fractions from spore extracts were assayed with this DNA. However, no activity was obtained in the fractions corresponding to RNA polymerase II. While none of these explanations for the inability to detect RNA polymerase II activity in the spore can be eliminated, the possibility remains that RNA polymerase II is not present in an active form in the dormant spore.

Properties of RNA polymerase III from both spore states were indistinguishable, whereas RNA polymerase I from the ungerminated spores eluted from the DEAE-cellulose column slightly earlier and responded to divalent cations differently than the corresponding fraction from germinated spores. Figure 5 shows when during germination RNA polymerase II appeared and when the characteristics of RNA polymerase I were altered so that they resembled the enzyme from the germinated spores. RNA polymerase II was detected after 3 h, just prior to germ tube formation. At the same time (3 h), RNA polymerase I attained the characteristics of the germinated spore enzyme.

In other eucaryotic organisms, RNA polymerases I and II direct the synthesis of rRNA and mRNA, respectively. The function of RNA polymerase III has not been established, but it has been reported that this enzyme transcribes small RNA species such as 5S RNA and tRNA [38] or that it is of mitochondrial origin [39]. Evidence that specific classes of RNA are synthesized by specific RNA polymerases in these eucaryotic organisms resulted mainly from studies with isolated nuclei and nucleoli. Similar studies, however, are difficult to conduct with *R. stolonifer* because of the harsh conditions required to disrupt the spores. Indirect evidence that *R. stolonifer* RNA polymerases I, II, and III correspond to those from other eucaryotic organisms is provided by the sensitivity of *R. stolonifer* RNA polymerase II to α-amanitin and the effect of increasing $(NH_4)_2SO_4$ concentrations on the activity of the three polymerases.

If the RNA polymerase data are valid and if our assumptions on the functions of RNA polymerases I and II are correct, the following predictions about RNA synthesis and *R. stolonifer* spore germination can be made: (1) the spore is incapable of synthesizing mRNA or its precursor [heterogeneous nuclear RNA (HnRNA)] until an active RNA polymerase II appears between 2 and 3 h into the germination process; (2) RNA polymerase II must be translated from preexisting mRNA in the spores if it is the result of de novo synthesis; (3) the initial synthesis of rRNA and probably tRNA must occur prior to mRNA, since uracil and adenine are incorporated into RNA almost immediately after the spores are placed in a germination medium (Fig. 4B).

The kinetics of synthesis of different RNA classes have been examined [32] in order to determine if the synthesis of all classes of RNA begin simultaneously or if rRNA and/or tRNA initiate synthesis prior to the initiation of mRNA synthesis [32]. Figure 6 shows the results of polyacrylamide gel electrophoresis of total RNA extracted from spores pulse-labeled with [^{3}H]adenine at 0 to 15 min, 15 to 30 min, 45 to 60 min, and 225 to 240 min. [^{14}C]rRNA and [^{14}C]tRNA were also included in each gel as internal markers. RNAs which comigrated with large and small rRNAs, 5S RNA, and tRNA were synthesized within the first 15 min of germination (Fig. 6A) and during all subsequent time periods. The two slower migrating radioactive RNA species present in Fig. 6A-D and especially prominent in Fig. 6A-B are characteristic of precursor rRNAs. They have mol wts of approximately 2.4×10^6 and 1.45×10^6, which are similar to the mol wts of precursor rRNAs reported for yeast [40,41]. Furthermore, the radioactivity in these fractions disappeared during a 30-min chase with unlabeled adenine, and a corresponding increase in

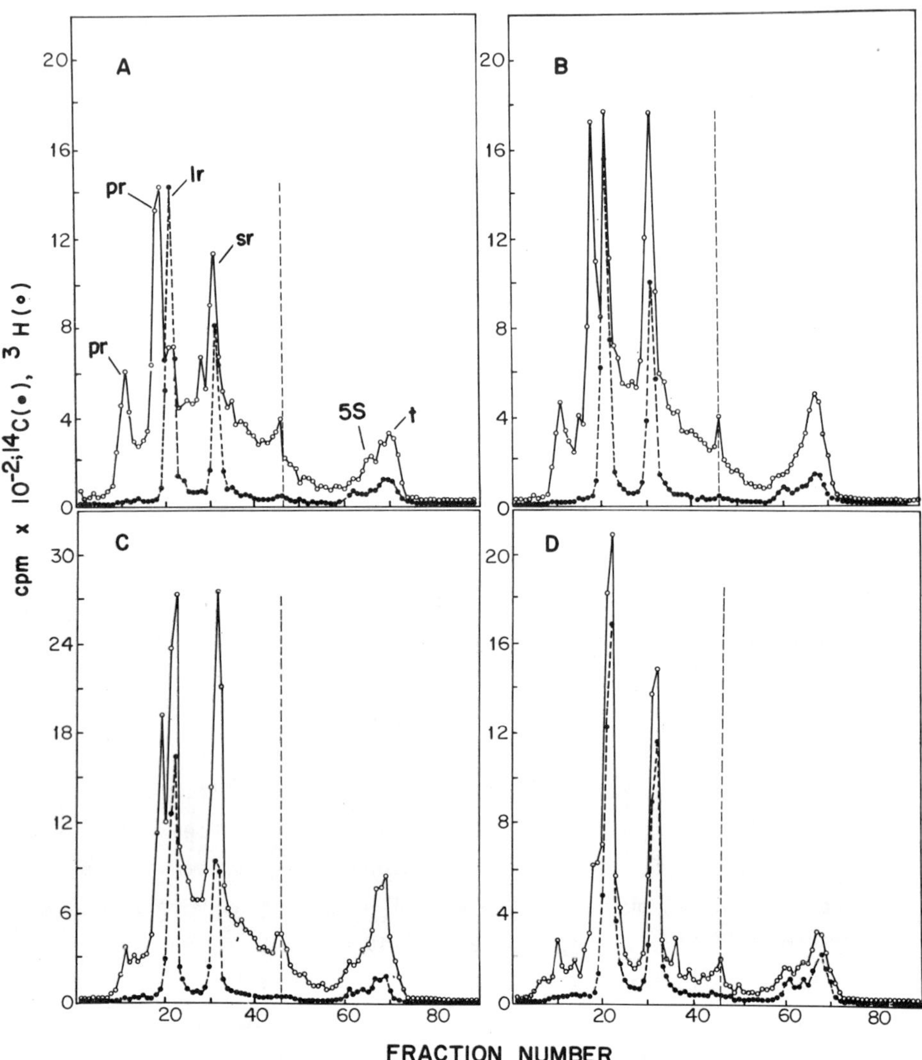

FIG. 6. Polyacrylamide gel electrophoresis of total RNA from germinating sporangiospores of *R. stolonifer* which were pulsed for 15-min periods with [^{3}H]adenine. Pulse-labeled from (A), 0 to 15 min; (B), 15 to 30 min; (C), 45 to 60 min; and (D), 225 to 240 min. Precursor rRNAs (pr), large rRNA (lr), small rRNA (sr), 5S RNA (5 S), and tRNA (t) are labeled in A. The open circles represent the cpm of the pulsed material and the solid circles represent the cpm of a [^{14}C]RNA marker. The gels were composed of 3% (left side of the dashed line) and 5% (right side of the dashed line) acrylamide. Reproduced with permission from Academic Press.

radioactivity appeared in the rRNA fractions so that the ratio of radioactivity in the large rRNA to small rRNA increased to about 2-2.5:1 after the chase period. Adenine incorporated into tRNA resulted from de novo synthesis and was not due to terminal addition at the 3'-hydroxy ends [32].

Recent reports indicate that many HnRNAs and mRNAs from eucaryotic organisms including *Achlya* and yeast [42-45] contain polyadenylic acid segments attached to their 3'-hydroxy ends [termed poly(A)+ RNA]. Preliminary experiments established that an RNA fraction from *R. stolonifer* spores pulse-labeled for 15 min at 4 h into germination eluted from a polydeoxythymidylic acid-cellulose column in a manner expected of poly(A)+ RNA. Further characterization of this fraction revealed that: (1) it comprised 3 to 5% of the adenine incorporated into total RNA; (2) it sedimented in a heterogeneous manner between 7 S and 25 S in sucrose density gradients; (3) it had a tendency to aggregate; (4) it contained few methylated bases; (5) it hybridized with about 8% of the *R. stolonifer* DNA; and (6) rRNA did not compete with it in DNA:RNA hybridization experiments [32]. Since all of these properties are characteristics of HnRNA and mRNA, we concluded that this poly(A)+ RNA fraction is enriched in HnRNA and/or mRNA.

To determine when the synthesis of poly(A)+ RNA began during germination, spores were pulse-labeled with [^{3}H]adenine at six time periods and the total RNA was separated into poly(A)+ RNA and poly-(A)- RNA fractions. As noted in Table 3, adenine was incorporated into both fractions within the first 15 min of germination. However, more adenine was incorporated into poly(A)+ RNA relative to poly(A)- RNA during the early stages of germination, and the ratio decreased with increasing germination time.

It is possible, however, that the [^{3}H]adenine incorporated into poly(A)+ RNA was added as a poly A tail to preexisting spore RNA and was not a result of de novo HnRNA and mRNA synthesis, since it is known that the polyadenylate segment is added after transcription of HnRNA in other organisms [46]. Results from two types of experiments eliminate this possibility: (1) poly(A)+ RNA was iso-

TABLE 3

Distribution of *R. stolonifer* RNA into Poly(A)+ RNA and Poly(A)- RNA after Chromatography on a Poly dT-cellulose column[a]

Time of pulse, min	% of Total radioactivity incorporated into RNA			
	[^{3}H]Adenine		[^{3}H]Uracil	
	Poly (A)+ RNA	Poly (A)- RNA	Poly (A)+ RNA	Poly (A)- RNA
0-15	13.4	86.6	12.1	87.9
15-30	13.9	86.1	13.4	86.6
45-60	10.8	89.2	7.4	92.6
105-120	6.0	94.0	6.0	94.0
165-180	4.9	95.1	4.4	95.6
225-240	4.8	95.2	4.3	95.7

[a]The RNA was isolated from cells which had been pulse-labeled with either [^{3}H]adenine or [^{3}H]uracil for 15-min periods during the germination process [32].

lated from cells pulse-labeled with [^{3}H]adenine during the first 15 min of germination and incubated with RNAases A and T_1. About 75% of the radioactivity incorporated into this poly(A)+ RNA fraction was digested with the nucleases; (2) spores were pulse-labeled for 15-min periods at six time intervals during germination with either [^{3}H]adenine or [^{3}H]uracil. As noted in Table 3 uracil was also incorporated into the poly(A)+ RNA fraction during the first 15 min of germination. Furthermore, the distribution of uracil into the poly(A)+ RNA and poly(A)- RNA fractions paralleled the distribution of adenine incorporated into these RNA fractions during germination (Table 3). Therefore, these data indicate that poly(A)+ RNA is synthesized de novo within the first 15 min of germination.

In summary, the RNA synthesis experiments indicate that all classes of RNA initiate synthesis within 15 min after placing *R. stolonifer* spores in a germination medium. In fact, synthesis of all classes of RNA can be detected within the first 8 min [47]. Although it is still possible that these RNAs initiate synthesis sequentially, synthesis of all of them begins within the first few minutes. Finally, the RNA synthesis data conflict with our interpretation of the qualitative and quantitative changes observed in DNA-directed RNA polymerases associated with germination. At present we have no satisfactory explanation for this paradox.

Recently we have examined the length of the poly(A) segment in poly(A)+ RNA synthesized at various times during germination. Table 4 shows the percentage of adenine incorporated into RNAase A- and RNAase T_1-resistant poly(A)+ RNA isolated from total RNA from spores pulse-labeled at several time periods during germination.

TABLE 4

The Percentages of Newly Synthesized Poly(A)+ RNA Isolated from Total RNA Which Were Resistant to RNAase A and RNAase T_1 Digestion[a]

Time of pulse, min	Percentage of radioactivity in poly(A)+ RNA resistant to RNAase digestion
0-15	26.5
15-30	24.8
45-60	23.8
105-120	16.9
165-180	13.8
225-240	15.6

[a]The poly(A)+ RNA was isolated from germinating spores of *R. stolonifer* pulse-labeled with [^{3}H]adenine.

About 26% of the adenine incorporated into poly(A)+ RNA during the first 15 min of germination was resistant to RNAase digestion; this percentage decreased to about 15 to 17% at 2 to 4 h. The poly(A) segments isolated from spores pulse-labeled at the earliest time periods migrate slightly slower on 12% polyacrylamide gels than those isolated from cells pulse-labeled at later times during germination (Fig. 7A). The average lengths of the poly(A) segments isolated from total RNA were determined directly by measuring the ratio of radioactivity in adenylic acid to that in adenosine after KOH hydrolysis (Table 5). The average length of the poly(A) segment synthesized during the first 15 min of germination was about 100 nucleotides; the length decreased with germination time until it was about 60 nucleotides at 4 h. Thus, the length of the newly synthesized poly(A) segments attached to poly(A)+ RNA isolated from total RNA decreased during germination. Although there are several explanations for these results, it is possible that the spore synthesizes poly(A)+ RNA with long poly(A) segments during the early stages of germination; with increased germination time shorter poly(A) segments are synthesized. Alternatively, the change in the length of the poly(A) segment might simply reflect the speed with which HnRNA is processed to mRNA.

Experiments were then conducted to determine: (1) if any newly synthesized poly(A)+ RNA immediately became associated with ribosomes (as polysomes), and (2) if the length of the poly(A) segment present in mRNA isolated from polysomes also decreased during germination. Spores were pulse-labeled with [^{3}H]adenine for 15-min periods throughout germination, polyribosomes were isolated, RNA was isolated from these fractions, and the RNA was fractionated into poly(A)+ RNA and poly(A)- RNA. [^{3}H]adenine was present in both poly(A)+ RNA and rRNA isolated from polysomes in spores pulse-labeled from 0 to 15 min. Thus, at least some mRNA and rRNA are synthesized de novo, processed, and assembled into functional polysomes during the first 15 min of *R. stolonifer* spore germination.

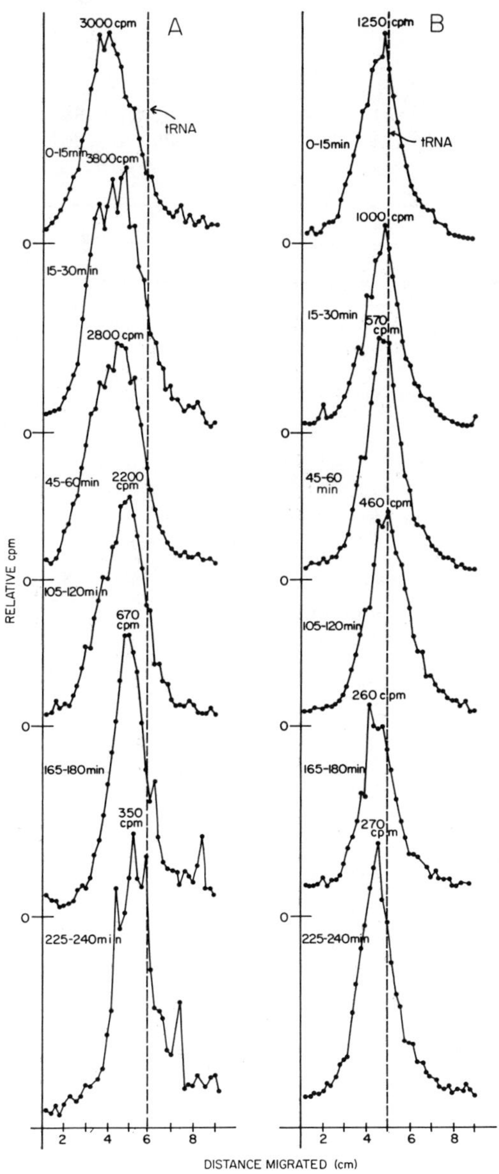

FIG. 7. Polyacrylamide gel electrophoresis of polyadenylate segments isolated from (A) total RNA or (B) polysomal RNA. The spores were pulsed for the time periods indicated, poly(A) segments isolated, and electrophoresed for 5 h on 12% polyacrylamide. The dashed line represents the migration of tRNA which was added as an internal marker in each of the gels. The cpms indicated on the figure are the highest cpm in the peak fraction for each of the samples.

TABLE 5

Length of Newly Synthesized Poly(A) Segments of Poly(A)+ RNA Isolated from Total RNA or Polysomal RNA[a]

Time of pulse, min	Ratio of cpm in AMP to cpm in adenosine	
	From total RNA	From polysomal RNA
0-15	98 ± 12	52
15-30	92 ± 6	54
45-60	85 ± 4	54
105-120	70 ± 12	41
165-180	72 ± 12	48
225-240	60 ± 4	44
Average	--	49

[a]Total RNA and polysomal RNA were isolated from spores pulsed with [^{3}H]adenine for 15-min periods. Poly(A) segments were isolated [48], hydrolyzed with 0.3 M KOH for 18 h at 37°, chromatographed [49], and the radioactivity in adenosine monophosphate and adenosine determined.

In contrast to the results with total poly(A)+ RNA the lengths of the poly(A) segments isolated from polysomal poly(A)+ RNA were nearly identical at all times during germination as judged by their migration on polyacrylamide gels (Fig. 7B) and each contained about 50 adenylic acid residues (Table 5). A logical interpretation of these results is that the majority of the poly(A)+ RNA synthesized during the early periods of germination might take longer to be processed to mRNA and, hence, is isolated as HnRNA. After the spores are incubated in a germination medium for longer periods of time, the poly(A)+ RNA may be synthesized, processed, and associated with ribosomes faster so that the majority of the poly(A) segments isolated at later times are associated with mature mRNA. If these interpretations are correct, at least part of the processing of HnRNA to mRNA in *R. stolonifer* may result in a loss of about 40 to 50 adenine residues from the poly(A) region of the HnRNA. Experiments to test this possibility are presently underway.

III. FINAL COMMENTS

Because fungal spores can be easily produced, collected, and synchronously germinated under well-defined conditions, they provide a useful system for studying cellular differentiation. In addition to the fundamental physiological and molecular information that such studies may yield, several practical and potentially applicable aspects can be derived. Most human and plant diseases incited by fungi are initiated by spores, and germination is the first event essential to disease establishment. Thus, a basic understanding of the biochemical processes associated with germination should eventually result in more effective control methods of fungal diseases.

ACKNOWLEDGMENTS

We thank Ken Nickerson, Les Lane, Jim Partridge, Wendell Gauger, and Robert Knight for critically reading the manuscript. The scanning electron micrographs were the work of Kit Lee. Research from our laboratory was supported by Public Health Service Grant AI 108057 from the National Institute of Allergy and Infectious Diseases.

REFERENCES

1. Walker, J. C., 1957. Plant Pathology, Third ed., McGraw-Hill, New York.
2. Austwick, P. K. C., 1966. *In* The Fungus Spore, M. F. Madelin (ed.), Butterworths, London, p. 321.
3. Daldorf, G., 1962. Fungi and Fungus Diseases, C. C Thomas, Springfield, Illinois.
4. Davis, B. D., R. Dulbecco, H. N. Eisen, H. S. Ginsberg, and W. B. Wood, 1967. Microbiology, Harper & Row, New York.
5. Alexopoulos, C. J., 1962. Introductory Mycology, Second ed., John Wiley and Sons, New York.
6. Merlo, D. J., H. Roker, and J. L. Van Etten, 1972. Can. J. Microbiol. 18: 949-956.

7. Flanagan, P. W., 1969. Can. J. Bot. 47: 2055-2059.

8. Hawker, L. E., and A. Beckett, 1971. Phil. Trans. Roy. Soc. London, Ser. B, 263: 71-100.

9. Gauger, W. L., 1961. Amer. J. Bot. 48: 427-429.

10. Gooday, G. W., 1973. *In* Microbial Differentiation. Twenty-third Symposium of the Society for General Microbiology, J. M. Ashworth and J. E. Smith (eds.), University Press, Cambridge, p. 269.

11. Van den Ende, H., and D. Stegwee, 1971. Bot. Rev. 37: 22-36.

12. Ekundayo, J. A., 1966. J. Gen. Microbiol. 42: 283-291.

13. Hawker, L. E., and P. M. V. Abbott, 1963. J. Gen. Microbiol. 32: 295-298.

14. Hess, W. M., and D. J. Weber, 1973. Protoplasma 77: 15-33.

15. Hawker, L. E., 1966. *In* The Fungus Spore, M. F. Madelin (ed.), Butterworths, London, p. 151.

16. Buckley, P. M., N. F. Sommer, and T. T. Matsumoto, 1968. J. Bacteriol. 95: 2365-2373.

17. Bartnicki-Garcia, S., 1968. Ann. Rev. Microbiol. 22: 87-108.

18. Ekundayo, J. A., and M. J. Carlile, 1964. J. Gen. Microbiol. 35: 261-269.

19. Van Etten, J. L., L. A. Bulla, and G. St. Julian, 1974. J. Bacteriol. 117: 882-887.

20. Bussel, J., N. F. Sommer, and T. Kosuge, 1969. Phytopathology 59: 946-952.

21. Dunkle, L. D., K. Nickerson, and J. L. Van Etten, unpublished results.

22. Dunkle, L. D., and J. L. Van Etten, 1972. *In* Spores, Vol. 5, H. O. Halvorson, R. Hanson, and L. L. Campbell (eds.), American Society for Microbiology, Washington, D.C., p. 283.

23. Van Etten, J. L., unpublished results.

24. Nickerson, K., and J. L. Van Etten, unpublished results.

25. Sussman, A. S., and H. A. Douthit, 1973. Ann. Rev. Plant Physiol. 24: 311-352.

26. Van Etten, J. L., 1969. Phytopathology 59: 1060-1064.

27. Van Etten, J. L., L. D. Dunkle, and R. H. Knight, 1976. *In* The Fungal Spore, Form and Function, D. J. Weber and W. M. Hess (eds.), Wiley-Interscience, New York, p. 243.

28. Raghu, K., and D. J. Weber, 1972. Mycologia 64: 81-91.

29. Van Etten, J. L., and R. Knight, unpublished results.

30. Freer, S., and J. L. Van Etten, unpublished results.

31. Van Etten, J. L., R. Koski, and M. M. El-Olemy, 1969. J. Bacteriol. 100: 1182-1186.

32. Roheim, J. R., R. H. Knight, and J. L. Van Etten, 1974. Dev. Biol. 41: 137-145.

33. Villa, V. D., and R. Storck, 1968. J. Bacteriol. 96: 184-190.

34. Gong, C.-S., L. D. Dunkle, and J. L. Van Etten, 1973. J. Bacteriol. 115: 762-768.

35. Gong, C.-S., and J. L. Van Etten, 1972. Biochim. Biophys. Acta 272: 44-52.

36. Gong, C.-S., and J. L. Van Etten, 1974. Can. J. Microbiol. 20: 1267-1272.

37. Biswas, B. B., A. Ganguly, and A. Das, 1975. *In* Progress in Nucleic Acid Research and Molecular Biology, Vol. 15, W. E. Cohn (ed.), Academic Press, New York, p. 145.

38. Weinmann, R., and R. G. Roeder, 1974. Proc. Nat. Acad. Sci. USA 71: 1790-1794.

39. Horgen, P. A., and D. H. Griffin, 1971. Nature New Biol. 234: 17-18.

40. Retel, J., and R. J. Planta, 1970. Biochim. Biophys. Acta 224: 458-469.

41. Udem, S. A., and J. R. Warner, 1972. J. Mol. Biol. 65: 227-242.

42. Edmonds, M., and D. W. Kopp, 1970. Biochem. Biophys. Res. Commun. 41: 1531-1537.

43. McLaughlin, C. S., J. R. Warner, M. Edmonds, H. Nakazato, and M. H. Vaughan, 1973. J. Biol. Chem. 248: 1466-1471.

44. Reed, J., and E. Wintersberger, 1973. F E B S Letters 32: 213-217.

45. Silver, J. C., and P. A. Horgen, 1974. Nature 249: 252-254.

46. Darnell, J. E., W. R. Jelinek, and G. R. Molloy, 1973. Science 181: 1215-1221.

47. Roheim, J. R., R. H. Knight, and J. L. Van Etten, unpublished results.

48. Freer, S., S. Mayama, and J. L. Van Etten, unpublished results.

49. Mendecki, J., S. Y. Lee, and G. Brawerman, 1972. Biochemistry 11: 792-798.

ZOOSPORE GERMINATION AND EARLY DEVELOPMENT OF *BLASTOCLADIELLA EMERSONII*

James S. Lovett, Cheng-Shung Gong, and Steven A. Johnson

Department of Biological Sciences
Purdue University
West Lafayette, Indiana

I. INTRODUCTION

Blastocladiella emersonii--a small, nonfilamentous water mold --was originally described by Dr. E. C. Cantino as an isolate from the botany garden pool at the University of Pennsylvania [1]. The versatility of *Blastocladiella* for experimental work was immediately recognized by Dr. Cantino and was exploited over several years in studies of a number of enzyme differences between cells developing along alternative developmental pathways in multigeneration cultures [2]. The ready experimental induction of these different developmental sequences presented a useful system for developmental analysis, which was further improved by the discovery that excellent synchrony could be obtained even with heavy zoospore inoculations, thus permitting the analysis of single generations of growth and development throughout either life cycle [3,4]. This has already been reviewed in detail [2,5], and we want to discuss work made possible by further refinements of the system. We found that even higher densities (1 to 4 × 10^6/ml) of spores inoculated into a chemically defined medium, or an inorganic salts solution, would undergo encystment and outgrowth (referred to collectively as germination) to produce a tubular rhizoid, rapidly and with rather good synchrony [6]. Soll and Sonneborn, using lower cell densities, further defined the conditions for achieving rapid and synchronous germination and have shown the potassium ion concentration to be critical for optimal synchrony in the process [7].

The zoospore is a motile, nongrowing, highly differentiated cell, whose primary function is presumed to be asexual propagation and dispersal of the organism. Particularly striking, even by light microscopy, are its large membrane-enclosed ribosomal nuclear cap, posterior flagellum, and characteristic side body composed of a large acentric mitochondrion and associated lipid droplets (Fig. 1). Under conditions appropriate for germination this highly differentiated state is rapidly lost during zoospore encystment, which in a given cell requires only about 1 min. During this brief interval the cell rounds up, retracts its flagellum, and deposits a thin but clearly distinguishable cell wall. After inoculation into appropriate medium, cells begin to encyst at about 10 min, and rhizoidal outgrowth begins around 25 min; by 60 min the rhizoid has elongated considerably and soon produces branches. The first nuclear division occurs between 140 and 190 min, but is not accompanied by cytokinesis [6]. Cells germinated in an inorganic medium follow an identical sequence, including considerable rhizoid elongation and branching, but fail to undergo mitosis.

At the ultrastructural level encystment involves not only the retraction of the flagellar axoneme and deposition of a thin cell wall, but also the spreading out of the mitochondrion. The anchoring microtubules and flagellar rootlets disappear, with partial disorganization of the previously intact nuclear cap membrane and breakdown of the zoospore gamma particles. Subsequently, the nuclear cap ribosomes are redistributed in the cytoplasm, the mito-

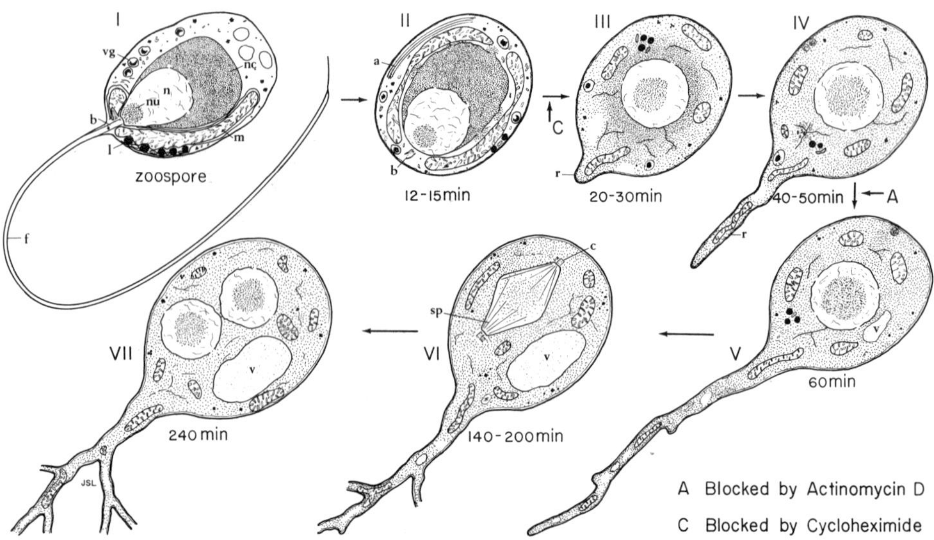

FIG. 1. Stages in zoospore germination and early growth. A series of diagrammatic sketches to illustrate the morphological and intracellular changes of cells germinating in complete medium: a, axoneme of retracted flagellum; b, basal body; c, centriole; f, flagellum; l, lipid; m, mitochondrion; nc, nuclear cap; nu, nucleolus; n, nucleus; r, rhizoid; sp, mitotic spindle; v, vacuole; vg, vesicle-enclosed gamma particle. Reprinted from [6] with permission of the American Society for Microbiology.

chondrion divides at least once [8], and the microtubules of the retracted flagellar axoneme are degraded or disassembled. These changes have been described in detail elsewhere [6,9-13]. Normal encystment occurs in the presence of either cycloheximide or actinomycin D, while production of the rhizoidal outgrowth requires new protein synthesis, but not new RNA synthesis [6,14]. Actinomycin D, however, does not cause developmental arrest until the early rhizoid elongation stage (40 to 50 min), regardless of when the antibiotic is added [6]. Cells encysted in cycloheximide do not break down the retracted flagellar axoneme but otherwise complete the known cytological events which precede rhizoid formation [11,14]. If this inhibitor is removed again, normal development follows after a delay [14], and if cells are exposed to cycloheximide at later time points rhizoid initiation escapes inhibition [15].

Other significant changes also occur during encystment [10]; there is a lag in dry wt increase, which does not begin until ca. 60 min, while net increases in RNA and protein content are detectable after 40 min and 80 min, respectively [6]; in the first 30 min the lipid content decreases by 60 to 70% [16] and the cellular glycogen by 90% [17]; the respiratory rate increases more than 100% [18]; and the high levels of zoospore cAMP and cGMP drop rapidly [19]. It has also been shown that prelabeled zoospore proteins are degraded at the rate of 5%/h through at least the first two hours [20].

From these and other experiments it is apparent that the rapid cytological changes (of the zoospore) during germination are accompanied by an equally swift activation of cellular metabolism and at least some biosynthetic systems. The effects of the inhibitors clearly indicate (1) that encystment requires neither RNA nor protein synthesis; (2) that rhizoid outgrowth and flagella axoneme disassembly both depend upon new protein synthesis; and (3) that the required protein synthesis is accomplished using the nuclear cap ribosomes and stored messenger RNA (mRNA) [6,14]. These conclusions further imply that a completely preformed but inactive system for protein synthesis is carried by the nonsynthetic zoospore even to the extent that the stored mRNA contains all the necessary information for the essential early protein synthesis. In discussing our experimental work we would like to concentrate first on our attempts to assess these assumptions concerning the zoospore, and then to describe experiments on the activation of new synthesis during zoospore germination.

II. CURRENT RESEARCH

A. Compartmentalization in Zoospores

In regard to the regulation of protein synthesis it seems to us reasonable to assume, a priori, that at least a portion of each essential protein factor (e.g., initiation and elongation factors, aminoacyl-tRNA synthetases, etc.) must be present in zoospores to enable protein synthesis to resume so quickly at germination. However, such factors could be either excluded from the ribosomes by the nuclear cap membrane or regulated by various mechanisms, such as inactivation. Both of these possibilities have been examined as potential modes for the regulation of protein synthesis.

To assess the role of compartmentalization in *Blastocladiella* we have taken advantage of procedures to isolate the ribosomal nuclear caps in reasonable yields [21-23]. The distributions of various components needed for protein synthesis are summarized in Table 1. Only the ribosomes and associated 5S ribosomal RNA (rRNA) have been found exclusively in the nuclear caps. The proportions of the total elongation factor and synthetase activities are less

TABLE 1

Distribution of Components for Protein Synthesis in the Nuclear Caps and Extra-cap Fractions of Zoospores

	Percentage of total	
Component	Cap	Extra-cap
Soluble protein	14	86
Ribosomes	100	0
Elongation factor-1	1.2 (8.6)[a]	98.8
Elongation factor-2	4.1 (29.3)	95.9
Transfer RNA	61	39
5S RNA	100	0
Messenger RNA [Poly(A)+ RNA]	80.2	19.8
Met-tRNA synthetase (purified)	5.8 (41.4)	94.2
Phenylalanine synthetase (purified)	8.1 (57.9)	91.9
Group synthetase (assayed with protein hydrolysate)	29.6	70.4

[a]Numbers in parentheses indicate the percentage of the expected activity found, based on the assumption that 14% of each soluble component should be present in the cap fraction as found for the soluble protein.

than would be expected from the distribution of the total soluble proteins, while mRNA [poly(A)+ RNA; see below] seems to be in a relative excess. The values in parentheses show the percentage of each protein component recovered in the cap fraction based on that expected if all proteins were evenly distributed in the total soluble fraction, of which 14% occurs in the cap. The recovery of 14% of the protein in the cap fractions seems reasonable since the caps comprise ca. 20% of the zoospore volume [24] and ca 28.5% of its dry wt [21]. In any case, the presence of the aminoacyl-tRNA synthetases, charged tRNA [22], and both elongation factors in the cap provides little direct support for a regulatory mechanism that acts to exclude these factors from the vicinity of the ribosomes. The presence and distributions of the initiation factors, initiation-specific met-$tRNA_F^{met}$, and nucleoside triphosphates have not yet been examined so this conclusion must remain tentative. Also, polyacrylamide gel electrophoresis of the nuclear cap and extra-cap protein fractions does suggest that all proteins may not be evenly distributed.

B. Ribosome Inhibition in Zoospores

We have recently found that the zoospores *do* contain a small fraction of polysome-like ribosomal aggregates. Of the ribosomes, 84% occur as free 80S monomers, with 4% subunits and about 10% polysomes. Evidence that these are true polysomes rather than random aggregates of ribosomes will be presented later, but the mere presence of polysomes might indicate that some protein synthesis takes place in zoospores (as suggested earlier by Soll and Sonneborn [25]). However, we have measured the incorporation of radioactive amino acids by motile zoospores and found that virtually all of the small amount of incorporation detectable, when the polysomes are fractionated on gradients after a 10-min pulse with [^{3}H]yeast protein hydrolysate, is solubilized by hot TCA treatment (Fig. 2). This radioactivity probably represents exchange reactions with the charged zoospore tRNA, but it has not been further analyzed.

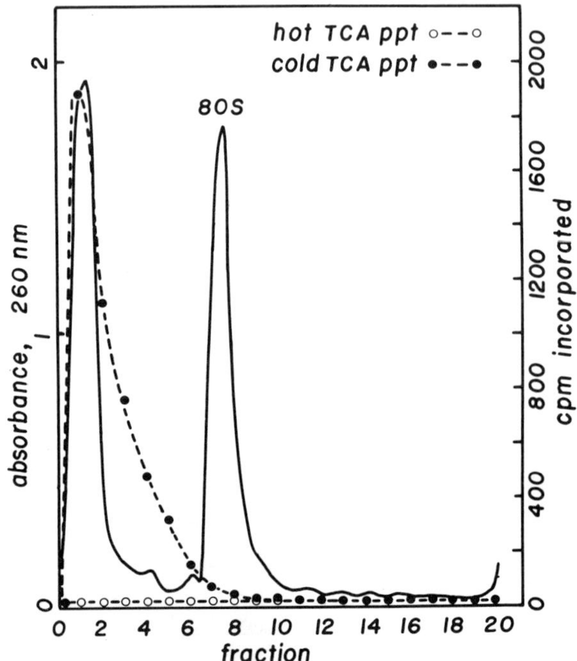

FIG. 2. In vivo incorporation of amino acids by motile zoospores. Zoospores were incubated in 1 mM Tris-maleate (pH 6.8) buffer containing 1 mM $CaCl_2$ and 1 μCi/ml [^{3}H]yeast protein hydrolysate for 10 min at 24°. The cells were collected by centrifugation and sonicated 15 sec in 20 mM Tris-HCl, pH 8.5, 250 mM NaCl, 50 mM $MgCl_2$, 25 mM EGTA, 2% Triton x-100. After a low-speed centrifugation (12K × g, 15 min) four A_{260} units of supernatant were centrifuged through a 0 to 32.5% sucrose gradient (in 10 mM Tris-HCl, pH 7.6, 10 mM KCl, 10 mM $MgCl_2$, 0.1 mM EDTA) for 35 min at 45,000 rpm (Spinco SW 50.1 rotor). The gradient was scanned at 260 nm, fractions were collected, and the radioactivity was determined as previously described [23]. Solid line, absorbance; dashed lines, radioactivity/fraction [from Ref. 35].

The failure to detect synthesis in vivo is consistent with our earlier reports that isolated nuclear cap ribosomes were inactive in vitro when isolated in low-salt buffers and assayed in poly(U)-dependent phenylalanine incorporation assays using supernatants from growth phase cells (Table 2) [22,23].

Furthermore, material removed from the nuclear cap ribosomes by 0.5 M KCl buffer inhibits in vitro amino acid incorporation if added to either active ribosomes from growth phase cells or the

TABLE 2

In Vitro Incorporation Activity of Nuclear Cap Ribosomes Washed in Buffers with Different KCl Concentrations

Molarity of KCL in buffer[a]		Ribosomal protein[b] cpm/mg
Exp. 1	0 M	0
Exp. 2	0.05 M	13,480
	0.5 M	23,120

[a]The molarity of KCl in the 10 mM Tris-HCl (pH 7.2), 10 mM $MgCl_2$ buffers used to wash the ribosomes prior to assay.

[b]Measured in a standard poly(U)-dependent phenylalanine incorporation assay using supernatants of growth phase cells as the source of enzymes (see Refs. 22, 23).

activated (high salt-washed) zoospore ribosomes [22,23]. This inhibitory activity also occurs in the extra-cap fraction of the zoospores and, as a result, our original assays were all run with high-speed supernatants from growth phase cells. The inhibitor is equally effective in blocking either poly(U)-directed phenylalanine incorporation in vitro with plant ribosomes or incorporation of amino acid mixture by isolated growth phase polysomes (Fig. 3). Although this suggests a primary effect on peptide chain elongation, normal initiation is not required in the poly(U) system, and the plateau in incorporation after 30 min also indicates a lack of initiation in the polysome system (Fig. 3, growth phase enzymes and polysomes). Thus, an additional effect of the inhibitor on initiation cannot be ruled out.

The ribosome inhibitor is resistant to heat, 10% TCA treatment, and pronase or RNAase; it is dializable and has fractionation characteristics consistent with a nucleotide or related compound [23,26]. However, we have not yet been able to purify and identify it. The inhibitor can be removed from zoospore preparations by dialysis against 0.1 to 0.5 M KCl, Tris-HCl, $MgCl_2$, buffers, and such cell-free preparations will then catalize amino acid incorporation

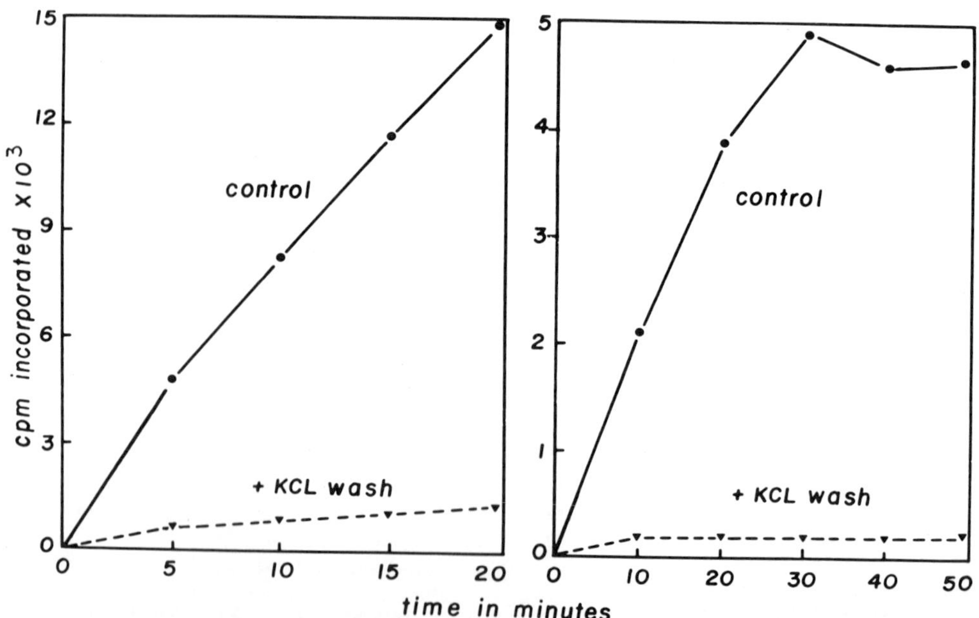

FIG. 3. Inhibition of amino acid incorporation by the inhibitor from zoospores. Nuclear caps of zoospores were isolated and the inhibitor extracted from the caps as described by Adelman and Lovett [23]. Left. Poly(U)-dependent [^{3}H]phenylalanine incorporation by high-salt-washed 80S ribosomes and high-speed supernatant from growth phase cells (see Ref. 23 for method). Right. Amino acid incorporation ([^{3}H]yeast protein hydrolysate) by growth phase polysomes with high-speed supernatant (see footnote, Table 3). The same amount of inhibitor preparation was used in each assay with appropriate controls for the KCl added.

in an entirely homologous zoospore system. In such inhibitor-free low-speed (S 23) zoospore preparations, 0.1 mM concentrations of the initiation inhibitors aurintricarboxylic acid and pactamycin reduce amino acid incorporation by 86% and 75%, respectively (Table 3). This inhibition clearly indicates that the zoospores do contain functional initiation factors in vivo, although they have not yet been isolated. The ability of dialized zoospore supernatants to support linear incorporation by plant polysomes for more than 50 min also supports our original hypothesis that both initiation and elongation factors should exist in the nonsynthetic zoospores [27].

TABLE 3

Amino Acid Incorporation
by Whole-zoospore Cell-free Systems (S 23)[a]

Assay mixture	Specific activity (cpm/mg protein)	% of Control
Complete	35843	100
Minus S 23 supernatant	0	0
Minus GTP	14105	39
Plus aurintricarboxylic acid (0.1 mM)	5147	14
Plus pactamycin (0.1 mM)	8961	25

[a] 1 ml packed cell volume of zoospores was mixed with 2 ml high-salt TKM buffer (20 μM Tris-HCl, pH 7.6, 500 mM KCl, 1 mM $MgCl_2$, 0.1 mM EDTA, and 5 mM dithiothreitol) and sonicated 15 sec. The homogenate was centrifuged at 23K × g for 15 min and the supernatant (S 23) dialized against several changes of TKM buffer with 100 mM KCl. Amino acid incorporation was measured in the following at a final volume of 0.5 ml: 100 mM Tris-HCl (pH 7.6); 2 mM ATP, 0.4 mM GTP, 10 mM $MgCl_2$, 100 mM KCl, 50 μg yeast tRNA, 5 mM phosphoenol pyruvate kinase, 1 μCi ^{3}H-reconstituted protein hydrolysate (suplemented with 60 μM of the lacking amino acids), 10 mM dithiothreitol, and 0.1 ml of dialized S 23 fraction. After incubation at 30° for 30 min the samples were worked up and the hot TCA radioactivity determined as described elsewhere [23]. [From Ref. 13.]

Although we do not yet know if initiation factors occur in the cap fraction, our evidence suggests that the ribosome inhibitor could be the primary regulator of protein synthesis in zoospores. Unequivocal proof of this hypothesis nevertheless requires more direct evidence for the identity and mode of action of the inhibitor as well as examination of the localization of initiation factors, met-$tRNA_F^{met}$, and the sizes of the nucleotide pools.

One obvious factor that could limit protein synthesis would be a deficiency of mRNA or its storage in a nontranslatable state. The inhibitor experiments mentioned earlier and the presence of a few inactive polysomes, which have been found to contain polyadenylic acid-associated RNA [poly(A)+ RNA], both indicate that the lack of synthesis may result from blocked translation rather than the absence of mRNA. To test this directly we have measured the amount of poly(A)+ RNA in zoospores and growth phase cells, and also the distribution of the poly(A)+ RNA between the nuclear caps and extra-cap regions of zoospores. Although all the mRNA does not necessarily carry poly(A) sequences, we have concentrated on the fraction which does because it can easily be separated from other RNAs by oligo(dT)-cellulose chromatography.

C. Poly(A)+ RNA in Zoospores

The total RNA of zoospores contains 2.6 ± 0.2% poly(A)+ RNA, which we assume includes the stored mRNA used during germination. Young growth phase cells (6 h) contain a virtually identical percentage of poly(A)+ RNA. The poly(A)+ RNA from zoospores is polydisperse on gradients or when separated by gel electrophoresis, and has a mean mol wt of between 350,000 and 500,000 daltons (Fig. 4). The nuclease-resistant (RNase A + T_1) poly(A) segment itself comigrates with 4S tRNA on 5% polyacrylamide gels (using either A_{260}, or radioactivity with [^{3}H]adenosine-labeled RNA) (Fig. 5). The adenosine/AMP ratio obtained following alkaline hydrolysis indicates an average length of 40 adenine residues for the poly(A) segment. This is about the size of yeast poly(A) [28]. The base composition of the zoospore poly(A)+ RNA differs only slightly from the whole-cell RNA (G + C = 48.4 vs 45.7), but is significantly different from the reported base composition of the nuclear DNA (65%) [29]. The reason for this difference is unknown.

When the intracellular distribution of the zoospore poly(A)+ RNA was examined we were surprised to discover 80.2% of this putative stored mRNA in the cap fraction, suggesting a possible direct association of the poly(A)+ RNA with cap ribosomes (Table 4). However, when nuclear caps were lysed and centrifuged to place the ribosomes near the bottom of cushioned 10 to 35% gradients, only about one-third of the total poly(U)-hybridizable material sedimented with monosomes or polysomes. We do not yet know if the

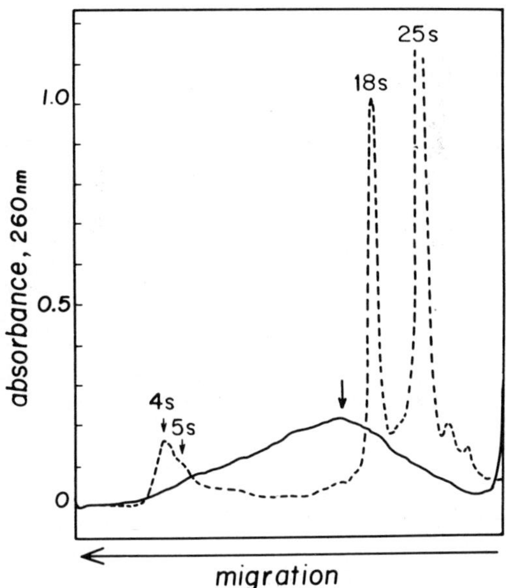

FIG. 4. Polyacrylamide gel electrophoresis of zoospore poly(A)+ RNA. Poly(A)+ RNA purified by phenol extraction and oligo(dT)-cellulose chromatography was heated 5 min at 70° in 0.3 M NaCl, pH 7.0, and quick-chilled to eliminate aggregation. It was then electrophoresed in 2.4% polyacrylamide-SDS gels for 2 h at 6 ma/gel. The solid line represents the poly(A)+ RNA scan and the dashed line marker rRNA run on a parallel gel. The arrow indicates the approximate mean mol wt of 450,000.

TABLE 4

The Distribution of Poly(A) in the Nuclear Cap and Extra-cap Fractions of Zoospores[a]

Fraction	% Total RNA	% Total poly(A)	% Total DNA
Zoospore homogenate	100	100	100
Low-speed supernatant (extra-cap fraction)	18.1	19.8	84.5
Nuclear caps	67.6	80.2	5.2

[a]Zoospores were homogenized and the nuclear caps isolated as described [23]. The RNA was isolated from each fraction, and the poly(A) content determined by ^{3}H-poly(U) hybridization using the procedures of Wilt [33]. DNA was estimated by the fluorometric method of Hinegardner [34], and served as a control for nuclear contamination of the nuclear cap fraction.

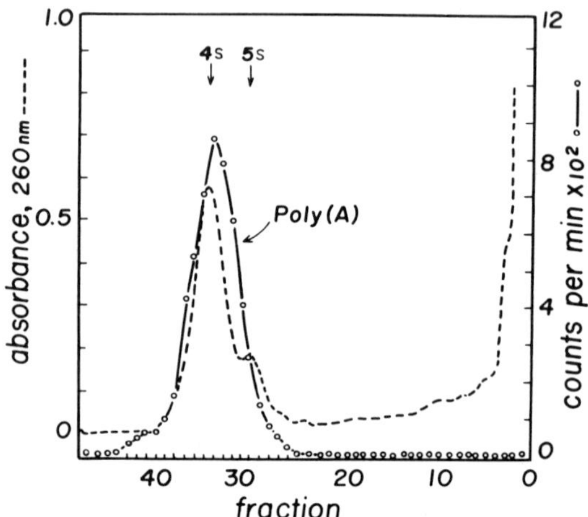

FIG. 5. Polyacrylamide electrophoresis of the poly(A) fragment from zoospore poly(A)+ RNA. [^{3}H]adenine-labeled zoospore poly(A)+ RNA, isolated as described in Fig. 4, was treated with pancreatic and T_1 ribonucleases (one unit each/20 µg RNA, in 0.3 M NaCl, pH 7.0) for 30 min at 37°. The enzyme-resistant material was recovered by oligo(dT)-cellulose chromatography, and precipitated with carrier poly(A) with 2 volumes of ethanol. The poly(A) was dissolved in running buffer and electrophoresed in 5% polyacrylamide-SDS gels for 2.5 h at 6 ma/gel. The gels were then scanned, sliced, and the slices counted by standard procedures [29]. Solid line, poly(A) radioactivity; dotted line, 4S and 5S RNA absorbancy on a parallel gel. [From Ref. 36.]

poly(A)+ RNA in the ribosome region of the gradients is actually associated with the ribosomes, occurs as aggregates that have a similar sedimentation coefficient, or was merely swept along with the bulk of material represented by the ribosomes. Since the small fraction of zoospore polysomes can only account for ca. 16% of this poly(A)+ RNA which sediments in the ribosome fraction of the gradient, the possible specific association of the remainder with inactive 80S ribosomes requires assessment. It might seem unlikely that mRNA is stored in the spore as a specific initiation complex, since initiation inhibitors greatly reduce in vitro amino acid incorporation and delay postencystment germination events. However, further repeated initiation of such complexes would be required to fully load the mRNA to make a normal sized polysome, and in a regulatory sense, it seems more efficient to prevent initiation rather than elongation; however, we cannot prove or disprove this at the present time. Certainly compartmentalization seems an unlikely mechanism for the control of mRNA function with 80% of the mRNA in the cap region.

D. Initiation of Protein Synthesis

It is obvious that our understanding of the regulatory mechanism in zoospores is incomplete. What we do know, on the other hand, is consistent with the presence of a completely functional preprogrammed but suppressed system for protein synthesis. The rapidity with which germination takes place is certainly consistent with a preprogrammed system, as is the ability of the cells to complete a significant degree of development in the absence of any nutrients [6,30]. A number of internal membrane rearrangements accompany encystment, but it is not known how new biosynthesis is triggered. Release of the ribosome inhibitor could be an important event and there is some evidence for its appearance in the extracellular medium. Following encystment, the appearance of new polysomes is remarkably rapid, reaching levels of around 50% in a 10-min interval [20]. This rapid, initial increase is only slightly reduced by actinomycin D, but significantly so by cycloheximide and is consistent with the effects seen on development and amino acid incorporation by the same inhibitors. The inhibitor effects also provide direct evidence for the utilization of the zoospore-stored mRNA, since no new mRNA should have been synthesized in the actinomycin D-treated cells. Even in untreated cells the entry of newly labeled RNA into the polysome fractions is too slow to provide significant new mRNA until after the initial polysome rise [30]. The initiation inhibitor D-MDMP--i.e., 2-(4-methyl-2,6-dinitroanilino)-N-methylpropionamide [31]--severely delays both the increase in amino acid incorporation and rhizoid production [30], indicating that rapid initiations with the stored mRNA and ribosomes give rise to these first polysomes (Fig. 6). From this curve it is also apparent that leucine incorporation does not increase as rapidly as would be expected from the kinetics of polysome formation. The explanations for this observation, as well as the fact that total protein remains constant until 70 to 80 min, are two-fold. Silverman et al. [15] have reported that the leucine pool does not reach a high specific activity rapidly during pulses at early time points, and this may in turn be the result of pool expansion due to the proteolysis (5%/h) found by Lodi and Sonneborn [20]. The second contributing factor is the persistence of the ribosome inhibitor when

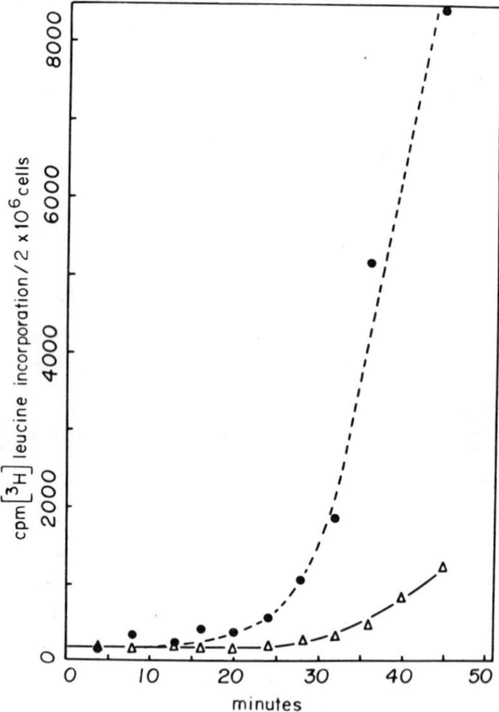

FIG. 6. The effect of the initiation inhibitor D-MDMP [31] on [^{3}H]leucine incorporation during zoospore germination. 2 × 10^6 cells/ml were incubated with 1 μCi/ml [^{3}H]leucine and either 5 × 10^{-5} M L-MDMP (control ●) or 5 × 10^{-5} M D-MDMP (inhibited △) in complete medium. At the times indicated 1 ml samples were removed and the hot TCA-insoluble radioactivity determined. [Modified from Ref. 30.]

using high cell densities for germination experiments [23,30]. Because of the residual inhibitor the actual polypeptide elongation rate (average ribosome transit time per mRNA) at 30 min is half of that measured for early growth phase cells (120 min) [30]. The reduced rates and turnover thus explain why so little net synthesis occurs despite a relatively high level of polysomes. Although we know that part of this new synthesis is essential we do not yet know what the critical proteins are.

E. Synthesis of Polysomal RNA During Germination

As indicated earlier actinomycin D fails to block development prior to 45 to 50 min, and this implies a specific requirement for new synthesis at this stage or possibly earlier. Even though it is apparently not essential initially, new RNA synthesis can be detected immediately after encystment and rapidly reaches a linear rate. Since the ATP pools can be labeled even earlier, there seems little question that the activation of synthesis occurs at or immediately after encystment. Pulse labeling with [^{3}H]uridine at different stages has shown that label enters both rRNA and polydisperse RNA as early as 22 to 23 min, and probably earlier [30]. Before 45 min, rRNA precursors are prominent after gel electrophoresis of RNA pulsed for 10 min, indicating sluggish processing. This is no longer true by 45 min, when a net accumulation of RNA per cell becomes measurable [6], and the rate of processing has reached growth phase levels. The accumulation of new rRNA in polysomes can be detected beginning at around 26 min [30].

If new ribosome synthesis is not important until 40 to 50 min, what about new mRNA? Does it contribute to the newly formed polysomes and to the continued increase in polysomes after 20 min? In an attempt to answer this question, zoospores were uniformly labeled by growth of zoosporangia on agar plates with $^{32}PO_4{}^{3-}$ for 12 h. These spores were then washed and germinated in medium containing 2.5 mM carrier phosphate and high specific activity [^{3}H]adenosine to label all newly synthesized RNA. Cells were collected and the polysomal RNA was extracted at 10-min intervals beginning at 25 min. Ideally the ^{32}P-radioactivity in the poly(A)+ RNA of polysomes should have represented stored zoospore mRNA and [^{3}H]adenosine radioactivity, new germling mRNA. ^{3}H-Radioactivity did enter the polysomes rapidly during germination and the poly(A)+ RNA fraction had already reached a high specific activity by 25 min (Fig. 7). The poly(A)- fraction (rRNA + tRNA) of polysomes, on the other hand, accumulated [^{3}H]adenosine label much more slowly. Thus, it is evident that newly labeled mRNA becomes rapidly associated with the old nuclear cap ribosomes and could potentially contribute to the increase in cellular polysome content after 20 min. If the post 20-min decline in polysomes observed in actinomycin D-treated cells [30] represents normal mRNA turnover, newly synthesized mRNA must normally replace it. The high percentage of the adenosine label in polysomal poly(A)+ RNA (24%) clearly suggests that new mRNA may be preferentially synthesized and transported to the cytoplasm at early time points (Fig. 8). Since the nuclear RNA polymerase responsible for the synthesis of mRNA is present and functional in *Blastocladiella* zoospores [32], early synthesis is entirely plausible. However, a portion of this early polysomal oligo(dT)-bound RNA may reflect polyadenylation of preexisting poly(A)- zoospore mRNA. The polysomal poly(A)+ RNA was therefore treated with pancreatic and T_1 ribonucleases and the hydrolysates fractionated on oligo(dT)-cellulose to recover the enzyme-resistant material [poly(A) base sequence]. In each successive 10-min fraction beginning at 25 min, about 20% of the ^{3}H-radioactivity in the total poly(A)+ RNA was nuclease-resistant, indicating a constant level of polyadenylation during the first hour of germination. This level may reflect (1) polyadenylation of cold preexisting spore messages which do or do not already have a poly(A) segment; (2) polyadenylation of newly synthesized mRNA; or, more likely, a combination of (1) and (2). Although we do not have sufficient data to further elaborate on these possibilities, it seems certain that at least a portion of the ^{3}H label in the oligo(dT)-cellulose-

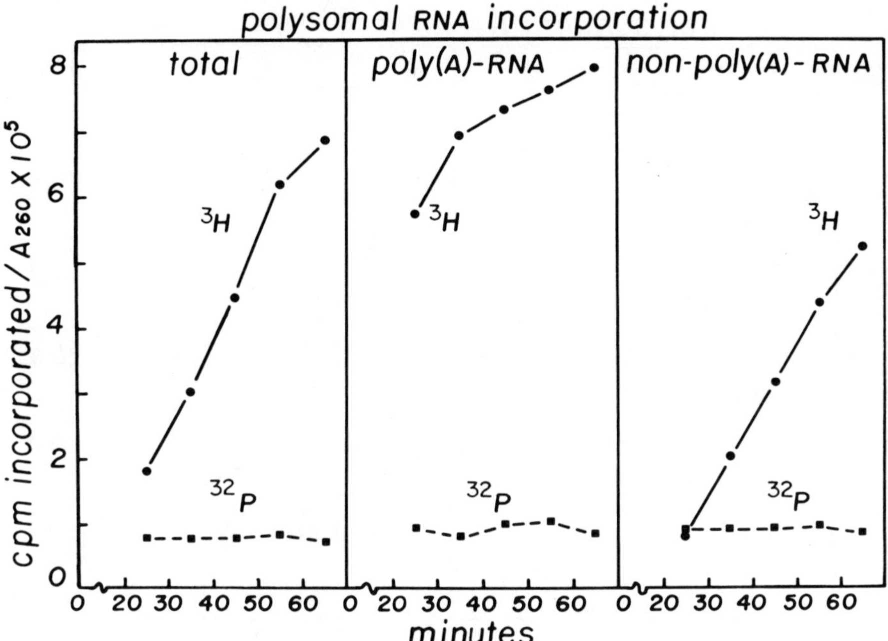

FIG. 7. Specific activity of different classes of polysomal RNA synthesized during early germination. Prelabeled zoospores were harvested from zoosporangia grown on PYG agar plates with $^{32}PO_4^{-3}$, washed, and inoculated at 2×10^6 cells/ml into synthetic medium [6] containing 1 μCi/ml [^{3}H]adenosine. Samples were removed at 25 min and subsequent 10-min intervals, the cells collected, washed, and homogenized in polysome extraction buffer (Fig. 2). The polysomes were pelleted through 50% sucrose by centrifugation at 45,000 rpm for 90 min (Spinco SW 50.1 rotor) and the RNA extracted from the pellet in 6% PAS-1% TNS in 0.01 M Tris-HCl, pH 7.5 [30] with phenol-chloroform-isoamyl alcohol. After precipitation the poly(A)+ RNA was separated from the poly(A)- RNA (rRNA, tRNA) by oligo(dT)-cellulose chromatography. Aliquots of the total RNA, poly(A)+ RNA, and poly(A)- RNA fractions were used to measure the A_{260} and counted in Aquasol to measure their ^{3}H and ^{32}P content by standard procedures.

bound polysomal RNA of germinating cells results from polyadenylation of preexisting mRNA. The decline in the percent cpm bound to oligo(dT)-cellulose after 25 min in Fig. 8 is largely due to the increasing synthesis of new [^{3}H]adenosine-labeled rRNA as newly produced ribosomes enter the polysome fraction. The ^{32}P data are

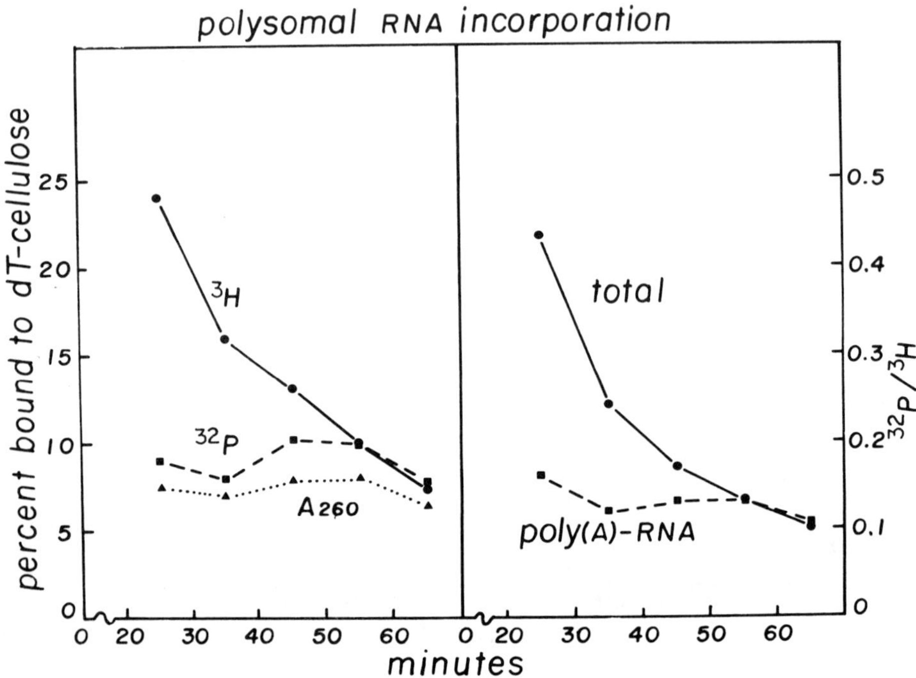

FIG. 8. The percentage of radioactivity or A_{260} from polysomal RNA which binds to oligo(dT)-cellulose vs germination time. See Fig. 7 for the details of polysome and RNA extraction, and oligo(dT)-cellulose chromatography. The percentage of the total radioactivity or A_{260} for each sample bound to oligo(dT)-cellulose represents the poly(A)+ RNA label (radioactivity) or amount (A_{260}).

more difficult to interpret, since there are indications that *Blastocladiella* may have a large polyphosphate pool, and the chase was clearly ineffective; as a result both old and new RNA were labeled to the same specific activity, and we cannot tell from this experiment whether the stored mRNA turned over after originally entering the polysomes.

III. FINAL COMMENTS

Our preliminary evidence for the contribution of new mRNA to polysome formation and protein synthesis raises an interesting point. One can calculate that the stored nuclear cap mRNA is in a two- to four-fold excess over that needed to form the initial actinomycin D-resistant polysomes [27]. If new synthesis supplements this fraction and also gradually replaces it as it turns over, it would be of interest to know the fate of the excess, unused poly(A)+ RNA messages. Unlike embryos of higher organisms, the asexual propagules of fungi and many other microorganisms may be required to undergo early development in widely varying environments. One intriguing possibility is that the excess stored mRNA in zoospores could represent a spectrum of potentially necessary information (e.g., for biosynthetic or degradative enzymes, etc.) from which a specific fraction, dictated by the sufficiency of the external environment, is selected. The remainder of the mRNA which was inappropriate for the particular situation could then be discarded via turnover without translation. In this sense the zoospore could be considered to be "preprogrammed" for a variety of different situations it might meet. In response to a specific environment one or more different subsets of the whole population of mRNA sequences might be selected in addition to the messages required for some common developmental program required primarily for rhizoid formation. In our view *Blastocladiella* is an excellent organism with which to examine such questions concerning the function of stored mRNA.

Despite a promising start it should go without saying, perhaps, that we are far from a complete understanding of the control of protein and RNA synthesis in zoospores and their reactivation at germination. Our work in the future will be aimed at filling some of the many gaps in our understanding and at answering many of the still unanswered questions. It will be important to complete our study of the protein and nonprotein components required for protein synthesis in zoospores, i.e., to be certain that a fully preformed system exists as we propose. It is equally important to identify the ribosome inhibitor and to establish its mode of action, as well as its synthesis during differentiation and inactivation at germination. We would also like to answer the following questions concerning the zoospore-stored mRNA. Is mRNA function specifically regulated in zoospores and, if so, how is this altered during germination? How much genetic information is carried by the stored mRNA? What fraction of this information is utilized during germination and does this depend upon the extracellular milieu? And, last but far from least, what *specific* gene products encoded in the stored mRNA are *essential* for germination?

ACKNOWLEDGMENTS

The authors are pleased to acknowledge the contributions of C. J. Leaver, F. H. Wilt, T. G. Adelman, and I. R. Schmoyer to different parts of the research described. This work was supported by Public Health Service Grant AI-04873 from the National Institute of Allergy and Infectious Diseases.

REFERENCES

1. Cantino, E. C., 1951. Antonie van Leeuwenhoek 17: 59-96.
2. Cantino, E. C., 1966. *In* The Fungi, Vol. 2, G. C. Ainsworth and A. S. Sussman (eds.), Academic Press, New York, p. 283.
3. Lovett, J. S., and E. C. Cantino, 1960. Amer. J. Bot. 47: 499-505.
4. Knouw, B. T., and H. D. McCurdy, 1969. J. Bacteriol. 99: 197-205.
5. Cantino, E. C., and J. S. Lovett, 1964. Adv. Morphogenesis 3: 33-93.
6. Lovett, J. S., 1968. J. Bacteriol. 96: 962-969.

7. Soll, D. R., and D. R. Sonneborn, 1969. Develop. Biol. 20: 218-235.

8. Bromberg, R., 1974. Develop. Biol. 36: 187-194.

9. Soll, D. R., R. Bromberg, and D. R. Sonneborn, 1969. Develop. Biol. 20: 183-217.

10. Truesdell, L. C., and E. C. Cantino, 1971. *In* Current Topics in Developmental Biology, Vol. 6, Academic Press, New York, p. 1.

11. Barstow, W. E., and J. S. Lovett, 1974. Protoplasma 82: 103-117.

12. Cantino, E. C., L. C. Truesdell, and D. S. Shaw, 1968. J. Elisha Mitchell Sci. Soc. 84: 125-146.

13. Lovett, J. S., 1975. Bact. Rev. 39: 345-404.

14. Soll, D. R., and D. R. Sonneborn, 1971. J. Cell Sci. 9: 679-699.

15. Silverman, P. M., M. M. O. Huh, and L. Sun, 1974. Develop. Biol. 40: 59-70.

16. Smith, J. D., and P. M. Silverman, 1973. Biochem. Biophys. Res. Comm. 54: 1191-1197.

17. Suberkropp, K. F., and E. C. Cantino, 1972. Trans. Brit. Mycol. Soc. 59: 463-475.

18. Cantino, E. C., K. F. Suberkropp, and L. C. Truesdell, 1969. Nova Hedwigia 18: 149-158.

19. Silverman, P. M., and P. M. Epstein, 1975. Proc. Nat. Acad. Sci. USA 72: 442-446.

20. Lodi, W. R., and D. R. Sonneborn, 1974. J. Bacteriol. 117: 1035-1042.

21. Lovett, J. S., 1963. J. Bacteriol. 85: 1235-1246.

22. Schmoyer, I. R., and J. S. Lovett, 1969. J. Bacteriol. 100: 854-864.

23. Adelman, T. G., and J. S. Lovett, 1974. Biochim. Biophys. Acta 335: 236-245.

24. Cantino, E. C., 1975. Biochem. Biophys. Res. Comm. 63: 343-348.

25. Soll, D. R., and D. R. Sonneborn, 1971. Proc. Nat. Acad. Sci. USA 68: 459-463.

26. Gong, C.-S., and J. S. Lovett, unpublished results.

27. Lovett, J. S., 1976. *In* The Fungal Spore: Form and Function, D. J. Weber and W. M. Hess (eds.), Wiley-Interscience, p. 189.

28. McLaughlin, C. S., J. R. Warner, M. Edmunds, H. Nakazato, and M. H. Vaughn, 1973. J. Biol. Chem. 248: 1466-1471.

29. Comb, D. G., R. Brown, and S. Katz, 1964. J. Mol. Biol. 8: 781-789.

30. Leaver, C. J., and J. S. Lovett, 1974. Cell Differentiation 3: 165-192.

31. Weeks, D. P., and R. Baxter, 1972. Biochemistry 11: 3060-3064.

32. Horgen, P. A., 1971. J. Bacteriol. 106: 281-282.

33. Wilt, F. H., 1973. Proc. Nat. Acad. Sci. USA 70: 2345-2349.

34. Hinegardner, R. T., 1971. Anal. Biochem. 39: 197-201.

35. Gong, G. S., and J. S. Lovett, 1977. Exp. Mycology (in press).

36. Johnson, S. A., J. S. Lovett and F. H. Wilt, 1977. Develop. Biol. (in press).

AUTHOR INDEX

SUBJECT INDEX